Small Molecule Drug Discovery

Small Molecule Drug Discovery

Methods, Molecules and Applications

Andrea Trabocchi

Department of Chemistry "Ugo Schiff"
University of Florence
Sesto Fiorentino, Florence
Italy

Elena Lenci

Department of Chemistry "Ugo Schiff"
University of Florence
Sesto Fiorentino, Florence
Italy

ELSEVIER

Elsevier
Radarweg 29, PO Box 211, 1000 AE Amsterdam, Netherlands
The Boulevard, Langford Lane, Kidlington, Oxford OX5 1GB, United Kingdom
50 Hampshire Street, 5th Floor, Cambridge, MA 02139, United States

Notices
Knowledge and best practice in this field are constantly changing. As new research and
experience broaden our understanding, changes in research methods, professional
practices, or medical treatment may become necessary.

Practitioners and researchers must always rely on their own experience and knowledge in
evaluating and using any information, methods, compounds, or experiments described
herein. In using such information or methods they should be mindful of their own safety
and the safety of others, including parties for whom they have a professional
responsibility.

To the fullest extent of the law, neither the Publisher nor the authors, contributors, or
editors, assume any liability for any injury and/or damage to persons or property as a
matter of products liability, negligence or otherwise, or from any use or operation of any
methods, products, instructions, or ideas contained in the material herein.

Library of Congress Cataloging-in-Publication Data
A catalog record for this book is available from the Library of Congress

British Library Cataloguing-in-Publication Data
A catalogue record for this book is available from the British Library

ISBN: 978-0-12-818349-6

For information on all Elsevier publications visit our website at
https://www.elsevier.com/books-and-journals

Publisher: Susan Dennis
Acquisition Editor: Emily M McCloskey
Editorial Project Manager: Kelsey Connors
Production Project Manager: Sreejith Viswanathan
Cover Designer: Alan Studholme

Typeset by TNQ Technologies

Working together
to grow libraries in
developing countries

www.elsevier.com • www.bookaid.org

Contents

Contributors

Michelle R. Arkin
Small Molecule Discovery Center and Department of Pharmaceutical Chemistry, University of California San Francisco, San Francisco, CA, United States; Buck Institute for Research on Aging, Novato, CA, United States

Andrea Basso
Department of Chemistry and Industrial Chemistry, University of Genova, Genova, Italy

Pietro Capurro
Department of Chemistry and Industrial Chemistry, University of Genova, Genova, Italy

Stuart J. Conway
Department of Chemistry, Chemistry Research Laboratory, University of Oxford, Oxford, United Kingdom

Margarida Espadinha
Research Institute for Medicines (iMed.ULisboa), Faculty of Pharmacy, Universidade de Lisboa, Lisbon, Portugal

M. Isabel García-Moreno
University of Seville, Seville, Spain

José M. García Fernández
Institute for Chemical Research (IIQ), CSIC – University of Seville, Seville, Spain

Kamal Kumar
Max Planck Institute of Molecular Physiology, Dortmund, Germany; Aicuris Antiinfective Cures GmbH, Wuppertal, Germany

Elena Lenci
Department of Chemistry "Ugo Schiff", University of Florence, Sesto Fiorentino, Florence, Italy

Qingliang Li
National Center for Biotechnology Information, National Library of Medicine, National Institutes of Health, Besthesda, MD 20894, United States

José L. Medina-Franco
Department of Pharmacy, School of Chemistry, Universidad Nacional Autónoma de México, Mexico City, Mexico

Carmen Ortiz Mellet
University of Seville, Seville, Spain

Amy Trinh Pham
School of Pharmacy, Health Sciences Campus, University of Waterloo, Waterloo, ON, Canada

Praveen P.N. Rao
School of Pharmacy, Health Sciences Campus, University of Waterloo, Waterloo, ON, Canada

Fernanda I. Saldívar-González
Department of Pharmacy, School of Chemistry, Universidad Nacional Autónoma de México, Mexico City, Mexico

Elena M. Sánchez-Fernández
University of Seville, Seville, Spain

Maria M.M. Santos
Research Institute for Medicines (iMed.ULisboa), Faculty of Pharmacy, Universidade de Lisboa, Lisbon, Portugal

Arash Shakeri
School of Pharmacy, Health Sciences Campus, University of Waterloo, Waterloo, ON, Canada

Andrea Trabocchi
Department of Chemistry "Ugo Schiff", University of Florence, Sesto Fiorentino, Florence, Italy

Chris G.M. Wilson
Small Molecule Discovery Center and Department of Pharmaceutical Chemistry, University of California San Francisco, San Francisco, CA, United States

Foreword

The discovery of new drugs is an endeavor of high scientific demand and societal relevance. It requires interdisciplinary research spanning the life sciences, chemistry, pharmcology, and even material science. It benefits mankind because the treatment of disease is one of societies' most urgent demands to science.

Among the pharmacopoeia available to us, small molecules historically are most prevalent, and they form the largest group of new chemical entities in medicinal chemistry research to this very day. Undoubtedly biologicals, in particular antibodies, have gained major importance and are here to stay, but it is also evident that small molecule drugs will remain to be of highest relevance in drug discovery in the foreseeable future.

Hence the science that underlies the discovery and development of new bioactive small molecules that can be considered drug candidates and that may inspire new approaches to the treatment of disease is of utmost importance and calls for continuous introduction of new methods and principles.

This necessity underlies the articles compiled in the book edited by Andrea Trabocchi and Elena Lenci. Collectively the authors shine light on a very impressive ensemble of some of the most relevant topics in contemporary medicinal chemistry and drug discovery. These include chemical synthesis, cheminformatics, and biophysical and computational methods and highlight individual case studies focusing on some of the greatest challenges in this science, as for instance Alzheimer disease.

The Editors have chosen the topics wisely and with deep insight into drug discovery. Thereby the book not only gives an overview of recent developments, it also guides the reader to the frontiers of medicinal chemistry research. It will be both entertaining to read and educative such that it will be of interest to the professional skilled in the art, as well as to newcomers to the field, in particular, advanced graduate and postdoctoral students.

I hope that this book will find widespread interest from practitioners in medicinal chemistry and simply curious scientists trying to get a glimpse at and an understanding of the science that drives small molecule drug discovery.

Herbert Waldmann
Max Plank Institute of Molecular Physiology
Dortmund, Germany
September 2019

Preface

The identification of novel molecular entities capable of specific interactions represents a significant challenge in early drug discovery. Despite the success of biopharmaceuticals, small molecules still dominate the market, being more than 95% of the top 200 most prescribed drugs in 2018. Small molecules are low-molecular-weight organic compounds that include natural products and metabolites, as well as drugs and other xenobiotics. The entire drug discovery process has changed a lot during the last decades due to the difficulties in finding new lead compounds for all those "undruggable" targets and for addressing complex oncology and CNS diseases. The rational design of ligands is still a powerful approach, especially in combination with computer-aided methods when the biological target is well-defined and structurally known. Nevertheless, new synthetic methods able to generate high-quality chemical libraries have been exploited over the last decades to meet the need of improving the quality and quantity of small molecules for biological screenings. Since the synthetic efforts characterized by the trial-and-error approach of the 1980s and combinatorial chemistry of the 1990s, new attitudes are now gaining wide attention in synthetic chemistry for small molecule drug discovery, in order to maximize the quality of libraries and reducing the waste of generating and screening random unnecessary compounds. New frontiers in the synthesis of small molecule libraries have been explored. Diversity-Oriented Synthesis has proven to be very effective to access large areas of the chemical space, primarily through the creation of many distinct molecular scaffolds. Also, Biology-Oriented Synthesis has been conceived with the purpose of taking inspiration from nature to select promising molecular scaffolds being related to natural products in terms of biological output. Today, an important part of modern medicinal chemistry is represented by computer-aided methods, for rational drug design (i.e, virtual screening), and for the smart design of small molecule libraries. As the number of publicly accessible biological data is rapidly increasing, chemoinformatics is gaining relevance as a tool for developing better chemical libraries.

The book is organized in three parts, exploring selected topics on small molecule drug discovery on key synthetic and screening methods, representative small molecule categories, and selected biomedical applications. The first part encompasses the methods for the synthesis, structure classification, and biological evaluation of small molecules. Specifically, Chapter 1 reports an in-depth overview of strategic approaches for the achievement of small molecules, and Chapter 2 gives a thorough account about most relevant chemical reactions for building small molecules. Chapters 3 and 4 report the chemoinformatic tools to assess chemical diversity of small molecule libraries and virtual screening methods, respectively. Chapter 5 concludes the first part on methods discussing screening approaches and biophysics of small molecules. In the second part, representative small molecule classes derived from natural products are reported. Chapter 6 describes the principles and applications of small molecule peptidomimetics, and Chapter 7 reports the chemistry of sp2-

iminosugars within the field of carbohydrates. Chapter 8 outlines the synthesis and structural features of small molecules characterized by spiroacetal moiety, and Chapter 9 reports the case study of centrocountins as nature inspired indoloquinolizines. The third part contains two selected case studies about the successful application of small molecules in biomedical research. Chapter 10 deals with PPIs as therapeutic targets for anticancer drug discovery and describes the case study of MDM2 and BET bromodomain inhibitors, and Chapter 11 is an account of the discovery of small molecules for the treatment of Alzheimer disease.

These presentations have been conceived for a broad readership and should interest not only those readers who currently work in the field of organic and medicinal chemistry addressing drug discovery, but also those who are considering this approach in the field of chemical biology, taking advantage of the use of small molecule as chemical probes for dynamically interrogating biological systems and for investigating potential drug targets. We hope these Chapters will stimulate further advances in the ever-developing field of small molecule drug discovery.

Andrea Trabocchi
Elena Lenci
Florence, September 2019

Abbreviations

(TR)-FRET	Time-resolved fluorescence resonance energy transfer
2D	Two-dimensional
3D	Three-dimensional
ACD	Available chemicals directory
ACE	Angiotensin-converting enzyme
ACh	Acetylcholine
AChE	Acetylcholinesterase
ACN	Acetonitrile
AcOH	Acetic acid
AD	Alzheimer disease
ADME	Absorption-distribution-metabolism-excretion
ADP	Adenosine diphosphate
AIDS	Acquired immunodeficiency syndrome
ALK	Anaplastic lymphoma kinase
AlphaScreen	Amplified luminescent proximity homogeneous assay
AMBER	Assisted Model Building with Energy Refinement
ANS	Anthocyanidin synthase
APDS/TRP	Alanine-proline-aspartate-serine/threonine-arginine-proline
ApoA1	Apolipoprotein A-1
APP	Amyloid precursor protein
APT1	Acyl protein thiosterase 1
APV	Amprenavir
AR	Androgen receptor
AS/MS	Affinity selection followed by mass spectrometry
ATP	Adenosine triphosphate
B/C/P	Build/couple/pair
BACE-1	Beta-site amyloid precursor protein cleaving enzyme 1
BAL	Backbone amide linker
BBB	Blood-brain barrier
BET	Bromodomain and extraterminal domain
Bcl-X$_L$	B-cell lymphoma
BCPs	Bromodomain-containing proteins
BEDROC	Boltzmann enhanced discrimination of ROC
BIOS	Biology-oriented synthesis
BLI	Bio-layer interferometry
BOP	Benzotriazol-1-yloxytris(dimethylamino)phosphonium hexafluorophosphate
BRCA	Breast cancer gene
BRD	Bromodomain
BRET	Bioluminescence resonance energy transfer
BSA	Bovine serum albumin
BZD	Benzodiazepine
CAN	Ceric ammonium nitrate

CAS	Catalytic active site
CBD	Condition-based divergence
CCR2	Chemokine Receptor type 2
CCR5	Chemokine Receptor type 5
CDC	Cross-dehydrogenative couplings
CDK	Chemistry Development Kit
CDKs	Cyclin-dependent kinases
CDP	Consensus diversity plot
CETP	Cholesteryl ester transfer protein
CETSA	Cellular thermal shift assay
ChE	Cholinesterase
CHI	Chalcone isomerase
CHS	Chalcone synthase
CLL	B-cell chronic lymphocytic leukemia
clogP	Calculated octanol/water partition coefficient
CMC	Critical micelle concentration
CNS	Central nervous system
COPD	Chronic obstructive pulmonary disease
COX	Cyclooxigenase
cAMP	Cyclic adenosine monophosphate
cGMP	Cyclic guanosine monophosphate
CPA	Chiral phosphoric acid
CPAs	Carboxypeptidases A
Crm1	Chromosome region maintenance 1
Cryo-EM	Cryogenic electron microscopy
CS	Castanospermine
CSA	Camphorsulphonic acid
CSP	Chemical shift perturbation
CSR	Cyclic system recovery curves
CuAAC	Cu-catalyzed Azide Alkyne Click
CYP3A4	Cytochrome P450 3A4
DIPEA	N,N-Diisopropylethylamine
DCE	Dichloroethane
DCM	Dichloromethane
DCN	1,4-Dicyanonaphthalene
DDQ	Dichlorodicyanobenzoquinone
DECLs	DNA-encoded chemical libraries
DEEP-STD	Differential epitope mapping-STD
DIAD	Diisopropyl azodicarboxylate
DLS	Dynamic light scattering
DMAP	4-Dimethylaminopyridine
DMEDA	1,2-Dimethylethylenediamine
DMF	N,N-Dimethylformamide
DMPU	N,N'-dimethyl-N,N'-propylene urea
DMSO	Dimethylsulfoxide
DNA	Deoxyribonucleic acid
DNJ	Deoxynojirimycin
DNP	Dictionary of Natural Products

DOS	Diversity-oriented synthesis
DPPH	Diphenyl-1-picrylhydrazyl
DR	Diabetic retinopathy
DRR	Double reactant replacement
DRV	Darunavir
DSF	Differential scanning fluorimetry
EC$_{50}$	Half maximal effective concentration values
ECFP	Extended-connectivity fingerprint
EeAChE	Electric eel acetylcholinesterase
EF	Enrichment factor
ELT	Encoded library technology
EMA	European Medicines Agency
ERα	Estrogen receptor α
ESIPT	Excited state intramolecular proton transfer
ET	Energy transfer
EYFP	Enhanced yellow fluorescent protein
FA	Fluorescence anisotropy
FACS	Fluorescence-activated cell sorting
FC	Fusicoccin
FDA	US Food and Drug Administration
FLIM	Fluorescence lifetime imaging microscopy
FP	Fluorescence polarization
FPV	Fosamprenavir
FRET	Fluorescence resonance energy transfer
GalNAc	*N*-Acetyl-d-galactosamine
GABA	Gamma-aminobutyric acid
GBSA	Generalized Born surface area
GFP	Green fluorescent proteins
GlcNAc	*N*-Acetyl-d-glucosamine
GluCl	Glutamate-gated chloride channel
GOLD	Genetic Optimisation for Ligand Docking
GPCRs	G protein–coupled receptors
GPx	Glutathione peroxidase
GSK3β	Glycogen synthase kinase 3β
GTM	Generative topographic mapping
H3R	Histamine H3 receptor
HAT	Hydrogen atom transfer
HATU	1-[Bis(dimethylamino)methylene]-1H-1,2,3-triazolo[4,5-b]pyridinium 3-oxid hexafluorophosphate
HBA	Hydrogen bond acceptors
HBD	Hydrogen bond donors
hBuChE	Human butyrylcholinesterase
HCV	Hepatitis C virus
HCV NS3	Hepatitis C Virus nonstructural protein 3
HDx	Hydrogen/deuterium exchange
HER2	Human epidermal growth factor receptor 2
Hh	Hedgehog
HIV	Human immunodeficiency virus

HLMs	Human liver microsomes
HMG-CoA	3-Hydroxy-3-methyl glutaryl coenzyme A
HMQC	Heteronuclear Multiple Quantum Coherence
HO-1	Heme oxygenase-1
HPLC	High-performance liquid chromatography
HRP	Horseradish peroxidase
HSQC	Heteronuclear Single Quantum Coherence
hTR	Human telomerase RNA
HTRF	Homogeneous time-resolved FRET
HTS	High-throughput screening
ICR	Institute of Cancer Research
icv	Intracerebroventricular
IDH1	Isocitrate dehydroganse type 1
IDV	Indinavir
IMAP-FP	Ion affinity-based fluorescence polarization
IMCRs	Isocyanide-based multicomponent reactions
iNOS	Inducible nitric oxide synthase
ISC	Intersystem crossing
ITC	Isothermal titration calorimetry
IUPAC	International Union of Pure and Applied Chemistry
JAK2	Janus kinase 2
KAc	Acetylated lysine residues
KAHA	α-KetoAcid-HydroxylAmine
KATs	Lysine acetyltransferases
KDACs	Deacetylated by lysine deacetylases
KNIME	Konstanz Information Miner
LBVS	Ligand-based virtual screening
LC-MS	Liquid chromatography-mass spectrometry
LD$_{50}$	Lethal dose, 50%
LED	Light-emitting diode
LPS	Lipopolysaccharide
LSDs	Lysosomal storage diseases
LSF	Late-stage functionalization
LTP	Long-term potentiation
mAb	Monoclonal antibody
MACCS	Molecular ACCess System
MAO	Monoamine oxidase
MAPK	Mitogen-activated protein kinase
MB	Methylene blue
MCF-7	Michigan Cancer Foundation-7
***m*CPBA**	*m*-Chloroperoxybenzoic acid
MCR	Multicomponent reaction
MCR^2	Combining multicomponent reactions
MCSS	Maximum Common Substructure
MDM2	Mouse double minute 2 homolog
MeCN	Acetonitrile
MEK1/2	MAP (mitogen-activated protein) kinase/ERK (extracellular signal-regulated kinase) Kinase 1/2

MFS	Multifusion similarity maps
MMP	Matrix metalloprotease
MptpA	Low-molecular-weight protein-tyrosine phosphatase A
MptpB	Low-molecular-weight protein-tyrosine phosphatase B
MRS	Modular reaction sequences
MS	Mass spectrometry
MST	Microscale thermophoresis
MCC	Matthews correlation coefficient
MOE	Molecular Operating Environment
MTDLs	Multi-target-directed ligands
MTT	3-(4,5-Dimethylthiazol-2-yl)-2,5-diphenyltetrazolium bromide
MUC1	Mucin 1
MW	Molecular weight
NADPH	Nicotinamide adenine dinucleotide phosphate hydrogen
NaN$_3$	Sodium azide
NF-κB	Nuclear factor-kappa B
NFTs	Neurofibrillary tangles
NHC	*N*-heterocyclic carbene
NK1	Neurokinin 1 receptor
NMDA	*N*-methyl-D-aspartate
NMDAR	NMDA receptor
NMP	N-Methyl-2-pyrrolidone
NMR	Nuclear magnetic resonance
NN	Neural network
NOR	Novel object recognition
NPM	Nucleophosmin
ORAC-FL	Oxygen radical absorbance capacity
ORTEP	Oak Ridge Thermal Ellipsoid Plot
P-3CR	Passerini reaction
PADAM	Passerini reaction/Amine Deprotection/Acyl Migration
PAINs	Pan-assay interference compounds
PAMPA	Parallel artificial membrane permeability assay
PBMC	Peripheral blood mononuclear cells
PBSA	Poisson-Boltzmann surface area
PCA	Principal component analysis
PCIs	Protein-chromatin interactions
PCR	Polymerase chain reaction
PD	Pharmacodynamics
PDB	Protein Data Bank
PDE	Phosphodiesterase
PDE5	Phosphodiesterase type 5
PET	Positron emission tomography
PHFs	Paired helical filaments
PIAs	Phosphatidylinositol ether lipid analogues
PI3K	Phosphoinositide-3-kinase
PK	Pharmacokinetics
PMI	Principal moment of inertia
PPI	Protein-protein interaction

PS	Polystyrene
PSSC	Protein structure similarity clustering
PTP1B	Protein-tyrosine phosphatase 1B
PUMA	Platform for Unified Molecular Analysis
PVDF	Polyvinylidene difluoride
QSAR	Quantitative structure—activity relationship
RB	Rose bengal
RBs	Rotatable bonds
RCM	Ring closing metathesis
RF	Random forest
RGD	Arg-Gly-Asp
RIfS	Interference spectroscopy
RNA	Ribonucleic acid
ROC	Receiver operating characteristics
ROCS	Rapid Overlay of Chemical Structures
ROM	Ring opening metathesis
ROS	Reactive oxygen species
RTV	Ritonavir
RU	Response units
RXR	Retinoid X receptor
SAR	Structure—activity relationship
SBS	Society for Biomolecular Sciences
SBVS	Structure-based virtual screening
ScFv	Single-chain variable fragment
SCONP	Structural classification of natural products
SDS-PAGE	Sodium dodecyl sulfate-polyacrylamide gel electrophoresis
SE	Shannon entropy
sEH	Soluble epoxide hydrolase
SET	Single electron transfer
SGLT2	Sodium-glucose linked transporter 2
SHG	Second harmonic generation
SIFt	Structural interaction fingerprint
SLL	Small lymphocytic lymphoma
SlogP	Octanol/water partition coefficient
SMM	Small molecule microarray
SOCE	Store-operated calcium entry
SOMs	Self-organizing maps
SPOS	Solid-phase organic synthesis
SPR	Surface plasmon resonance
SPRs	Structure—properties relationships
SPS	Solid-phase synthesis
SRR	Single reactant replacement
SQV	Saquinavir
STAT3	Signal transducers and activators of transcription 3
STD	Saturation transfer difference
SVM	Support vector machines
t-SNE	Distributed stochastic neighbor embedding
TASK3	TWIK-related acid-sensitive K^+ channel 3

TBAF	Tetrabutylammonium fluoride
TCM	Traditional Chinese medicine
TEM	Transmission electron microscopy
TFA	Trifluoroacetic acid
THF	Tetrahydrofuran
ThT	Thioflavin T
TINS	Target immobilized NMR screening
TNF-α	Tumor necrosis factor-α
TPP	Tetraphenylporphirine
TPSA	Topological polar surface area
TOS	Target-oriented synthesis
TRH	Thyrotropin-releasing hormone
TRK	Tropomyosin receptor kinase
TrxR	Thioredoxin reductase
U-5C-4CR	Ugi 5-center-4-component reaction
UDC	Ugi/deBoc/cyclization
Ugi-4CC	Ugi-4 component reaction
UNPD	Universal Natural Product Database
USR	Ultrafast shape recognition
UV-B	Ultraviolet B-rays
VE-PTP	Vascular endothelial-protein-tyrosine phosphatase
VHR	Vaccinia H1-related
WHO	World Health Organization
YFP	Yellow fluorescent protein

Synthetic approaches toward small molecule libraries

Elena Lenci, Andrea Trabocchi
Department of Chemistry "Ugo Schiff", University of Florence, Sesto Fiorentino, Florence, Italy

1.1 Introduction

Drug discovery is the long and arduous process that can eventually bring molecules from the laboratories to the market. Although the number of new approved drugs showed about a 30% increase over 2017, marking a new record after 1996 [1], in general only 1 molecule out of 5000 hit candidates can reach the market [2].

The process of discovering, testing, and gaining approval for a new drug has changed a lot during the last century. From the isolation of active ingredients from traditional remedies and natural products, drug discovery has evolved into a multidisciplinary and complex process that brings together the efforts of biologists, pharmacologists, and chemists. Many different approaches nowadays can be applied in drug discovery. From one hand, the rational design of ligands remains the "gold standard" in medicinal chemistry, especially when the biological target is well defined and structurally known (Fig. 1.1, top) [3]. On the other hand, a parallel new approach has emerged, especially in those fields, such as cancer and neurodegenerative disorders, where the biological target or the mode of binding is not well known [4,5], or difficult to study in traditional drug discovery programs [6].

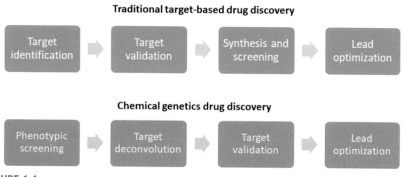

FIGURE 1.1

Comparison between conventional target-based and chemical genetics drug discovery approaches.

Small Molecule Drug Discovery. https://doi.org/10.1016/B978-0-12-818349-6.00001-7

When researchers are experiencing this impasse, one alternative, for the discovery of new targets and new lead compounds, is the application of large small molecules libraries in high-throughput screening (HTS), phenotypic assays, and chemical genetics studies (Fig. 1.1, bottom) [2,7−10]. The relevance of this approach is also highlighted by the emergence of international screening initiatives, such as EU-OPENSCREEN [11] or the European Lead Factory [12,13].

In both approaches, synthetic chemistry plays a key role in generating high-quality small molecules collections. In fact, despite the vast success of the biological drugs (monoclonal antibodies or recombinant proteins), the favorable pharmacokinetic properties of small molecules libraries allowed them to remain as the gold standard for the development of new medications, especially in the case of enzyme inhibitors. In fact, among the 59 new drugs approved by the FDA in 2018, 42 are small molecules and only 17 are biologic drugs [1]. In Table 1.1 are reported, for example, the 11 small molecules approved by the FDA as new drugs for cancer therapy in 2018.

Table 1.1 Small molecules approved by the FDA as new drugs for cancer therapy in 2018.

Name	Structure	Company	Biological effect
Encorafenib		Array	BRAF inhibitor. Used in combination with binimetinib for the treatment of BRAF-mutated melanoma
Binimetinib		Array	MEK1/2 inhibitor. Used in combination with Encorafenib for the treatment of BRAF mutated melanoma
Talazoparib		Pfizer	Poly (ADP-ribose) polymerase type 1 and 2 inhibitor. Used in the treatment of BRCA-mutated HER2-negative breast cancer
Ivosidenib		Agios	Isocitrate dehydrogenase type 1 (IDH1) inhibitor. Used in the treatment of acute myeloid melanoma
Gilteritinib		Astellas	FLT3, AXL, and ALK kinases inhibitor. Used in the treatment of acute myeloid melanoma

Table 1.1 Small molecules approved by the FDA as new drugs for cancer therapy in 2018.—*cont'd*

Name	Structure	Company	Biological effect
Glasdegib		Pfizer	Hedgehog (hh) signaling pathway inhibitor. Used in the treatment of acute myeloid melanoma
Duvelisib		Verastam	Phosphoinositide-3-kinase (PI3K) inhibitor. Used in the treatment of chronic lymphocytic leukemia or small lymphocytic lymphoma
Larotrectinib		Bayer and Loxo	Tropomyosin receptor kinase (TRK) A/B/C inhibitor. Used in the treatment of solid tumors that have the neurotrophic receptor tyrosine kinase gene fusion
Lorlatinib		Pfizer	ATP-competitive inhibitor of anaplastic lymphoma kinase (ALK) and c-Ros oncogene 1 (Ros)1. Used in the treatment of ALK-positive metastatic non—small cell lung cancer
Dacomitinib		Pfizer	Covalent ligand of human epidermal growth factor receptors Her-1, Her-2, and Her-4. Used in the treatment of metastatic non—small cell lung cancer
Apalutamide		Janssen	Androgen receptor (AR) antagonist. Used in the treatment of prostate cancer

Thus, to address this demand, very powerful synthetic methods are necessary for the generation of large small molecules libraries. Several efforts have been devoted to improve the quality and quantity of small molecules representing a library. In particular, during last decades, organic chemists have taken advantage of high-throughput synthesis methods, such as solid-phase techniques [14–17], and combinatorial chemistry [18,19]. Unfortunately, despite the apparent success, these chemistry approaches have not fulfilled the desired expectations as the automation of discovery processes has proven to be inefficient [20,21]. Thus, new frontiers in the synthesis of small molecules libraries are being explored, with the aim of improving the quality of the small molecules representing a library, where the synthetic efforts are not guided by a specific core structure, but rather by concepts like molecular diversity (i.e., diversity-oriented synthesis) and bioactivity or biosynthetic pathway (i.e., biology-oriented synthesis). This chapter focuses on main synthetic approaches for the generation of large, high-quality small molecule collections, with an emphasis on organic synthesis and technical methods rather than assay results.

1.2 What is a small molecule?

Considering that there is no strict definition, the term small molecule can be referred to any organic compound with a molecular weight of less than 1500 Da [22]. The cutoff limit of 1500 Da is arbitrary, as in the literature it is possible also to find this limit fixed on 900−1000 Da, but it is correlated to the ability of small molecules to rapidly diffuse across cell membranes and reach the intracellular sites of action [22]. Small molecules are compounds that alter the activity or the function of a biological target, by interacting with a biological macromolecule, such as DNA, RNA, and proteins [23], often in a selective and dose-dependent manner, showing a beneficial effect against a disease, or a detrimental one (such as in the case of teratogens and carcinogens). Small molecules can be naturally occurring or of synthetic origin and can have a variety of different applications beyond drugs, as pesticides [24] or as probes and research tools to perturb biological systems in order to identify and discover novel biological targets, such as in the field of chemical genetics [2,7−10,25,26]. In fact, they work rapidly, reversibly, and in tunable conditions depending on the concentration, in contrast with genetic approaches, so they are better probes to analyze complex biological networks. In pharmacology, the term "small molecule" is used to differentiate drugs below 1000 Da from all the other classes of larger and complex biologic drugs that include antibodies, peptides, nucleic acid-based compounds, cytokines, replacement enzymes, polysaccharides, and recombinant proteins.

Biologic drugs have been increasing over the last decade, thanks to the advances of biotechnology and analytical techniques. Although they have some advantages over small molecules, such as their high specificity and biocompatibility, they often suffer of poor Absorption, Distribution, Metabolism, and Excretion (ADME) properties, and the oral delivery route remains practically unattainable, as most of them are still delivered using subcutaneous injections. Also, they are much more expensive as compared to low-molecular-weight drugs, and their structural characterization and quality control is more challenging.

Small molecules still dominate the market, as more than 95% among the top 200 most prescribed drugs in 2018 are small chemical entities [27]. In Table 1.2, the first 15 small molecules of this list are reported. Despite that, in the list of 15 top selling drugs of 2018, only five are small molecules (Table 1.3), whereas all the rest are

Table 1.2 First 15 small molecules of top 200 most prescribed drugs in 2018.

Name	Structure	Biological effect
Lisinopril		ACE inhibitor, used in the treatment of hypertension and symptomatic congestive heart failure
Levothyroxine		Natural thyroxine analogue, used in the treatment of hypothyroidism
Atorvastatin		3-hydroxy-3-methyl glutaryl coenzyme A (HMG-CoA) reductase inhibitor, used in the treatment of hyperlipidemic diseases
Metformin		Antihyperglycemic agent, used in the treatment of type II diabete
Simvastatin		3-hydroxy-3-methyl glutaryl coenzyme A (HMG-CoA) reductase inhibitor, used in the treatment of hypercholesterolemia
Omeprazole		proton-pump (potassium-transporting ATPase alpha chain 1) inhibitor, used for the treatment of gastric acid-related disorders
Amlodipine		calcium channel blocker, used in the treatment of high blood pressure and angina
Metoprolol		Beta-1 blocker, used in the treatment of angina, heart failure, and hypertension
Paracetamol		Antipyretic and analgesic
Salbutamol		Beta-2 adrenergic receptor agonist, bronchodilator, used in the treatment of asthma and Chronic Obstructive Pulmonary Disease (COPD)
Hydrochlorothiazide		Diuretic, used in the treatment of edema, hypertension, and hypoparathyroidism
Losartan		Angiotensin-receptor blocker, used in the treatment of hypertension

Continued

Table 1.2 First 15 small molecules of top 200 most prescribed drugs in 2018.—*cont'd*

Name	Structure	Biological effect
Gabapentin		Increases the synaptic concentration of Gamma-AminoButyric Acid (GABA), anticonvulsant, used in the treatment of epilepsy
Sertraline		Serotonin-reuptake inhibitor, antidepressant
Furosemide		Diuretic, used in the treatment of edema

Table 1.3 Small molecules in the 15 top selling drugs of 2018.

Name	Structure	Company	Biological effect
Apixaban		Bristol-Myers Squibb and Pfizer	Anticoagulant, used for the treatment and the prevention of strokes and blood clots
Lenalidomide		Celgene	Chemotherapeutic agent, used for the treatment of multiple myeloma and other cancers
Nimodipine		Merck	Calcium channel blocker, used for the treatment of high blood pressure
Rivaroxaban		Bayer and Johnson & Johnson	Anticoagulant, used for the treatment of blood clots
Pregabalin		Pfizer	Inhibitor of neuronal activity used for the treatment of epilepsy and generalized anxiety disorder

biologic drugs, mainly because the high cost of producing and evaluating bio-pharmaceuticals reflects their high price of sales in the market and their high consumer cost [28].

On the other hand, small molecule drugs usually follow the Lipinski rule-of-five criteria or their variants, which allows for transcellular transport through intestinal epithelial cells. Usually they have no more than five hydrogen bond donors and no more than 10 hydrogen bond acceptors, an octanol-water partition coefficient (log P) below 5 and a molecular mass less than 500 Dalton. Such molecular weight cutoff is not a strict condition for oral bioavailability, as some antibiotics with a molecular weight up to 900 Da have been found to show digestive absorption ability. Also, in recent years, there has been an average increase of the molecular weight and of the number of nitrogen atoms of new chemical entities approved for marketing, so that scientists are claiming that conducting drug discovery campaigns in the "beyond rule of 5" (bRo5) chemical space can increase the chance of finding new medications [29]. All these properties allow small molecules to have better drug-like properties (solubility, permeability, metabolic stability, transportation, pharmacodynamic and pharmacokinetic properties), and more importantly higher stability (which reflects in an easier and longer conservation), cheaper preparation, and oral availability, although some of them are only absorbed if given as prodrugs.

1.3 Historical perspective

The roots of drug discovery go back thousands of years ago, as natural extracts and their preparations have been used for medicinal purposes for millennia (Fig. 1.2) [30]. However, the beginning of the modern era of drug discovery as "the traditional scientific discipline concerning the discovery, development, identification and interpretation of the mode of action of biologically active compounds at the molecular and cellular level" [31] can be defined only after 1930, when Dr. Paul Ehrlich demonstrated that infectious diseases can be treated with pure chemicals, instead of crude extracts of medicinal plants, and postulated the receptor concept.

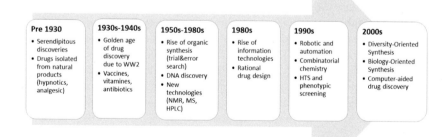

FIGURE 1.2

Schematic representation of the history of drug discovery.

After that, both pharmaceutical industry and academic laboratories started a run of discovery projects that went in parallel with the rise of synthetic organic chemistry, moved by the aspiration of finding a medication for all possible diseases, as synthesis offered the promise that if a drug could be envisioned, it could be made. This idea remained the dominant positive feeling for several decades, at least until the 1980s. In the same decades, some technologies, such as the Nuclear Magnetic Resonance (NMR) spectroscopy, mass spectrometry, High-Performance Liquid Chromatography (HPLC), and other purification methods, became available, thus giving a great impulse to organic synthesis, which assumed a leading role in refining and optimizing the activity of known drugs. Different analogues of known drugs were synthesized randomly and empirically, by adding different functional groups and appendages, thus providing a first empirical but successful way of the so-called lead optimization.

Fortunately, as soon as this positive spiral started giving signs of failure, advances in the field of biochemistry and molecular biology allowed the researchers to start thinking about the molecular targets behind some diseases and how to rationally design a molecular candidate for a specific target [32]. Starting from the 1980s, the attention began to shift away from trial-and-error searches to a more rational and predictive computer-aided drug design, thanks to the growth of computer processing power and the advances in structural biology [33,34].

Organic chemists began working in multidisciplinary teams, collaborating with many scientists from different fields, in order to synthesize only molecules with higher chance of possessing biological potency, good pharmacokinetic behavior, and low prediction of cytotoxic effects. However, despite some remarkable success of this rational approach (as explained in details in paragraph 1.3), this model showed to be not as productive as expected, and after an impressive growth at the end of the last century, the number of new molecular entities launched on the marked rapidly decreased in the recent past [35,36]. Thus, the research community experienced soon a return to empirical methods, with the so-called reverse pharmacology [37] or phenotypic drug discovery, that are the most frequently used approaches still today [32]. The renaissance of phenotypic drug discovery was driven by the advent of robotic and miniaturized techniques for the high-throughput screening of many compounds, and also by the development of new synthetic approaches for the generation of large compound collections, such as combinatorial chemistry [18,19] and solid-phase synthesis [14–17]. Strategies like the *split-and-pool* techniques made it possible to generate up to millions of compounds in a short period of time. Despite that, a new "ice age" of drug discovery was approached at the beginning of this century, mainly due to the lack of real chemical diversity in the industrial libraries that have been synthesized and screened [38]. In fact, combinatorial libraries often showed limited diversity, as most of them only vary for substituents around a common scaffold, which was also "flat" and structurally simple, whereas it has been demonstrated how a higher scaffold complexity and a more three-dimensional character are generally associated with a more successful outcome in drug discovery and development [39,40].

New strategies and new ideas were necessary to develop better high-quality chemical libraries, and in this view, diversity-oriented synthesis, biology-oriented synthesis, and activity-directed synthesis are providing new tools to explore underpinned and underrepresented areas of the chemical space. Also, developing the therapeutic agents of the future may require to keep up to date the landscape and the techniques involved in the chemistry aspects, also including entirely new approaches, by introducing new disciplines and new technologies in the process.

Today, an important part of modern medicinal chemistry is represented by computer-aided drug design, where the structure of the potential drug candidate is predicted *de novo* by a virtual screening of potential chemical substructures interacting with a target protein [41,42]. Also, computational methods, such as molecular modeling and dynamics simulations [43,44], are used to improve the potency and the drug-likeness, in replacement of experimental SAR studies [45,46].

One remarkable example of the evolution of drug discovery is presented by the history of the development of angiotensin-converting enzyme (ACE) inhibitors [47]. All the story started when a Brazilian postdoc joined the group of Dr. Vane, who was actively studying hypertension, bringing with him an extract obtained from a Brazilian viper venom, that contained a mixture of peptides, including teprotide (Fig. 1.3), that was found to be a potent inhibitor of the ACE enzyme, thus blocking the conversion of angiotensin I to angiotensin II and ultimately reducing the blood pressure. However, this peptide was not orally available and had to be administered intravenously. Thus, Dr. Laragh and Dr. Squibb started a long drug discovery project made of structure-activity studies that took 10 years to identify the essential pharmacophore responsible for the activity against ACE enzyme. In order to develop one low-molecular-weight small molecule inhibitor, they studied this zinc-dependent metalloprotease, in analogy to the already known carboxypeptidases A (CPA) enzyme. Considering that the Zn^{2+} cation coordinates the carbonyl group next to the second proline starting from the *C*-terminus, they synthesized and assayed several *N*-acylated tripeptides containing the proline as the last amino acid. From this SAR studies they discovered that:

- the molecule should end with a COOH to mimic the *C*-terminus of the natural substrate;
- the thiol group is a better zinc-binding group as compared to amides, hydroxamates, or carboxylates; and
- bulky groups between the proline ring and the zinc-binding group are not well tolerated.

From these studies, captopril was developed in the early 1980s as one of the most active ACE inhibitor. However, this drug has some side effects related to cutaneous eruption and loss of taste sensation, due to the presence of the thiol group. Thus, advance in the development of ACE inhibitors were obtained by replacing the thiol with a carboxylate and balancing this affinity drop by introducing hydrophobic and aromatic groups to address the S1 subsite of the catalytic cleft. Thus, some of the

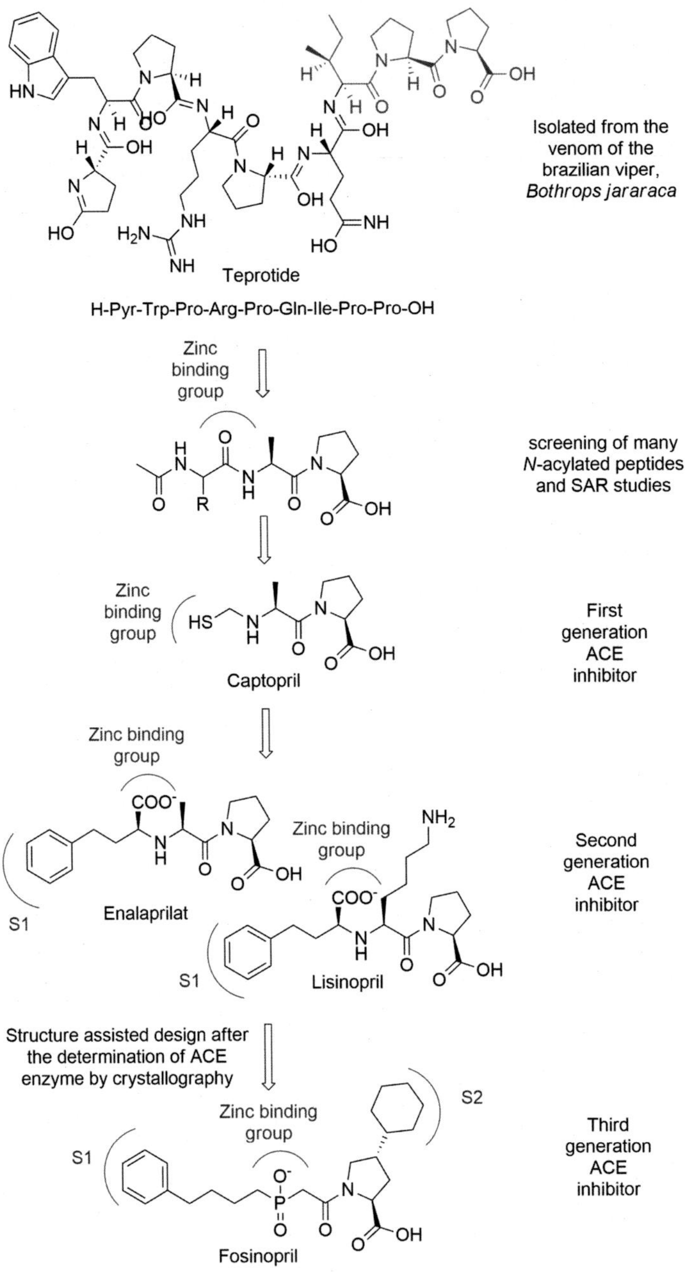

Isolated from the venom of the brazilian viper, *Bothrops jararaca*

Teprotide

H-Pyr-Trp-Pro-Arg-Pro-Gln-Ile-Pro-Pro-OH

Zinc binding group

screening of many *N*-acylated peptides and SAR studies

Zinc binding group

Captopril

First generation ACE inhibitor

Zinc binding group

Zinc binding group

Enalaprilat

Lisinopril

S1

S1

Second generation ACE inhibitor

Structure assisted design after the determination of ACE enzyme by crystallography

Zinc binding group

S1

S2

Fosinopril

Third generation ACE inhibitor

FIGURE 1.3

History of the development of angiotensin-converting enzyme (ACE) inhibitors.

most prescribed drugs around the world, including enalaprilat and lisinopril, were developed.

In 2003, the structure of the ACE enzyme was resolved by X-ray crystallography. The rational design of novel ACE inhibitors was then possible, taking advantage of the structural features of this enzyme. Just to give an example of this third generation of inhibitors, fosinopril contains a phosphinic acid group that coordinates the zinc cation strongly as the thiol group, but without provoking the same adverse effects.

1.4 Drugs from natural products

Nature has always been an extraordinary source of therapeutic agents, belonging to many different compounds classes, including glycosides, alkaloids, peptides, lipids, phenols, flavonoids, polyketides, and steroids. Medical preparations, obtained by drying or extracting active components from herbs and plants, have been used since the beginning and are used still today in the Ayurvedic medicine. But more importantly, nature has provided along these centuries a rich heritage of lead compounds for the development of drugs [48—51]. Historically, these compounds were discovered by serendipity, such as in the famous case of the antibiotic penicillin, discovered by Fleming in 1928 when working with the fungus *Penicillium notatum* [52]. Today, natural product research takes advantage of high-throughput screening, purification technologies, as well as chemoinformatic analysis [53,54]. Natural products can be used directly as drugs, as in the case of morphine, paclitaxel, and doxorubicin (isolated, respectively, from *Papaver somniferum,* from *Taxus brevifolia*, and from *Streptomyces peucetius*) (Fig. 1.4A). More often, they serve as lead compounds for the development of more complex (or more simplified) drugs, as in the case of aspirin, the world's best known analgesic agent, which is the acylated analogue of salicylic acid (isolated from *Salix alba)* [55], or ixabepilone, an orally available modified version of the natural product epothilone B (isolated from the myxobacterium *Sorangium cellulosum)*, which can be used as antitumor compound (Fig. 1.4B) [56]. Lastly, natural products can be used as tools to study biological mechanisms and identify novel molecular targets by the inhibition of key steps in biochemical processes in the so-called chemical genetics [57]. Just to give some examples, the abuse of nicotine by cigarette smoke opened the way to the studies of acetylcholine receptors, whereas the application of natural product-derived libraries in chemical genetics studies have recently revealed how the inhibition of yeast DNA topoisomerases by camptothecin (isolated from the tree *Camptotheca acuminate)* was fundamental for the correct eukaryotic DNA synthesis and repair (Fig. 1.4C) [58]. Also studies with artemisin, isolated from *Artemisia annua*, have helped in assessing the mode of action of this antimalarial compound by interfering with mitochondria function (Fig. 1.4C) [59].

These are the reasons why, still today, a third of FDA-approved drugs are natural products or their derivatives [50,60—62]. In fact, despite a period in which pharmaceutical industries decided to move back from the use of natural products (mainly

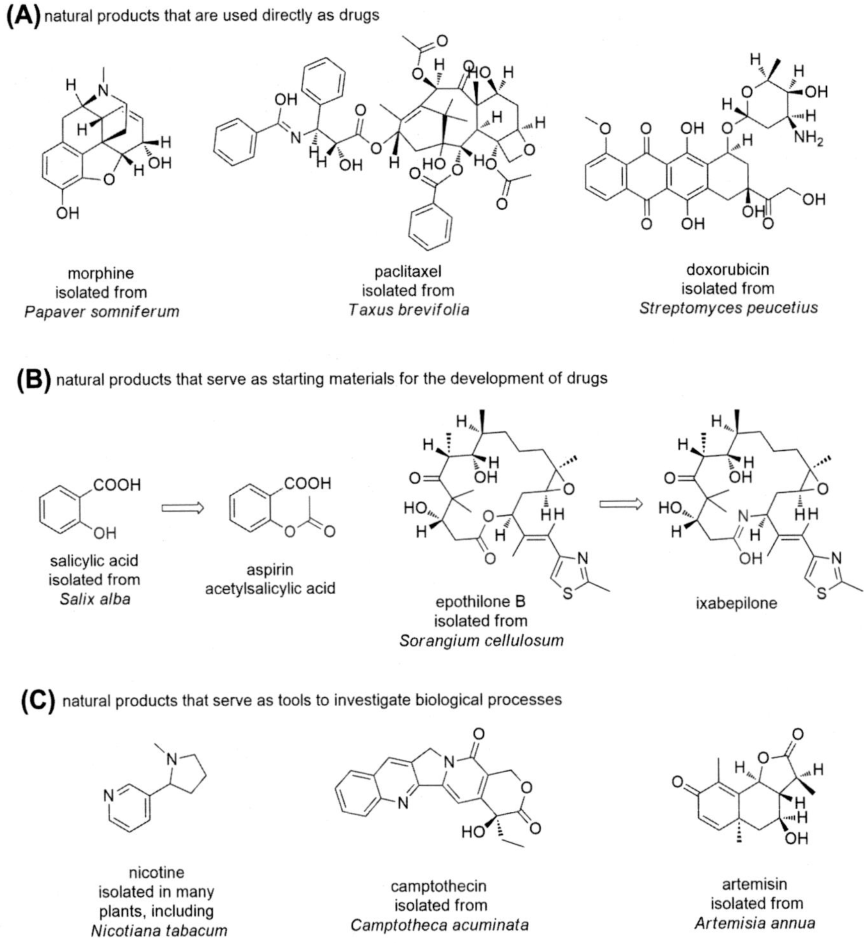

(A) natural products that are used directly as drugs

morphine
isolated from
Papaver somniferum

paclitaxel
isolated from
Taxus brevifolia

doxorubicin
isolated from
Streptomyces peucetius

(B) natural products that serve as starting materials for the development of drugs

salicylic acid
isolated from
Salix alba

aspirin
acetylsalicylic acid

epothilone B
isolated from
Sorangium cellulosum

ixabepilone

(C) natural products that serve as tools to investigate biological processes

nicotine
isolated in many
plants, including
Nicotiana tabacum

camptothecin
isolated from
Camptotheca acuminata

artemisin
isolated from
Artemisia annua

FIGURE 1.4

Representative examples of natural products that serve directly as drugs (A), as starting materials for the development of drugs (B) or as tools to investigate biological processes (C).

because of the difficulties in their access and purification, their inherent structural complexity, and the concerns about intellectual properties), natural products have been re-established in recent years as one the most valuable source for drug discovery projects. In fact, they show an intrinsic chemical diversity (which is correlated to the large variety of biological outcomes), a good compliance to Lipinski's rule of five for orally available compounds (which is accepted for most of natural compounds, although some of them possess higher molecular weights, more rotatable bonds, and more stereogenic centers). Typically, they have scaffolds with two to four rings, with resulting Van der Waals volumes that match the size of binding cavities of proteins, and possess several functional groups that can be used for the attachment of further substituents.

Most of the technical drawbacks associated with natural product research have been now solved, thanks to newest synthetic approaches for the development of chemical libraries (such as solid-phase techniques) and novel biotech methods that have increased the availability of natural product compounds, by conveniently producing them from bacteria or yeasts. Thus, many papers have appeared in the literature regarding the combinatorial chemistry starting from natural product-derived templates. Specifically, biology-oriented synthesis has been conceived for the purpose of taking inspiration from the conservative evolution of nature to select the most promising molecular scaffolds for the development of high-quality chemical libraries. In the future, there is hope that natural products, or their simplified core scaffolds, will be more often taken into account for the development of high-quality chemical libraries.

1.5 Rational design of small molecule drugs

In contrast to traditional trial-and-error methods of drug discovery, rational drug design starts from the identification of the cellular or molecular structure that is involved in the development of the disease or pathologic condition [63]. Although it is estimated that there are more than 5000 possible targets, the whole drug discovery research is focused only on less than 500 targets [64,65]. This is because "druggable" targets, such as protein kinases, G Protein-Coupled Receptors (GPCRs), and enzymes, are easy to be modulated with the use of small molecules [66], and they have gained a wide experience in the pharmaceutical industry, whereas more challenging targets, such as protein-protein interactions, transcription factors, or epigenetic targets, have proven to be very difficult to study with traditional approaches. Also, the identification of targets in complex disorders such as cancer or infectious diseases is challenging, as in these systems the pathological status is often defined by a complex interaction between diverse biomolecules [67].

Once a target is identified and validated, a huge number of compounds are prepared following the principles complementarity in shape and charge to the biomolecular target and then tested by the use of high-throughput screening campaigns [68]. Often, many compounds show similar activity, as they share common chemical features. Starting from their structures, it is possible to define a common pharmacophore, and by applying iterative structure-activity relationship (SAR) studies, the activity and the selectivity against the chosen target, as well as the drug-likeness of the molecule, can be improved to move them toward *in vitro* and *in vivo* testing. This process is called "lead optimization."

The rational design approach has been reevaluated with the advent of chemoinformatic methods. Thus, such approach nowadays is identified by the term "computer-aided drug design" [69] or "structure-based drug design" [70]. In fact, virtual screening, both in ligand- and in receptor-based approaches, can be used

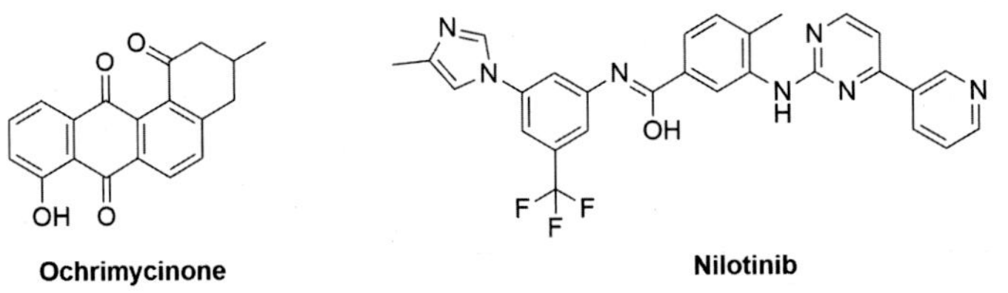

Ochrimycinone   **Nilotinib**

FIGURE 1.5

Representative examples of small molecule drugs identified by virtual screening against known biomolecular targets, specifically ochrimycinone (ligand of the Stat3 protein) and nilotinib (inhibitor of Bcr-Abl tyrosine kinase enzyme).

as a tool to improve high-throughput screening to increase the chance of discovering new lead candidates [71]. Also, computational methods are used to predict the drug-likeness of compounds and select only those that show good ADME and toxicological profiles [72,73].

Recent success of this approach is represented by the identification of ochrimy-cinone by virtual screening against the signal transducers and activators of transcription 3 (STAT3), a target that play an important role in human breast carcinoma [74], as well as of nilotinib as a potential kinase inhibitors in treatment of chronic myeloid leukemia (Fig. 1.5) [75].

1.6 Combinatorial chemistry and DNA-encoded libraries

The term combinatorial chemistry appeared in the literature in the early 1990s when, in two separate papers, Lam [76] and Houghten [77] described the preparation of large combinatorial collections of peptides. However, early examples of large peptide libraries prepared with the general principles of combinatorial chemistry can be found since the 1980s, although these methods were called differently, with the terms multipin or tea-bag technology [78]. Although combinatorial chemistry has been applied predominantly for the preparation of peptide libraries, this concept was also applied to the synthesis of small molecules libraries [15−20] and used for both lead discovery and optimization [79−81]. In fact, thanks to the high grade of automation, thousands different structures per day can be generated, enabling the rapid concurrent high-throughput screening of huge libraries against specific drug targets.

The basic principle of combinatorial chemistry is to allow reacting a large number of different *building blocks* in a mixture, by using versatile and efficient coupling reactions, such as amide bond formation or multicomponent reactions. The application of multicomponent reactions, such as the Ugi four-component reaction (Fig. 1.6), is particularly useful in this context, as starting from only 10 different carboxylic acids, 10 different amines, 10 different aldehydes, and 10 different nitriles, up to 10,000 diverse products can be obtained.

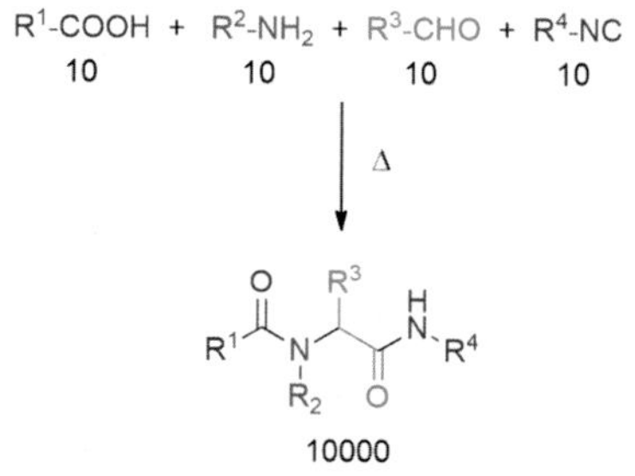

R¹-COOH + R²-NH₂ + R³-CHO + R⁴-NC
   10   10   10   10

$$R^1\text{-COOH} + R^2\text{-NH}_2 + R^3\text{-CHO} + R^4\text{-NC}$$

Δ

10000

FIGURE 1.6

Potential different products that can be achieved using a Ugi-4CC reaction starting from 10 carboxylic acids, 10 amines, 10 aldehydes, and 10 nitriles.

Parallel synthesis has been extremely helpful in developing fast methods to get a great variety of compounds with simple and automatic purification processes [15,82]. It can be achieved manually or through automation, in solution or using solid supports. The structures that are obtained are known in advance, and this technique is particularly relevant for the lead optimization process, especially if applied in combination with computational methods and prediction tools. In the *one-bead/ one-compound* approach, the compounds are prepared by using the *split-and-pool* technique [15,76,77]: starting materials are assembled on microbeads and divided in different batches (*split*); each bead is made reacting with the subsequent building block and the products are mixed again (*pool*). The resulting mixture is again split, different reagents are added, and then mixed again (Fig. 1.7). In this way, the number

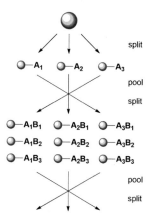

FIGURE 1.7

Solid-phase *split-pool* combinatorial technique.

of products increases exponentially with the number of building blocks and the number of steps, affording thousands of structures in a single mixture. However, after the cleavage, a complex mixture containing a large number of different products is obtained. This mixture of products is submitted to high-throughput screenings. Once a positive bead is found, the mixture need to be decoded in order to identify the hit compound. The identification of hit structures can be obtained by using affinity selection, indexed libraries, deconvolution methods, and positional screenings. Subsequently, the synthesis of the hit compounds can be optimized and the structures modified to get better properties in a hit-to-lead process.

In Fig. 1.8, it is reported a recent example of a combinatorial library of peptide-like compounds obtained starting from small building blocks and using a "synthetic fermentation" method in water that exploits the α-KetoAcid-HydroxylAmine (KAHA) amide-forming ligation, without the need of organisms, enzymes, or other chemical reagents [83]. The α-ketoacid initiator (I) reacts with the isoxazolidine elongation monomer (M) to produce oligomers that are accomplished by a terminator (T). By adjusting the addition order of the monomers and the reaction conditions, the authors obtained up to 6000 structurally different products, starting from only 23 building blocks that were screened directly for biological activity without the need of isolation or workup purification. From this assay, a hepatitis C virus NS3/4A protease inhibitor was identified and characterized.

Combinatorial chemistry includes a wide variety of techniques, including phage-display, yeast-display, and bacteria-display methods [84] and mRNA-display

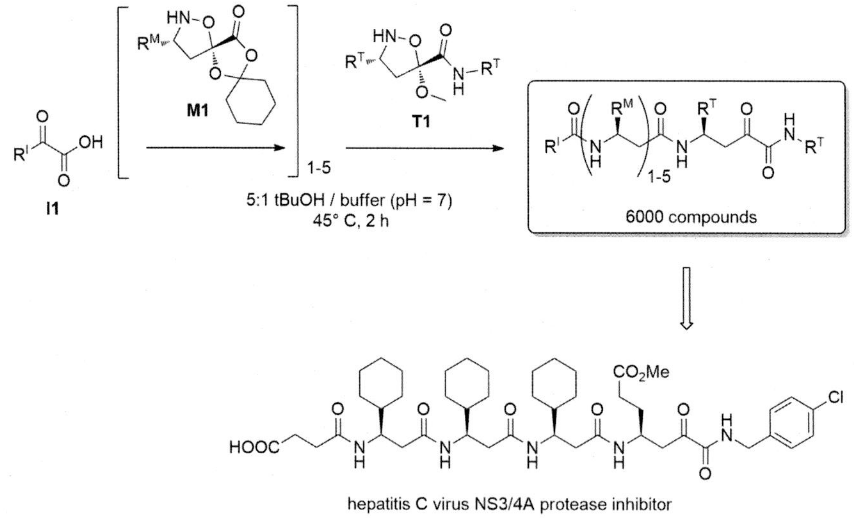

hepatitis C virus NS3/4A protease inhibitor

FIGURE 1.8

Schematic representation of a combinatorial library of peptide-like compounds obtained from small building blocks and using a "synthetic fermentation" method exploiting the α-KetoAcid-HydroxylAmine (KAHA) amide-forming ligation.

methods [85,86] that are more relevant for the preparation of peptide libraries. DNA-encoded chemical libraries (DECLs) are a combinatorial chemistry approach that exploits molecular biology technologies to create large collections of small molecules, peptides, or macrocyclic compounds. This approach consists of coupling building blocks to short DNA fragments that serve as identification codes. Then, the *split-and-pool* technique can be applied, although the coupling chemistry that can be used is limited, due to the mild reaction conditions required by the oligonucleotide tags. In this context, Litovchick and coworkers have recently made some advances, as they developed a chemical ligation based on the ability of the Klenow fragment of DNA polymerase to accept templates with triazole linkages in place of phosphodiesters, thus expanding the scope and diversity of the chemistry suitable for DECLs (Fig. 1.9) [87]. Specifically, in the first cycle, the starting material **S3,** obtained by the click reaction of **S1** and **S2,** was split and made reacting with 2259 primary amines by reductive amination. After removing the TIPS group, the material was split into 666 wells, tagged by CuAAC with different azido-derived compounds, and then acylated with 666 bromoarylcarboxylates. Finally, in the fourth cycle, 667 boronic acid/boronate ester building blocks were installed by Suzuki cross-coupling, thus obtaining a collection of 334 million compounds. The library was subjected to a high-throughput screening assay against the epoxide hydrolase (sEH) enzyme, a target involved in the pathophysiological development of diabetes and several cardiovascular diseases, identifying the building blocks responsible for the higher affinity toward this enzyme.

1.7 Diversity-oriented synthesis

As mentioned above, combinatorial chemistry has not fulfilled the expectations that were upon its shoulders [21,22,88]. Despite the huge investments that pharmaceutical industries put to equip their labs with automatic and robotic apparatus for this technique, the number of new molecular entities launched on the market did not increase. This was mainly due to the lack of diversity within the combinatorial libraries [89] and also to the flatness and the modest complexity of the resulting compounds. These collections can access only a limited area of the chemical space, as they are characterized by a relative narrow range of properties, whereas it has been pointed how expanding the region of the chemical space explored by a compound collection is essential for the discovery of both new ligands and new targets outside the so-called druggable genome [4]. The chemical space, the n-dimensional space in which each molecule is represented as a spot where the coordinates are its physicochemical properties, is the most reliable chemoinformatic way to measure the chemical diversity. The synthetic, combinatorial library compounds seem to cover only a limited and quite uniform area of the chemical space, whereas existing drugs and particularly natural products exhibit much greater chemical diversity, distributing more evenly to the chemical space. Library size is not as important as the quality of it, in terms of structural complexity and diversity, to increase the

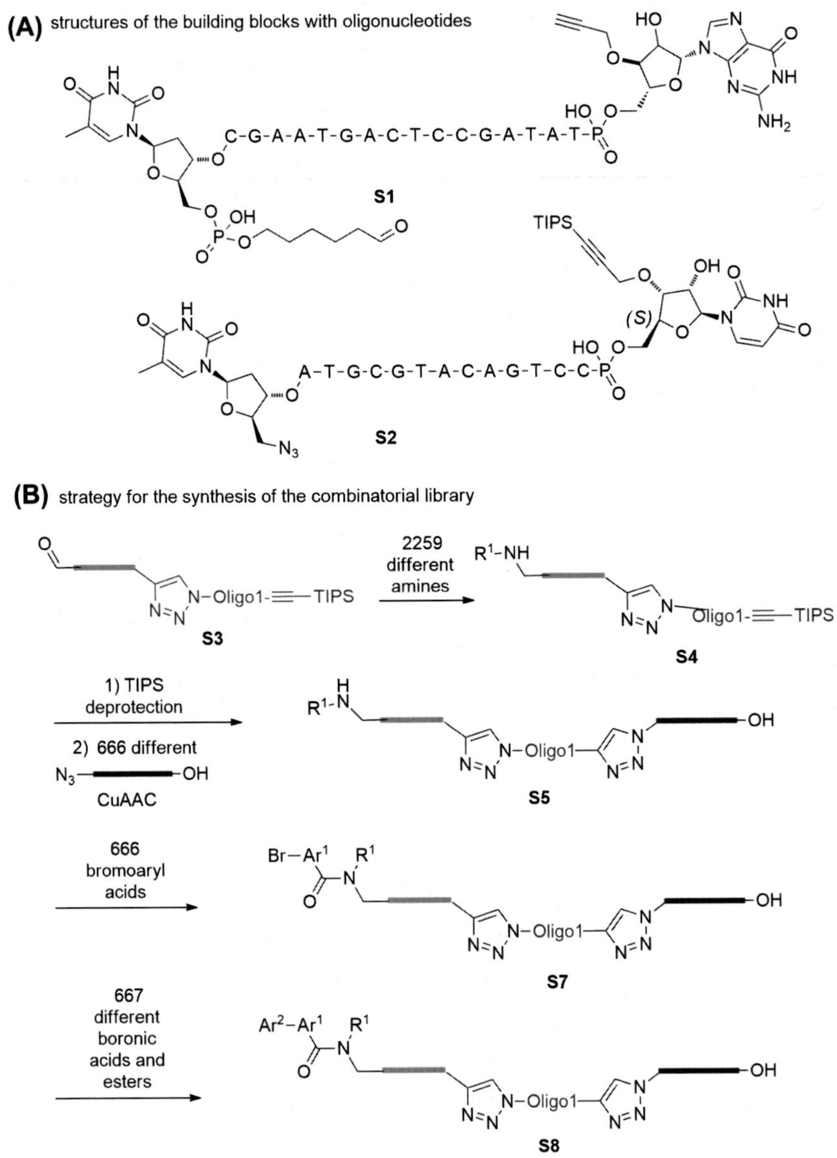

FIGURE 1.9

Oligonucleotide-derived building blocks S1 and S2 (A) used for the combinatorial synthesis of a DNA-encoded chemical library of 334 million members (B)

likelihood of discovering small molecule modulators for a broad range of biological targets [90–93].

In this context, diversity-oriented synthesis (DOS) has proven to be very effective to explore a large area of the chemical space, as it is an efficient synthetic strategy for the creation of functionally diverse small molecule libraries, primarily through the creation of novel molecular scaffolds [94,95]. This term was coined by Stuart Schreiber in 2000 [96], although some of the idea behind this approach were already present in the literature since the 1990s. The definition of diversity-oriented synthesis as the "deliberate, simultaneous and efficient synthesis of more than one target compound in a diversity-driven approach to answer a complex problem" was subsequently introduced by Spring [97], where complex problems were primarily referred to the discovery of both novel small molecules modulators and targets, but they can also be extended to the discovery of novel ligands, catalysts, or monomers for new materials.

DOS works in the opposite direction to the more traditional target-oriented synthesis and combinatorial chemistry. In fact, while in both these two approaches a target structure is in mind at the beginning of the synthesis and the synthetic pathways are designed with the use of retrosynthetic analyses, in DOS the syntheses are planned in forward direction, with the aim of obtaining as more compounds as possible, divergently, starting from simple starting materials and by using no more than five synthetic steps. The ideal DOS strategy is aimed to reach the chemical diversity in the most efficient manner possible [10], as the synthetic efficiency and the shortness of the procedures are required for two main reasons: first, the biological potential of the synthesized compounds is not known in advance, so the synthesis should be cheap and effective, as most of the compounds probably do not have a future in drug discovery; secondly, once interesting activities of any compounds are ascertained, it should be easy and convenient generating focused libraries around the most interesting hit structures. For this reason, DOS takes advantage of starting materials and intermediates that has to be chosen in order to maximize the potential diverse reactivity in the synthetic sequence. Also, it exploits multicomponent reactions, complexity-generating reactions, and product-substrate relationship in order to build quickly diverse and complex molecular skeletons.

With the term diversity, three different types of diversity are defined:

1) *Appendage diversity (or building block diversity)*: it is the simplest way to generate diversity, and it consists of decorating functional groups and reactive centers all around a common skeleton (Fig. 1.10A) [98];
2) *Stereochemical diversity*: it is the variation in the spatial orientation of the functional groups and potential interacting elements on the skeleton. It is an important type of diversity, as different stereoisomers have often different interaction toward biological systems (Fig. 1.10B) [99];
3) *Skeletal diversity*: it is the presence in the library of diverse and complex skeletons. This is the most important type of diversity as the overall shape of the molecule is a key factor in controlling the biological outcome of the entire

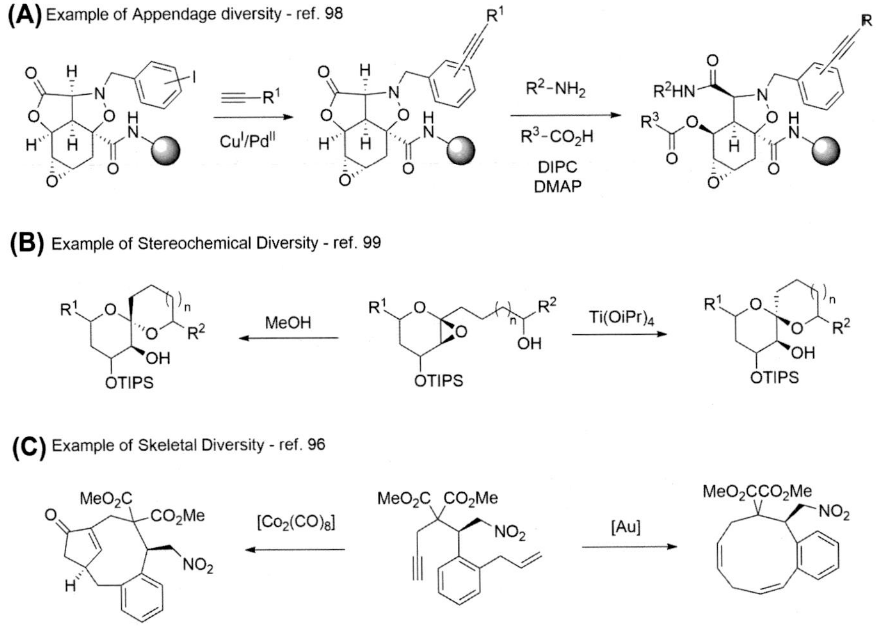

(A) Example of Appendage diversity - ref. 98

(B) Example of Stereochemical Diversity - ref. 99

(C) Example of Skeletal Diversity - ref. 96

FIGURE 1.10

Representative examples of the generation of appendage (A), stereochemical (B), and skeletal (C) diversity.

molecule. Biological targets will interact with a certain molecule only if it has a complementary three-dimensional binding surface [39,40]; thus, the molecular shape diversity and the structural complexity are the most important factors in determining the quality of the resulting library and its potential success in subsequent biological screenings (Fig. 1.10C) [96].

Skeletal diversity can be achieved in two main approaches:

1) a *reagent-based approach (or branching approach)*: this is the application of different reagents or different reaction conditions on the same substrate to the synthesis of skeletally distinct compounds. Such approach is usually obtained by exploiting two methods: (1) the use of densely functionalized molecules where orthogonal functional groups can be transformed selectively; (2) the use of pluripotent functional groups that can take part to diverse reactions under different experimental conditions.

2) a *substrate-based approach (or folding approach)*: where different starting materials lead to different products under the same reaction conditions thanks to the presence of σ-elements, that bring distinct "preencoded" skeletal information. Such method is usually based on intramolecular reactions that combine functional groups strategically positioned on different points of the starting materials.

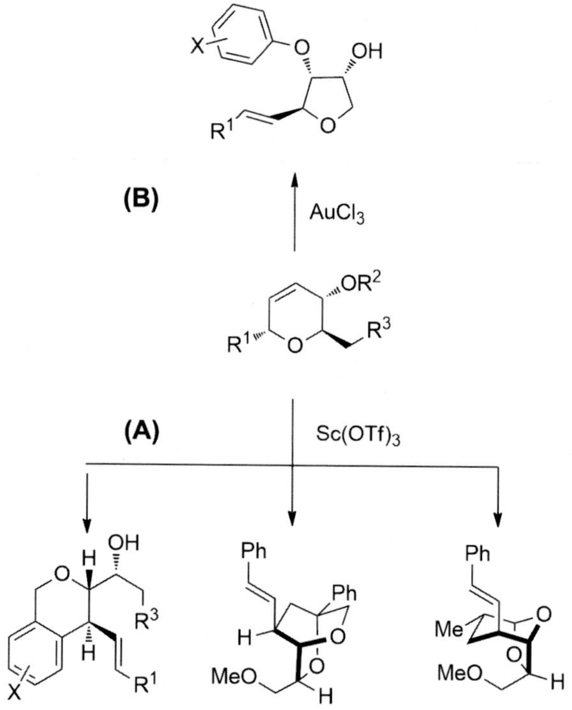

FIGURE 1.11

(A) Substrate-based and (B) reagent-based approach for the generation of skeletal diversity starting from a glycal derivative.

In Fig. 1.11, one example for each approach is reported, specifically applied to glycal-type carbohydrate-derived building blocks [100,101].

A DOS campaign can incorporate both approaches, and usually substrate-based approaches are applied at later stages in order to *fold* compounds into different molecular skeletons. This is the main idea behind the build/couple/pair strategy, developed by Schreiber in 2008 [102]. Such concept consists to *build* the starting materials, with several orthogonal protecting groups suitable for the subsequent steps; *couple* them to a "multipotent" polyfunctional intermediate and finally *pair* the appropriate functional groups by intramolecular reactions in order to generate different skeletons. All the process should count from three to five steps, avoiding unnecessary protection/deprotection steps and exploiting highly selective, versatile, and efficient reactions, both in terms of yield and atom economy. In Fig. 1.12, one example of a build/couple/pair strategy is reported starting from a modified tripeptide obtained from the assembly of three building blocks, which was subjected to different pairing reactions involving the nucleophilic addition to an iminium ion [103–105].

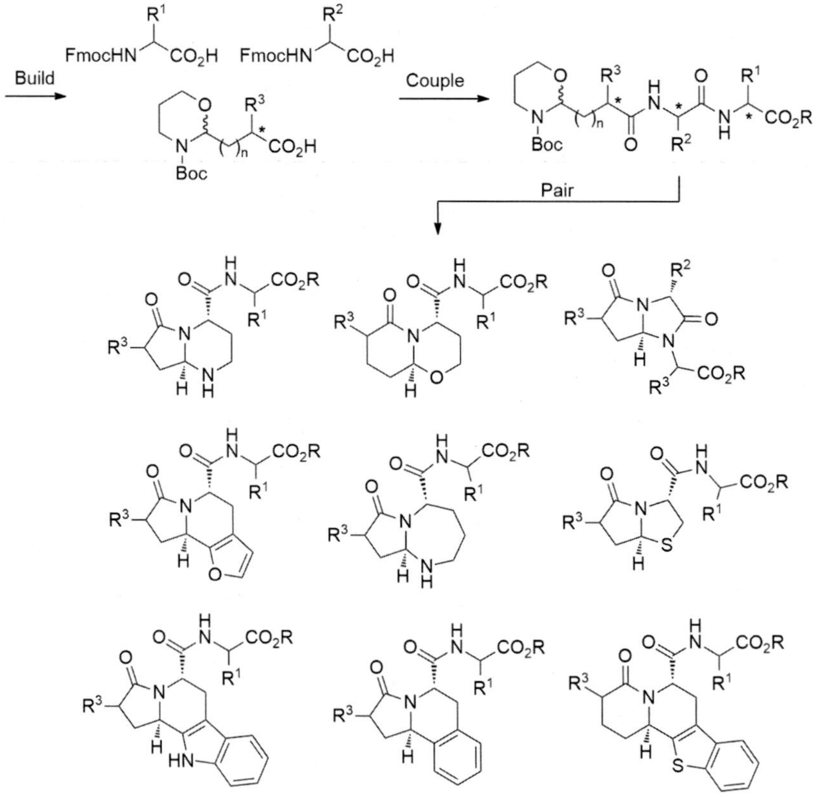

FIGURE 1.12

Build/couple/pair strategy starting from amino acid-derived building blocks.

The synthesis of high-quality chemical libraries (or "prospecting libraries," as defined in 1997 by Spaller) [106], which address novel and vast areas of the chemical space and contain high levels of structural diversity, has proven to be very effective in drug discovery [107,108]. A number of DOS libraries have already given as output some interesting novel small molecule modulators. Just to give some examples, robotnikinin (Fig. 1.13) is a potent Sonic Hedgehog inhibitor, discovered from the screening of a DOS library of 2070 amino alcohol-derived macrocycles with the small molecule microarray (SMM) technology [109], and emmacin is an antibacterial compound found upon the high-throughput phenotypic assay of a DOS library of 223 compounds [110,111]. As an evidence of the importance of the quality of the library above the size, the application of a small DOS collection of only 35 compounds but characterized by the presence of 10 distinct molecular scaffolds, resulted in the identification of dosabulin, an inducer of the cell mitosis arrest able to demolish the tubulin network, in a microscopy-based phenotypic screen [112].

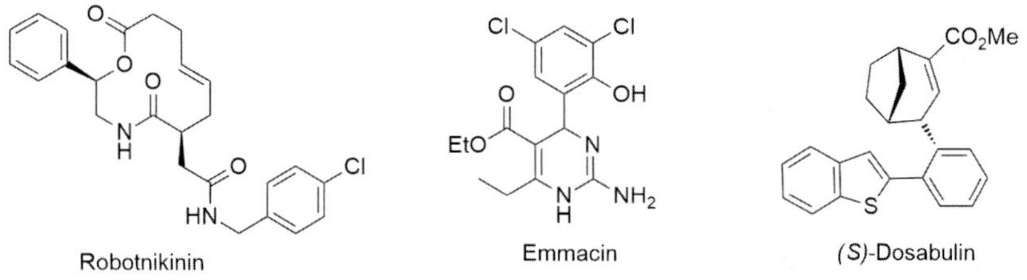

Robotnikinin   Emmacin   (S)-Dosabulin

FIGURE 1.13

Representative examples of small molecule modulators identified by the application of DOS libraries into phenotypic or high-throughput screening campaigns: robotnikinin, a Sonic Hedgehog inhibitor; emmacin, an antibacterial compound; and dosabulin, an antimitotic compound.

However, since it is estimated that the number of small molecules populating all the drug-like chemical space is more than 10^{60}, it will be physically impossible to achieve and study all the possible small molecules present in the chemical universe [113−115]. Thus, novel approaches are gaining wide attention to address only those areas of the chemical space that are enriched with biologically relevant compounds. The most relevant ones, biology-oriented synthesis [116,117] and activity-directed synthesis [118,119], aim to generate focused libraries based, respectively, around the core structures of natural products and of known biologically active molecules.

1.8 Biology-oriented synthesis

Biology-oriented synthesis (BIOS) [116,117] was introduced in the scientific community by Waldmann in the late 1990s, in response to the failure of pharmaceutical industry drug discovery campaigns that had removed natural products from their research. The main idea behind BIOS is that natural products possess molecular scaffolds that are enriched in biochemical and biological activity, considering how conservative has been the nature in the evolution of the protein binding sites. Thus, a driving force is offered to develop new small molecule modulators by the analysis and the chemoinformatic classification of the diversity of both protein binding sites and natural products. Two main complimentary approaches have been developed in BIOS by Waldmann's research group:

1) *Structural classification of natural products (SCONP)* [120]: it is the application of the scaffold tree concept (i.e., the hierarchical classification of molecular frameworks as defined by Murcko) [121] to all the compounds contained in the Dictionary of Natural Products (DNP). This classification allows to identify scaffolds that are already validated by nature in evolution and may be more promising for the development of small molecules able to target specific enzymes. As an example, in Fig. 1.14, it is reported the solid-phase BIOS synthesis of a library of 250 compounds based on the tetracyclic indolo [2,3]

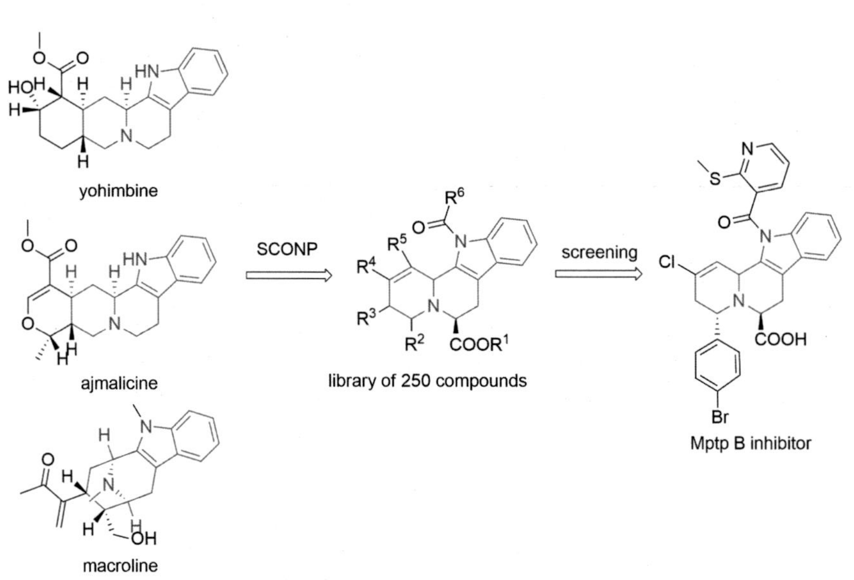

FIGURE 1.14

Development of a BIOS library based on the tetracyclic indolo [2,3]quinolizidines scaffold, identified by a structural classification of natural products (SCONP) analysis of different alkaloids.

quinolizidines scaffold, identified by an SCONP analysis of different alkaloids, including yohimbine, ajmalicine and macroline. From this library, novel inhibitors of several protein phosphatases (including VE-PTP, Shp-2, PTP1B, MptpA, MptpB, Cdc25A and VHR) were identified, showing higher potency as compared to the natural products itself [122,123].

2) *Protein structure similarity clustering (PSSC)*: this approach consists in analyzing the structural similarity and the evolutionary conservation of protein structures, with a focus on their binding sites, in order to identify the most promising ligand structures for a protein of interest. For example, the PSSC analysis performed to search enzymes with a catalytic site analogue of those of acyl protein thioesterase 1 (APT1) revealed that the dog gastric lipase has a high structural similarity to this enzyme, especially regarding the spatial arrangement of the catalytic amino acid residues. Starting from this observation, the authors generated a small library of lactone compounds similar to tetrahydrolipstatin (Orlistat), a known lipase inhibitor, and screened these compounds against APT1, finding that palmostatin B competitively inhibits APT1 with submicromolar activity (Fig. 1.15).

Both these two approaches can inspire the selection of compound library scaffolds on the basis of their relevance to nature and can be applied either separately or synergistically, by showing the similarity of proteins involved in the biosynthesis

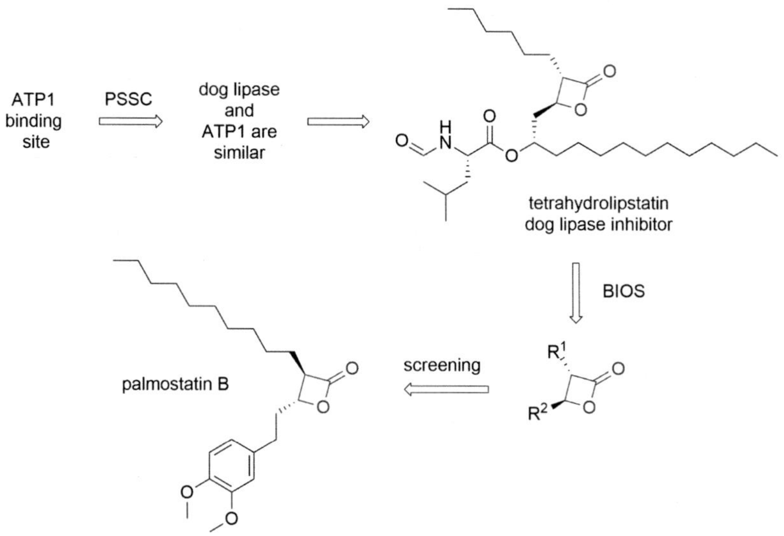

FIGURE 1.15

Identification of a novel acyl protein thioesterase 1 (ATP1) inhibitor, from a DOS library developed by the protein structure similarity clustering (PSSC) analysis of enzymes similar to ATP1.

of natural products [124]. A representative entry of the latter approach, the comparison of the protein fold topology of different enzymes involved in the biosynthesis of chalcones and flavonoids, namely chalcone synthase (CHS), chalcone isomerase (CHI), and anthocyanidin synthase (ANS) showed the similarity of these three enzymes with the catalytic site of phosphoinositide-3-kinase (PI3K), thus revealing why chalcones act also as kinase inhibitors [125].

BIOS concepts have been extended beyond natural products and protein ligands, to the screening of all those databases that include publicly available bioactivity data [126], such as ChEMBL and PubChem, in order to use these data to drive the synthetic efforts toward areas of the chemical space related to difficult targets. In the years to come, chemoinformatics and bioinformatics will become more and more important for the development of small molecule collections, because the systematic mining of currently available bioactive compounds possessing a precise chemotype can be used to discover novel underpinned biologically relevant regions of a defined space. Although the synthesis of complex natural products requires the application of more challenging chemical methods, in terms of synthetic steps and reaction feasibility, the investments in the chemical synthesis is perfectly balanced by the smaller size of the library needed and also by the highest rate of chance of finding hit compounds. Also, to overcome these challenges, novel methods are based on the synthesis and screening of natural products−derived fragments, that are easier to be prepared and purified. Fragments, as defined by Congreve's rule of three [127], have molecular weight below 300, less than three hydrogen bond donors and acceptors, and ClogP below 3. These limits act as guidelines, because the lead optimization process usually results in an increase in size, hydrophobic content, and stereochemical complexity. Also, fragment-based screening campaigns often give ligands that only weakly bind to the biological target, so that researchers often need to combine two or three fragments in order to produce a lead with a higher affinity [128]. In this context, Over and coworkers fragmented more than 200,000 natural products, achieving 2000 clusters of natural product—derived fragments [129]. Besides all the other properties in common with fragments [130], these NP-derived fragments are stereochemically complex and sp^3-rich, thus showing a higher three-dimensional complexity, a crucial aspect for binding to biomacromolecules. The relevance of this NP-derived fragments library was confirmed by the identification of numerous ligands of p38α MAP kinase and of several phosphatases, when the library was subjected to high-throughput screening assays.

1.9 Conclusions and future outlook

Despite the vast success of the so-called biologic drugs (monoclonal antibodies or recombinant proteins), the favorable pharmacokinetic properties of small molecules allow to keep them as the gold standard for the development of new drugs. Small molecules still dominate the market, being more than 95% of the top 200 most prescribed drugs in 2018, mostly because they are orally available and less expensive in terms of industrial preparation and purification. However, considering the low number of new molecular entities launched on the market in recent years and the difficulties in finding new lead compounds for all those "undruggable" targets, the entire drug discovery process has had to change a lot during the last decades. From one hand, the rational design of ligands is still a powerful approach, especially in combination with computer-aided methods when the biological target is well defined and structurally known. On the other hand, new synthetic methods able to generate high-quality chemical libraries are being developed in order to meet the need of improving the quality and quantity of small molecules for biological screenings. The relevance of this need is also highlighted by the rise of international initiatives, such as EU-OPENSCREEN or the European Lead Factory, that are continuously asking for new molecules. From the massive synthetic efforts characterized by the trial-and-error approach of the 1980s and the combinatorial chemistry of the 1990s, new attitudes are now gaining wide attention in the organic chemistry community, in order to maximize the quality of libraries and reducing the waste of generating and screening random unnecessary compounds. New frontiers in the synthesis of small molecules libraries have been explored. In particular, diversity-oriented synthesis has proven to be very effective to access large areas of the chemical space, primarily through the creation of many distinct molecular scaffolds. Also, biology-oriented synthesis has been conceived with the purpose of taking inspiration from nature to select promising molecular scaffolds being related to natural products in terms of biological output. Today, an important part of modern medicinal chemistry is represented by computer-aided methods, not only for rational drug design (i.e., virtual screening), but also for the smart design of small molecules libraries [131]. As the number of publicly accessible biological data is rapidly increasing, chemoinformatics represent a real opportunity for developing better chemical libraries. Today, we have access to millions of data about the biological activity of thousands of compounds and this big amount of data represents an incredible resource, although still underexplored. As the random and uneven exploration of the chemical space not always resulted in improving the success of screening campaigns, in the future, there is hope that organic chemists will take advantage more often of bioactivity databases and chemoinformatics tools, to drive the synthetic efforts toward those areas of the chemical space that have not yet been explored or that can be more biologically relevant. We are experiencing exciting new trends at the interface between computational and organic chemistry [132]. Chemoinformatics will become increasingly important for assisting organic synthesis, as novel tools are developed to assess the lead-likeness of molecular scaffolds, to systematic charting bioactive compounds possessing a precise chemotype, and to design new molecules by artificial intelligence. Chemoinformatic methods can also help organic chemists in rethinking synthetic pathways, to minimize the need of protecting groups, and to maximize the compound diversity, as a step forward in trying to explore the vast universe of small molecules, hoping that the number of new molecular entities launched on the market will continue to increase in the next years, as for 2018.

References

[1] A. Mullard, 2018 FDA drug approvals, Nat. Rev. Drug Discov. 18 (2019) 85–89.

[2] F. Cong, A.K. Cheung, S.-M.A. Huang, Chemical genetics-based target identification in drug discovery, Annu. Rev. Pharmacol. Toxicol. 52 (2012) 57–78.

[3] D.S. Tan, Diversity-oriented synthesis: exploring the intersections between chemistry and biology, Nat. Chem. Biol. 1 (2) (2005) 74–84.

[4] K.H. Altmann, J. Buchner, H. Kessler, F. Diederich, B. Krautler, S. Lippard, R. Liskamp, K. Muller, E.M. Nolan, B. Samorì, G. Schneider, S.L. Schreiber, H. Schwalbe, C. Toniolo, C.A.A. van Boeckel, H. Waldmann, C.T. Walsh, The state of the art of chemical biology, Chembiochem 10 (1) (2009) 16–29.

[5] W.P. Walters, M. Namchuk, Designing screens: how to make your hits a hit, Nat. Rev. Drug Discov. 2 (4) (2003) 259–266.

[6] D.P. Ryan, J.M. Matthews, Protein-protein interactions in human disease, Curr. Opin. Struct. Biol. 15 (4) (2005) 441–446.

[7] B.R. Stockwell, Chemical genetics: ligand-based discovery of gene function, Nat. Rev. Genet. 1 (2) (2000) 116–125.

[8] D.P. Walsh, Y.-T. Chang, Chemical genetics, Chem. Rev. 106 (6) (2006) 2476–2530.

[9] R.J. Spandl, A. Bender, D.R. Spring, Diversity-oriented synthesis; a spectrum of approaches and results, Org. Biomol. Chem. 6 (7) (2008) 1149–1158.

[10] S.L. Schreiber, The small-molecule approach to biology, Chem. Eng. News 81 (2003) 51−61.

[11] http://www.eu-openscreen.eu/.

[12] https://www.europeanleadfactory.eu.

[13] A. Karawajczyk, F. Giordanetto, J. Benningshof, D. Hamza, T. Kalliokoski, K. Pouwer, R. Morgentin, A. Nelson, G. Müller, A. Piechot, D. Tzalis, Expansion of chemical space for collaborative lead generation and drug discovery: the European Lead Factory Perspective, Drug Discov. Today 20 (11) (2015) 1310−1316.

[14] B.A. Bunin, J.A. Ellman, A general and expedient method for the solid-phase synthesis of 1,4-benzodiazepine derivatives, J. Am. Chem. Soc. 114 (27) (1992) 10997−10998.

[15] S.H. De Witt, J.S. Kiely, C.J. Stankovic, M.C. Schroeder, D.M. Reynolds Cody, M.R. Pavia, Diversomers": an approach to nonpeptide, nonoligomeric chemical diversity, Proc. Natl. Acad. Sci. USA 90 (15) (1993) 6909−6913.

[16] R.B. Merrifield, Solid phase peptide synthesis. I. The synthesis of a tetrapeptide, J. Am. Chem. Soc. 85 (14) (1963) 2149−2154.

[17] F. Camps, J. Castells, M.J. Ferrando, J. Font, Organic syntheses with functionalized polymers: I. Preparation of polymeric substrates and alkylation of esters, Tetrahedron Lett. 12 (1971) 1713−1714.

[18] F. Balkenholh, C. von dem Bussche-Hunnefeld, A. Lansky, C. Zechel, Combinatorial synthesis of small organic molecules, Angew. Chem. Int. Ed. Engl. 35 (1996) 2288−2337.

[19] M.J. Plumkett, J.A. Ellman, Solid-phase synthesis of structurally diverse 1,4-benzodiazepine derivatives using the stille coupling reaction, J. Am. Chem. Soc. 117 (1995) 3306−3307.

[20] C.J. O'Connor, L. Laraia, D.R. Spring, Chemical genetics, Chem. Soc. Rev. 40 (2011) 4332−4345.

[21] P. Landers, Drug industry's big push into technology falls short, Wall Street J. 24 (2004).

[22] M.R. Arkin, J.A. Wells, Small-molecule inhibitors of protein-protein interactions: progressing towards the dream, Nat. Rev. Drug Discov. 3 (4) (2004) 301−317.

[23] S.L. Schreiber, Small molecules: the missing link in the central dogma, Nat. Chem. Biol. 1 (2005) 64−66.

[24] S.K. Das, Screening of bioactive compounds for development of new pesticides: a mini review, Univ. J. Agric. Res. 4 (1) (2016) 15−20.

[25] D.R. Spring, Chemical genetics to chemical genomics: small molecules offer big insights, Chem. Soc. Rev. 34 (6) (2005) 472−482.

[26] B.R. Stockwell, Exploring biology with small organic molecules, Nature 432 (7019) (2004) 846−854.

[27] A.V. Fuentes, M.D. Pineda, K.C. Nagulapalli Venkata, Comprehension of top 200 prescribed drugs in the US as a resource for pharmacy teaching, training and practice, Pharmacy 6 (2) (2018) 43.

[28] T. Park, S.K. Griggs, D.C. Suh, Cost effectiveness of monoclonal antibody therapy for rare diseases: a systematic review, Biodrugs 29 (4) (2015) 259−274.

[29] D. DeGoey, H.-J. Chen, P.B. Cox, M.D. Wendt, Beyond the rule of 5: lessons learned from AbbVie's drugs and compound collection, J. Med. Chem. 61 (7) (2018) 2636−2651.

[30] P.W. Erhardt, Medicinal chemistry in the new millennium. A glance into the future, Pure Appl. Chem. 74 (5) (2002) 703−785.

[31] R. Brenk, D. Rauh, Change or be changed: reflections of the workshop 'Future in Medicinal Chemistry, Bioorg. Med. Chem. 20 (12) (2012) 3695–3697.

[32] F.K. Brown, Chemoinformatics: what is it and how does it impact drug discovery, Annu. Rep. Med. Chem. 33 (1998) 375–384.

[33] A. Ruiz-Garcia, M. Bermejo, A. Moss, V.G. Casabo, Pharmacokinetics in drug discovery, J. Pharm. Sci. 97 (2) (2008) 654–690.

[34] B. Munos, Lessons from 60 years of pharmaceutical innovation, Nat. Rev. Drug Discov. 8 (2009) 959–968.

[35] S.M. Paul, D.S. Mytelka, C.T. Dunwiddie, C.C. Persinger, B.H. Munos, S.R. Lindborg, A.L. Schacht, How to improve R&D productivity: the pharmaceutical industry's grand challenge, Nat. Rev. Drug Discov. 9 (3) (2010) 203–214.

[36] M.M. Hann, G.M. Keserü, Finding the sweet spot: the role of nature and nurture in medicinal chemistry, Nat. Rev. Drug Discov. 11 (5) (2012) 355–365.

[37] D.C. Swinney, J. Anthony, How were new medicines discovered? Nat. Rev. Drug Discov. 10 (7) (2011) 507–519.

[38] G. Krauss (Ed.), Biochemistry of Signal Transduction and Regulation, Wiley and Sons, 2008.

[39] T. Flagstad, G. Min, K. Bonnet, R. Morgentin, D. Roche, M.H. Clausen, T.E. Nielsen, Synthesis of sp(3)-rich scaffolds for molecular libraries through complexity-generating cascade reactions, Org. Biomol. Chem. 14 (21) (2016) 4943–4946.

[40] S. Stotani, C. Lorenz, M. Winkler, F. Medda, E. Picazo, R.O. Martinez, A. Karawajczyk, J. Sanchez-Quesada, F. Giordanetto, Design and synthesis of fsp(3)-rich, bis-spirocyclic-based compound libraries for biological screening, ACS Comb. Sci. 18 (6) (2016) 330–336.

[41] U. Rester, From virtuality to reality – virtual screening in lead discovery and lead optimization: a medicinal chemistry perspective, Cur. Opin. Drug Discov. 11 (4) (2008) 559–568.

[42] J.M. Rollinger, H. Stuppner, T. Langer, Virtual screening for the discovery of bioactive natural products, Prog. Drug Res. 65 (2008) 213–249.

[43] G.B. Barcellos, I. Pauli, R.A. Caceres, L.F. Timmers, R. Dias, W.F. de Azevedo, Molecular modeling as a tool for drug discovery, Curr. Drug Targets 9 (12) (2008) 1084–1091.

[44] J.D. Durrant, J.A. McCammon, Molecular dynamics simulations and drug discovery, BMC Biol. 9 (2011) 71.

[45] D.W. Borhani, D.E. Shaw, The future of molecular dynamics simulations in drug discovery, J. Comput. Aided Mol. Des. 26 (1) (2012) 15–26.

[46] M. Ciemny, M. Kurcinski, K. Kamel, A. Kolinski, N. Alam, O. Schueler-Furman, S. Kmiecik, Protein-peptide docking: opportunities and challenges, Drug Discov. Today 23 (8) (2018) 1530–1537.

[47] C.G. Smith, J.R. Vane, The discovery of captopril, FASEB J. 17 (8) (2003) 788–789.

[48] M.S. Butler, Natural products to drugs: natural product-derived compounds in clinical trials, Nat. Prod. Rep. 25 (2008) 475–516.

[49] A.L. Harvey, Natural products as a screening resource, Curr. Opin. Chem. Biol. 11 (5) (2007) 480–484.

[50] D.J. Newman, G.M. Cragg, Natural products as sources of new drugs over the last 25 years, J. Nat. Prod. 70 (3) (2007) 461–477.

[51] K.S. Lam, New aspects of natural products in drug discovery, Trends Microbiol. 15 (2007) 279–289.

[52] A. Fleming, On the antibacterial action of cultures of a penicillium with special reference to their use in the isolation of B. influenza, Br. J. Exp. Pathol. 10 (1929) 226—236.

[53] G.M. Rishton, Natural products as a robust source of new drugs and drug leads: past successes and present day issues, Am. J. Cardiol. 101 (2008) 43—49.

[54] A. Ganesan, The impact of natural products upon modern drug discovery, Curr. Opin. Chem. Biol. 12 (2008) 306—317.

[55] W. Sneader, The discovery of aspirin: a reappraisal, BMJ 321 (2000) 1591—1594.

[56] M.V. Cobham, D. Donovan, Ixabepilone: a new treatment option for the management of taxane-resistant metastatic breast cancer, Cancer Manag. Res. 1 (2009) 69—77.

[57] A. Lopez, A.B. Parsons, C. Nislow, G. Giaever, C. Boone, Chemical-genetic approaches for exploring the mode of action of natural products, Prog. Drug Res. 66 (237) (2008) 239—271.

[58] Y. Pommier, Topoisomerase I inhibitors: camptothecins and beyond, Nat. Rev. Cancer 6 (10) (2006) 789—802.

[59] W. Li, W. Mo, D. Shen, L. Sun, J. Wang, S. Lu, J.M. Gitschier, B. Zhou, PLoS Genet. 1 (3) (2005) e36.

[60] G.T. Carter, Natural products and Pharma 2011: strategic changes spur new opportunities, Nat. Prod. Rep. 28 (2011) 1783—1789.

[61] N.E. Thomford, D.A. Senthebane, A. Rowe, D. Munro, P. Seele, A. Maroyi, K. Dzobo, Natural products for drug discovery in the 21st century: innovations for novel drug discovery, Int. J. Mol. Sci. 19 (6) (2018) E1578.

[62] A.L. Harvey, Natural product in Drug discovery, Drug Discov. Today 13 (19) (2008) 894—901.

[63] C.R. Ganellin, R. Jefferis, S.M. Roberts, in: S.M. Roberts (Ed.), The Small Molecule Drug Discovery Process — from Target Selection to Candidate Selection in "Introduction to Biological and Small Molecule Drug Research and Development: Theory and Case Studies, Elsevier, 2013.

[64] J.W. Scanell, A. Blanckley, H. Boldon, B. Warrington, Diagnosing the decline in pharmaceutical R&D efficiency, Nat. Rev. Drug Discov. 11 (3) (2012) 191—200.

[65] M. Rask-Andersen, M.S. Almén, H.B. Schiöth, Trends in the exploitation of novel drug targets, Nat. Rev. Drug Discov. 10 (8) (2011) 579—590.

[66] Y. Yuan, J. Pei, L. Lai, Binding site detection and druggability prediction of protein targets for structure-based drug design, Curr. Pharmaceut. Des. 19 (12) (2013) 2326—2333.

[67] P.D. Leeson, B. Springthorpe, The influence of drug-like concepts on decision-making in medicinal chemistry, Nat. Rev. Drug Discov. 6 (11) (2007) 881—890.

[68] M. Entzeroth, H. Flotow, P. Condron, Overview of high-throughput screening, Curr. Protoc. Pharmacol. 44 (1) (2009) 1—27.

[69] P.-S. Hasemi, F. Mehri, The role of different sampling methods in improving biological activity prediction using deep belief network, J. Comput. Chem. 38 (10) (2016) 1—8.

[70] Drug Design, in: C.H. Reynolds, K.M. Merz, D. Ringe D (Eds.), Structure- and Ligand-Based Approaches, first ed., Cambridge, Cambridge, UK, 2010.

[71] S. Ghosh, A. Nie, J. An, Z. Huang, Structure-based virtual screening of chemical libraries for drug discovery, Curr. Opin. Chem. Biol. 10 (3) (2006) 194—202.

[72] J. Bajorath, Integration of virtual and high-throughput screening, Nat. Rev. Drug Discov. 1 (11) (2002) 882—894.

[73] H. Gohlke, G. Klebe, Approaches to the description and prediction of the binding affinity of small-molecule ligands to macromolecular receptors, Angew Chem. Int. Ed. Engl. 41 (15) (2002) 2644–2676.

[74] H. Song, R. Wang, S. Wang, J. Lin, A low-molecular-weight compound discovered through virtual database screening inhibits Stat3 function in breast cancer cells, Proc. Natl. Acad. Sci. USA 102 (13) (2005) 4700–4705.

[75] A.Y.S. Malkhasiana, B.J. Howlin, Automated drug design of kinase inhibitors to treat Chronic Myeloid Leukemia, J. Mol. Graph. Model. 91 (2019) 52–60.

[76] K.S. Lam, S.E. Salmon, E.M. Hersh, V.J. Hruby, W.M. Kazmierski, R.J. Knapp, A new type of synthetic peptide library for identifying ligand-binding activity, Nature 354 (1991) 82–84.

[77] R.A. Houghten, C. Pinilla, S.E. Blondelle, J.R. Appel, C.T. Dooley, J.H. Cuervo, Generation and use of synthetic peptide combinatorial libraries for basic research and drug discovery, Nature 354 (1991) 84–86.

[78] H.M. Geysen, R.H. Meloen, S.J. Barteling, Use of peptide synthesis to probe viral antigens for epitopes to a resolution of a single amino acid, Proc. Natl. Acad. Sci. USA 81 (1984) 3998–4002.

[79] J.P. Kennedy, L. Williams, T.M. Bridges, R.N. Daniels, D. Weaver, C.W. Lindsley, Application of combinatorial chemistry science on modern drug discovery, J. Comb. Chem. 10 (2008) 345–354.

[80] A. Rasheed, R. Farhat, Combinatorial chemistry: a review, Int. J. Pharm. Sci. Res. 4 (2013) 2502–2516.

[81] R. Liu, X. Li, K.S. Lam, Combinatorial chemistry in drug discovery, Curr. Opin. Chem. Biol. 38 (2017) 117–126.

[82] R.J. Simon, R.S. Kania, R.N. Zuckermann, V.D. Huebner, D.A. Jewell, S. Banville, S. Ng, L. Wang, S. Rosenberg, C.K. Marlowe CK, Peptoids: a modular approach to drug discovery, Proc. Natl. Acad. Sci. USA 89 (20) (1992) 9367–9371.

[83] Y.L. Huang, J.W. Bode, Synthetic fermentation of bioactive non-ribosomal peptides without organisms, enzymes or reagents, Nat. Chem. 6 (10) (2014) 877–884.

[84] G.P. Smith, Filamentous fusion phage: novel expression vectors that display cloned antigens on the virion surface, Science 228 (1985) 1315–1317.

[85] K. Josephson, M.C.T. Hartman, J.W. Szostak, Ribosomal synthesis of unnatural peptides, J. Am. Chem. Soc. 127 (33) (2005) 11727–11735.

[86] H. Murakami, A. Ohta, H. Ashigai, H. Suga, A highly flexible tRNA acylation method for non-natural polypeptide synthesis, Nat. Methods 3 (5) (2006) 357–359.

[87] A. Litovchick, C.E. Dumelin, S. Habeshian, D. Gikunju, M.A. Guie, P. Centrella, Y. Zhang, E.A. Sigel, J.W. Cuozzo, A.D. Keefe, M.A. Clark, Encoded library synthesis using chemical ligation and the discovery of sEH inhibitors from a 334-Million Member Library, Sci. Rep. 10 (2015) 10916.

[88] T. Kodadek, The rise, fall and reinvention of combinatorial chemistry, Chem. Commun. 47 (2011) 9757–9763.

[89] P. Seneci, G. Fassina, V. Frecer, S. Miertus, The effects of combinatorial chemistry and technologies on drug discovery and biotechnology—a mini review, Nova Biotechnol. Chim. 13 (2014) 87–108.

[90] W.R.J.D. Galloway, A. Bender, M. Welch, D.R. Spring, The discovery of antibacterial agents using diversity-oriented synthesis, Chem. Commun. (2009) 2446–2462.

[91] W.R.J.D. Galloway, D.R. Spring, Is synthesis the main hurdle for the generation of diversity in compound libraries for screening? Expert Opin. Drug Discov. 4 (5) (2009) 467–472.

[92] S.J. Haggarty, The principle of complementarity: chemical versus biological space, Curr. Opin. Chem. Biol. 9 (2005) 296–303.

[93] M.K. Schawrz, W.H.B. Sauer, Molecular shape diversity of combinatorial libraries: a prerequisite for broad bioactivity, J. Chem. Inf. Comput. Sci. 43 (3) (2003) 987–1003.

[94] M.D. Burke, S.L. Schreiber, A planning strategy for diversity-oriented synthesis, Angew. Chem. Int. Ed. 43 (1) (2004) 46–58.

[95] A. Trabocchi (Ed.), Diversity-Oriented Synthesis: Basics and Applications in Organic Synthesis, Drug Discovery, and Chemical Biology, Wiley and Sons, 2013.

[96] S.L. Schreiber, Target-oriented and diversity-oriented organic synthesis in drug discovery, Science 287 (5460) (2000) 1964–1969.

[97] D.R. Spring, Diversity-oriented synthesis; a challenge for synthetic chemists, Org. Biomol. Chem. 1 (22) (2003) 3867–3870.

[98] D.S. Tan, M.A. Foley, M.D. Shair, S.L. Schreiber, Stereoselective synthesis of over two million compounds having structural features both reminiscent of natural products and compatible with miniaturized cell-based assays, J. Am. Chem. Soc. 120 (33) (1998) 8565–8566.

[99] S.B. Moilanen, J.S. Potuzak, D.S. Tan, Stereocontrolled Synthesis of Spiroketals via Ti(O*i*-Pr)$_4$-Mediated Kinetic Spirocyclization of Glycal Epoxides with Retention of Configuration, J. Am. Chem. Soc. 128 (6) (2006) 1792–1793.

[100] A.R. Yeager, G.K. Min, J.A. Porco Jr., S.E. Schaus, Exploring skeletal diversity via ring contraction of glycal-derived scaffolds, Org. Lett. 8 (22) (2006) 5065–5068.

[101] M.R. Medeiros, R.S. Narayan, N.T. McDougal, S.E. Schaus, J.A. Porco Jr., Skeletal diversity via cationic rearrangements of substituted dihydropyrans, Org. Lett. 12 (14) (2010) 3222–3225.

[102] T.E. Nielsen, S.L. Schreiber, Towards the optimal screening collection: a synthesis strategy, Angew. Chem. Int. Ed. 47 (1) (2008) 48–56.

[103] T.E. Nielsen, M. Meldal, Solid-phase intramolecular N-acyliminium Pictet–Spengler reactions as crossroads to scaffold diversity, J. Org. Chem. 69 (11) (2004) 3765–3773.

[104] T.E. Nielsen, M. Meldal, Solid-phase synthesis of pyrroloisoquinolines via the intramolecular N-acyliminium Pictet–Spengler reaction, J. Comb. Chem. 7 (2005) 599–610.

[105] T.E. Nielsen, S.L. Quement, M. Meldal, Solid-phase synthesis of bicyclic dipeptide mimetics by intramolecular cyclization of alcohols, thiols, amines, and amides with N-acyliminium intermediates, Org. Lett. 7 (17) (2005) 3601–3604.

[106] M.R. Spaller, M.T. Burger, M. Fardis, P.A. Bartlett, Synthetic strategies in combinatorial chemistry, Curr. Opin. Chem. Biol. 1 (1) (1997) 47–53.

[107] E. Lenci, A. Guarna, A. Trabocchi, Diversity-oriented synthesis as a tool for chemical genetics, Molecules 19 (10) (2014) 16506–16528.

[108] W.R.J.D. Galloway, A. Isidro-Llobet, D.R. Spring, Diversity-oriented synthesis as a tool for the discovery of novel biologically active small molecules, Nat. Commun. 1 (2010) 80–87.

[109] B.Z. Stanton, L.F. Peng, N. Maloof, K. Nakai, X. Wang, J.L. Duffner, K.M. Taveras, J.M. Hyman, S.W. Lee, A.N. Koehler, J.K. Chen, J.L. Fox, A. Manodinova, S. Schreiber, A small molecule that binds Hedgehog and block its signaling in human cells, Nat. Chem. Biol. 5 (2009) 154–156.

[110] E.E. Wyatt, S. Fergus, W.R.J.D. Galloway, A. Bender, D.J. Fox, A.T. Plowright, A.S. Jessiman, M. Welch, D.R. Spring, Skeletal diversity construction via a branching synthetic strategy, Chem Commun. 31 (2006) 3296−3298.

[111] E.E. Wyatt, W.R.J.D. Galloway, G.L. Thomas, M. Welch, O. Loiseleur, T. Plowright, D.R. Spring, Identification of an anti-MRSA dihydrofolate reductase inhibitor from a diversity-oriented synthesis, Chem. Commun. 40 (2008) 4962−4964.

[112] B.M. Ibbeson, L. Laraia, E. Alza, C.J. O'Connor, Y.S. Tan, H.M.L. Davies, G. McKenzie, A.R. Venkitaraman, D.R. Spring, Diversity-oriented synthesis as a tool for identifying new modulators of mitosis, Nat. Commun. 5 (2014) 3155−3163.

[113] C.M. Dobson, Chemical space and biology, Nature 432 (2004) 824−828.

[114] R.S. Bohacek, C. McMartin, W.C. Guida, The art and practice of structure-based drug design: a molecular modeling perspective, Med. Res. Rev. 16 (1) (1996) 3−50.

[115] R.S. Bon, H. Waldmann, Bioactivity-guided navigation of chemical space, Acc. Chem. Res. 43 (8) (2010) 1103−1114.

[116] S. Wetzel, R.S. Bon, K. Kumar, H. Waldmann, Biology-oriented synthesis, Angew. Chem. Int. Ed. 50 (46) (2011) 10800−10826.

[117] H. van Hattum, H. Waldmann, Biology-oriented synthesis: harnessing the power of evolution, J. Am. Chem. Soc. 136 (34) (2014) 11853−11859.

[118] K. Liu, C. Zhu, J. Min, S. Peng, G. Xu, J. Sun, Stereodivergent synthesis of N-heterocycles by catalyst-controlled, activity-directed tandem annulation of diazo compounds with amino alkynes, Angew Chem. Int. Ed. Engl. 54 (44) (2015) 12962−12967.

[119] G. Karageorgis, M. Dow, A. Aimon, S. Warriner, A. Nelson, Activity-directed synthesis with intermolecular reactions: development of a fragment into a range of androgen receptor agonists, Angew Chem. Int. Ed. Engl. 54 (46) (2015) 13538−13544.

[120] M.A. Koch, A. Schuffenhauer, M. Scheck, S. Wetzel, M. Casaulta, A. Odermatt, P. Ertl, H. Waldmann, Charting biologically relevant chemical space: a structural classification of natural products (SCONP), Proc. Natl. Acad. Sci. USA 102 (48) (2005) 17272−17277.

[121] G.W. Bemis, M.A. Murcko, The properties of known drugs. 1. Molecular frameworks, J. Med. Chem. 39 (1996) 2887−2893.

[122] A. Noren-Muller, I. Reis-Corrþa, H. Prinz, C. Rosenbaum, K. Saxena, H.J. Schwalbe, D. Vestweber, G. Cagna, S. Schunk, O. Schwarz, H. Schiewe, H. Waldmann, Discovery of protein phosphatase inhibitor classes by biology-oriented synthesis, Proc. Natl. Acad. Sci. USA 103 (28) (2006) 10606−10611.

[123] I.R. Corrþa Jr., A. Noren-Muller, H.D. Ambrosi, S. Jakupovic, K. Saxena, H. Schwalbe, M. Kaiser, H. Waldmann, Identification of inhibitors for mycobacterial protein tyrosine Phosphatase   B (MptpB) by biology-oriented synthesis (BIOS),, Chem. Asian J. 2 (9) (2007) 1109−1126.

[124] B.M. McArdle, R.J. Quinn, Identification of protein fold topology shared between different folds inhibited by natural products, Chembiochem 8 (7) (2007) 788−798.

[125] B.M. McArdle, M.R. Campitelli, R.J. Quinn, A common protein fold topology shared by flavonoid biosynthetic enzymes and therapeutic targets, J. Nat. Prod. 69 (1) (2006) 14−17.

[126] Y. Wang, T. Suzek, J. Zhang, J. Wang, S. He, T. Cheng, B.A. Shoemaker, A. Gindulyte, S.H. Bryant, PubChem BioAssay: 2014 update, Nucleic Acids Res. 42 (2014) 1075−1082.

[127] M. Congreve, R. Carr, C. Murray, H. Jhoti, A 'rule of three' for fragment-based lead discovery? Drug Discov. Today 8 (19) (2003) 876–877.

[128] A.J. Price, S. Howard, B.D. Cons, Fragment-based drug discovery and its application to challenging drug targets, Essays Biochem. 61 (5) (2007) 475–484.

[129] B. Over, S. Wetzel, C. Grütter, Y. Nakai, S. Renner, D. Rauh, H. Waldmann, Natural-product-derived fragments for fragment-based ligand discovery, Nat. Chem. 5 (1) (2013) 21–28.

[130] P.D. Leeson, S.A. St-Gallay, The influence of the 'organisational factor' on compound quality in drug discovery, Nat. Rev. Drug Discov. 10 (2011) 749–765.

[131] E. Lenci, A. Trabocchi, Smart design of small-molecule libraries: when organic synthesis meets cheminformatics, Chembiochem 20 (9) (2019) 1115–1123.

[132] F. Häse, L. M Roch, A. Aspuru-Guzik, Next generation experimentation with self-driving laboratories, Trends Chem. 1 (3) (2019) 282–291.

Chemical reactions for building small molecules

2

Pietro Capurro, Andrea Basso

Department of Chemistry and Industrial Chemistry, University of Genova, Genova, Italy

2.1 Introduction

The history of drug discovery and medicinal chemistry has always entwined that of organic chemistry. Drug discovery has been existing since no longer than a hundred years, and its roots plunge into serendipitous discoveries made by chemists who were working with natural products. Over the years, the improvements achieved in chemistry, biology, and pharmacology caused drug research to rapidly evolve, becoming an interdisciplinary topic characterized by a constant dialogue between scientists from different research fields. While the diversification of known structures was once the most prominent process for developing new drugs for different therapeutic areas, today the advancements in molecular biology and pharmacology allow the molecular modelling of scaffolds tailored on a specific therapeutic target. Nevertheless, it is worth noting how current therapies are designed around a pool of less than 500 targets, despite the estimations suggesting that there could be around 5000—10000 molecular targets potentially exploitable [1]. It needs to be considered that such targets must undergo a strict validation process, i.e., the role they play in relation to a disease or condition is to be understood. To reach a high degree of validation, a connection between the reversal of the clinical conditions and the interaction of a certain drug with the target (e.g., the interaction of an antagonist with a receptor) must be established. Therefore, it is easy to understand why the efficient synthesis of libraries of diverse small molecules is still pivotal in modern drug discovery, despite the vast success of the so called "biotech" drugs (mostly monoclonal antibodies and recombinant proteins). The testing and screening of small molecules offer a dual approach in drug discovery. On one hand, they can be used to validate a therapeutic target. Small molecules play a prominent role in the endeavor of identifying a biological target and its working mechanism on a molecular level, as previously stated. On the other hand, the testing of libraries (i.e., arrays of small molecules sharing a common scaffold and differentiated by its substituents) is extremely useful to elucidate the structure-activity relationships (SARs) and structure-properties relationships (SPRs) against a therapeutic target. Furthermore, it allows to gather valuable information regarding the binding properties and the modulation potential of these molecules against such a specific target [2]. As a consequence, drug research evolved implementing new methodologies, such as cell-based assays and automated high-throughput screening (HTS) that allow parallel testing of a large number of molecules. Organic synthesis and medicinal chemistry evolved according to these requirements: solid phase synthesis (SPS), B/C/P (*build/couple/pair*) strategies [3], and combinatorial approaches have been developed to quickly supply large libraries of diverse, small molecules for screening. With these premises, it may seem odd that the vast majority of the reactions used in medicinal chemistry and drug research actually falls into few reaction classes. Recent reviews [4,5] pointed out that 89% of reactions used in modern medicinal chemistry can be categorized in just 10 classes: protecting group manipulations, alkylations, condensations, Pd-mediated cross-couplings (63% of the aforementioned reactions belong to these four categories), oxidations, reduction, heterocycle formation, halogenations, hydrolyzes, and metallations. Moreover, when we limit our interest to the reactions used in combinatorial chemistry, we notice that 87% of these falls either into alkylations, or condensations, or cross-couplings. Entire classes of reactions are scarcely found in most papers. Cycloadditions, click chemistry, radical reactions, photochemical processes, and multicomponent reactions are narrowly used when compared to the other classes cited before. This is due to several reasons. Firstly, robust synthetic pathways are usually preferred, to ensure a fast and reliable synthesis. Secondly, new methodologies are usually less tested toward functional group tolerance, and medicinal chemists tend to avoid the use of protective strategies whenever possible. Lastly, some processes introduce moieties that either increase excessively the product lipophilicity or represent a greater risk of attrition during the development process with respect to others; thus, these reactions are usually avoided. However, over time, the overwhelming abundance of amide bonds and para-substitution patterns, that are the most frequent motifs in drug-like products, led to the synthesis of molecules with increased linear shape and sp^2 character, leaving a surprisingly vast area of chemical space almost unexplored. Also, nitrogen-containing products seem to have been favored over oxygen-containing ones. It is questionable whether the predominance of the former in the current scenario was caused by the complexity of synthesis of the latter, or rather because nitrogen-containing compounds have a higher biological activity than oxygen-containing ones. Nonetheless, the large area of chemical space left empty holds the potential for a quantity of biologically active molecules. With all these considerations in mind, in the chapter herein we focused both on classical reactions (e.g., cross-couplings or cycloadditions) and recent strategies (such as photochemical processes or late-stage functionalization reactions) in organic synthesis to give a comprehensive outlook on the synthetic tools available today for the creation of small molecules libraries. Accordingly to our intent, we reported recent advances in cross-coupling chemistry, multicomponent reactions, photochemistry, and late-stage functionalization methods; with this perspective, we hope to offer a representative overview on the strategies available for diverse-oriented synthesis and modern drug research from a synthetic point of view (Fig. 2.1).

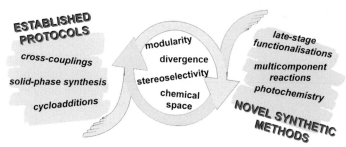

FIGURE 2.1

Different synthetic approaches for the synthesis of small molecules.

2.2 Cross-coupling reactions

Cross-couplings are, by definition, reactions in which an organometallic reagent (or rarely an unactivated one) combines with an organic electrophile in the presence of a metal catalyst to form a new C−C, C−H, C−N, C−P, C−O, C−S, or C-M bond [6]. A lot of metals can catalyze a cross-coupling reaction; however, palladium-mediated cross-couplings are by far the most famous ones (Scheme 2.1), and their development had such a huge impact on modern organic synthesis that, as a recognition for their ground-breaking work in the field, Richard Heck, Ei-ichi Negishi, and Akira Suzuki were awarded the Nobel Prize in Chemistry in 2010. Among these, Suzuki-Miyaura coupling proved to be the most popular one, followed by Heck-Mizoroki and Sonogashira couplings. The ascent of palladium complexes as catalysts for

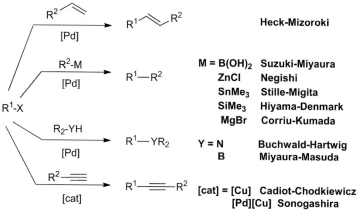

SCHEME 2.1

Most famous cross-coupling reactions. Palladium-mediated coupling proved to be the most robust and reliable in all fields of organic synthesis; other metal-catalyzed processes were efficiently converted into palladium-mediated ones (*e.g.* Corriu-Kumada reaction). X is a halide or pseudohalide (e.g., TfO); R^2 is usually a C(sp) or C(sp^2) partner.

cross-coupling reactions is due to several factors, but one above all. By mid-1970s, copper, nickel, and palladium proved to be efficient catalyst for a broad spectrum of reactions, each one with its unique features. Unsurprisingly, in the decade to come the research interest steered toward the investigation of palladium-mediated processes because of the possibility of modulating the catalyst reactivity through a plethora of different ligands while retaining its remarkable selectivity [7]. However, despite the powerful tool that cross-coupling reactions proved to be, they still have limitations. Only partial success has been achieved when one or both the coupling partners are C(sp^3), and the lack of detailed studies on enantioselective processes has hampered the diffusion of cross-couplings in asymmetrical syntheses. Furthermore, an electrophilic partner is required: despite being simple, the introduction of a suitable group must still be included in the planning of a synthesis. Thus, the interest toward direct C—H activation processes (catalyzed by metals such as ruthenium, iridium, copper, gold, iron, and nickel) and late-stage functionalization methods to overcome the aforementioned problems is rising constantly. Accordingly, we herein report a brief depiction of what palladium-mediated processes can offer to diversity-oriented and combinatorial synthesis, and then we will focus on emerging processes.

2.2.1 Palladium-mediated cross-couplings: applications in DOS, SPOS, and combinatorial synthesis

Solid-phase organic synthesis (SPOS) is a powerful tool in organic medicinal chemistry, and it found vast application in the synthesis of peptide and oligonucleotide libraries. However, the scope of polymer supported chemistry has been expanded, allowing the efficient synthesis of drug-like small molecules. With respect to traditional chemistry, SPOS offers a simple purification by washing and filtration of the polymer beads. Chromatographic separations are unnecessary, high-boiling solvents do not need to be evaporated, and reagents can be used in large excess to increase reaction yields. Moreover, functional groups on different molecules are less likely to interact, favoring intramolecular cyclizations that would otherwise be disadvantaged (*"pseudodilution effect"*) [8].

Heck, Sonogashira, Suzuki, and Stille couplings have been most effectively adapted to solid phase synthesis. Pseudodilution effect has been effectively exploited and difficult macrocyclizations have been achieved. Heck coupling occurs abundantly in SPOS, as it makes possible to diversify molecules in combinatorial libraries exploiting an unactivated alkene or alkyne moiety [9]. Both coupling partners can be supported, though supported halides tend to work better. Intramolecular reactions can also be achieved, as illustrated in Scheme 2.2: a seven-membered ring **2** as synthesized from an iodo alkene/alkyne supported on Wang resin **1** via Heck coupling. The subsequent cleavage and esterification with diazomethane gave the final product in 60% yield from the alkene and in 39%—73% yield from internal alkynes [10]. Benzodiazepine scaffold can be achieved in solid phase also by Stille coupling, as reported by Ellman and coworkers [11].

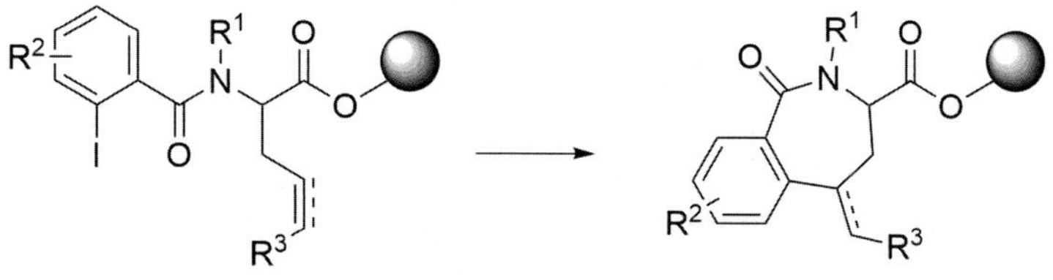

SCHEME 2.2

Solid-phase synthesis of an ε-caprolactam moiety fused to a benzene ring through an intramolecular Heck cross-coupling. Conditions: $Pd(OAc)_2$, PPh_3, KOAc, DMF, 70°C.

An interesting approach is the *cleavage-cross-coupling strategy*, in which the cleavage from the resin either releases one of the coupling partners or is a consequence of the coupling reaction itself. Anilines, for example, can be linked to an amine resin **5** as triazenes **3** (Scheme 2.3). Upon cleavage with TFA in MeOH, the diazonium ion formed in situ participates an Heck coupling in presence of catalytic amounts of $Pd(OAc)_2$ with very good yields of alkenes/alkynes **4**. This allows the simultaneous generation of the coupling partner and an efficient cleavage from the resin. Moreover, the resin can be regenerated and recycled [12].

A similar example of this strategy has been reported by Nicolaou and coworkers, that achieved an intramolecular macrocyclization via Stille cross-coupling within the total synthesis of (S)-zearalenone (Scheme 2.4). In this particular case, both the stannane and the aryl iodide are supported to the polystyrene resin **6**, and the organostannane moiety also plays the role of the linker. When the Stille reaction occurs, 14-membered ring **7** closes and, as a consequence, the product is cleaved from the resin and recovered (53% yield). The subsequent deprotection reaction led to the final product. This approach has been defined *cyclorelease* [13].

Considering the environmental concern for tin byproducts, Maleczka et al. reported an interesting option to avoid contamination of the products. A substoichiometric quantity of resin-bound organotin reagent (0.3 eq.) can be used in a catalytic Stille reaction, and tin by-products can be easily discarded by filtration [14].

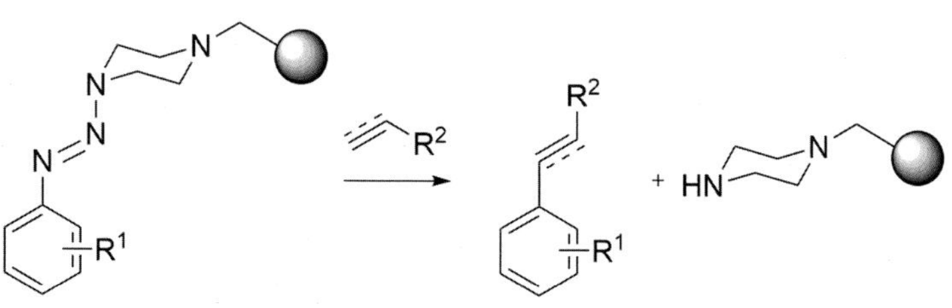

SCHEME 2.3

Cleavage-cross-coupling strategy with triazenes. Conditions: TFA, $Pd(OAc)_2$, MeOH.

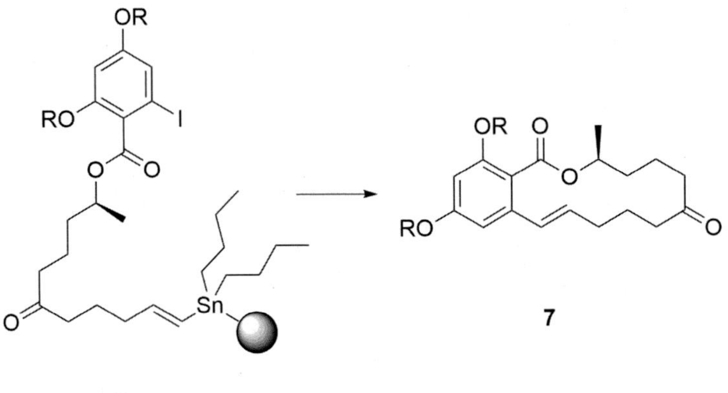

SCHEME 2.4

Cyclorelease through intramolecular Stille coupling, pseudodilution effect allows a smooth macrocyclization, achieving a 14-membered ring. Conditions: Pd(PPh$_3$)$_4$, toluene, 100°C.

Suzuki-Miyaura coupling found vast application in solid phase synthesis aimed toward diversity, as it occurs in mild condition and it is tolerant to a plethora of functional groups commonly found in combinatorial libraries. It can be carried out in biphasic solutions and water, and it can be exploited to attach linkers to a resin bearing alkene or halide moieties [9]. Among all the reports of efficient Suzuki couplings that current literature offers, it is worth mentioning Ellman's group work on prostaglandins analogues as a good example of combinatorial synthesis of active molecules (Scheme 2.5). Suzuki coupling on the immobilized bromide **8** occurred smoothly with various borane partners **9**, giving 37 different E$_1$-, E$_2$-, F$_1$-, F$_2$-prostaglandin analogues **10** with yields up to 60% [15,16]. Similarly, a combinatorial synthesis of a vitamin D$_3$ analogue and its derivatives has been optimized via solid phase Suzuki-Miyaura coupling [17]; also, a cleavage-cross-coupling strategy on supported enol phosphonates has been reported recently [18].

Similarly to Suzuki reaction, the mild conditions and wide functional group tolerance of Sonogashira coupling made the reaction ideal for combinatorial synthesis. Furthermore, the alkyne moiety can be easily converted into others. Schreiber

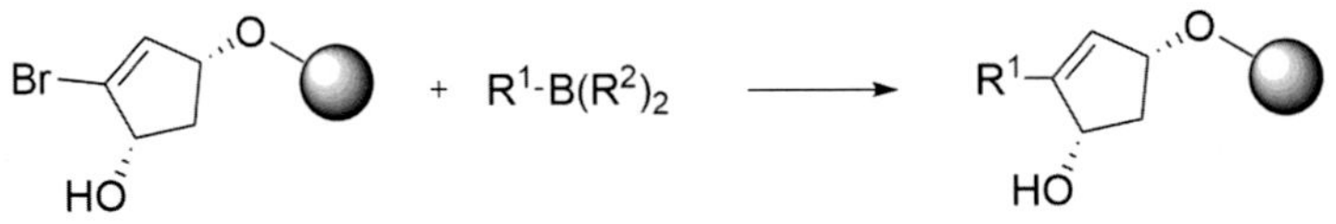

SCHEME 2.5

Combinatorial synthesis of E$_1$-, E$_2$-, F$_1$-, F$_2$-prostaglandin analogues via solid phase Suzuki-Miyaura coupling. Conditions: Pd(PPh$_3$)$_4$, Na$_2$CO$_3$ (2M in THF), 65°C.

and coworkers explored the possibility of applying Sonogashira couplings to DOS. Starting from 18 different tetracyclic structures bound to a resin, the consecutive reaction of these cores with 30 different alkynes (under standard Sonogashira conditions), 62 different amines and 62 different carboxylic acids led to the generation of a library of over two million diverse compounds [19]. Another interesting approach has been reported by Nilsson et al. Propargyl alcohol can be easily supported onto Merrifield resin **12**, and then an aryl halide **11** can be linked to the resin via Sonogashira coupling. Through a Nicholas reaction, simultaneous cleavage from the resin **13** and formation of a new C−C or C−O bond occurs, introducing additional diversity; this approach has been employed in the synthesis of a library of galactin inhibitors (Scheme 2.6) [20].

Other cross-couplings have been successfully adapted to SPOS but will not be discussed here in detail. Negishi coupling found multiple applications, though organozinc partners, being synthetized from the relative Grignard or organolithium compounds, are not as readily available as other coupling partners. As reported in Scheme 2.1, Kumada coupling uses Grignard reagents directly; however, no examples of its application in SPOS has been reported so far to the best of our knowledge. In contrast, immobilized bromides, iodides, and triflates coupled smoothly with (hetero)aryl- and alkylzinc halides [21]. Buchwald-Hartwig amination [22] and α-arylation of ketones [23] have been used in SPOS only recently and very few examples are known, in spite of their synthetic importance. Few examples of Hiyama-Denmark couplings are reported, despite organosilicon reagents being cheap, nontoxic, tolerant of many functional groups, and easily prepared from other reagents. This is mainly due to the low nucleophilicity of the silicon partners (C−Si bond is rather strong) that must be activated with a fluoride source, e.g., TBAF (*i.e.* tetrabutylammonium fluoride) or a base, limiting their use [8]. Only recently the introduction of aryloxysilanes as coupling partners has broadened the scope of

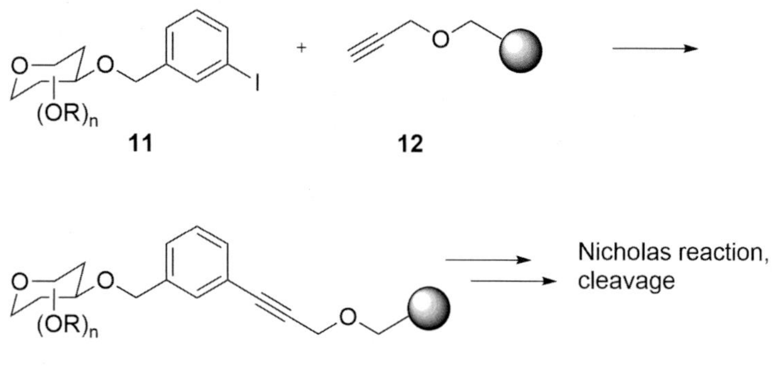

SCHEME 2.6

Sonogashira coupling—Nicholas reaction cleavage strategy for the synthesis of a library of potential galactin inhibitors. Conditions: Pd(PPh₃)₄, CuI, Et₃N.

this methodology [24]. Last, but not least, it is worth reminding that new proceedings in ligand technology, despite having not being used in DOS or SPOS yet, may help exploring some unpopulated regions of chemical space. Modern NHC (*N*-heterocyclic carbene) ligands, for example, proved to be useful to achieve C(sp^3)-C(sp^3) Negishi couplings in excellent yields [25].

Concluding, it is worth remarking that the possibility of modulating the reactivity of Pd (and other metals) through ligands opened the way to the so called *ligand-directed divergent synthesis* [26]. This approach should be highly considered when planning a diversity-oriented or a complex to diversity synthesis. While different scaffolds can be achieved from a common intermediate through different catalysts, with ligand-directed approaches it is possible to accomplish the same challenging task with different ligands to a same metal under different conditions. Other than Pd, many other metals offer this possibility (e.g., Au, see 2.2.3, but also Ni, Rh, Co, *etc.*). A remarkable example is provided by Buchwald et al. in their divergent synthesis of dibenzazepines **15**, carbazoles **16**, acridines **17**, and indoles **18**, through Pd-mediated processes from common substrates (Scheme 2.7). The variation of reaction conditions allowed to convert 2-chloro-N-(2-vinyl)anilines **14** in five-, six-, and seven-membered ring systems, all with potential pharmacological activity [27].

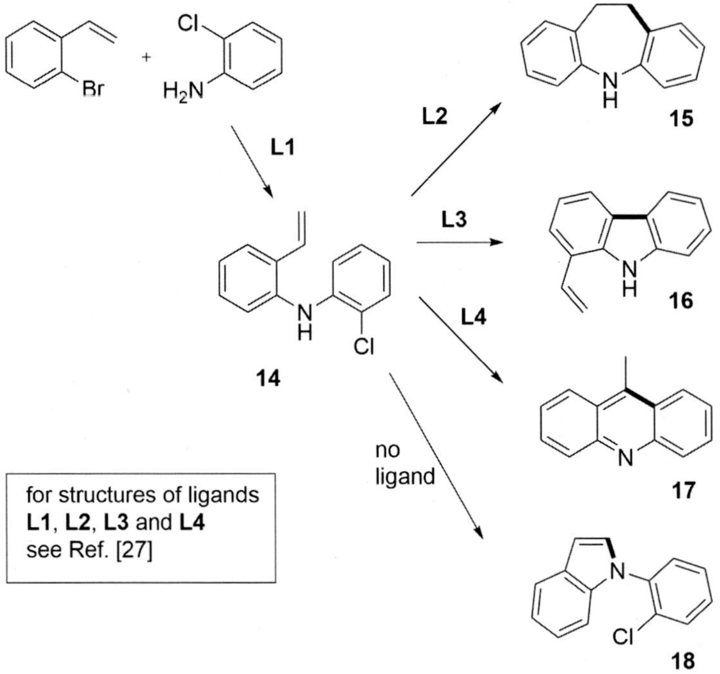
Cl
+
Br
H2N
L1
L2
L3
L4
no
ligand
N
H
Cl
14
15
16
17
18
N
for structures of ligands
L1, L2, L3 and L4
see Ref. [27]

SCHEME 2.7

Ligand-directed divergent synthesis of carbazoles, indoles, dibenzazepines, and acridines from 2-chloro-*N*-(2-vinyl)anilines. All reactions are catalyzed by Pd(0) species.

2.2.2 Other metal-mediated cross-couplings and their applications

2.2.2.1 Olefin metathesis in SPOS and DOS

Olefin metathesis is a powerful tool in synthetic organic chemistry and found valuable applications in ring closing reactions (RCM—ring closing metathesis), ring opening reactions (ROM—ring opening metathesis), and also polymerization reactions. Basically, olefin metathesis allows an exchange of moieties between two alkenes or an alkene and an alkyne, under ruthenium-carbene complexes catalysis. Despite its synthetic utility, only few examples are reported in diversity-oriented syntheses and SPOS. However, it can be used to exploit the synthetic potential of unactivated alkene and alkyne moieties, similarly to Heck reaction. Furthermore, the introduction of Hoveyda-Grubbs and second generation Grubbs catalysts broadened the scope of this reaction and mitigated its requirements; cross-metathesis efficiently occurs in mild conditions and is tolerant to a wide variety of unprotected functional groups [28]. Owing to its utility and potential, a wider use of olefin metathesis in modern high-throughput syntheses is highly desirable.

Cleavage-ring closing metathesis strategies have been reported and efficiently exploited. Piscopio et al. paved the way to access unsaturated Freidinger lactams (that play a key role in the development of conformationally constrained peptidomimetics drugs for different therapeutic areas) via a cleavage-RCM approach [29]. A similar approach has been reported by Nicolaou and coworkers, that took advantage of the pseudodilution effect to achieve the formation of a 16-membered ring with concomitant cyclorelease within the total synthesis of natural anticancer epothilone A [30].

Hoveyda and coworkers also reported an efficient way to support ruthenium catalysts onto monolithic sol-gel glass beads [31]. The supported catalyst can be recycled many times (>15 times per bead) retaining its activity and the removal of ruthenium impurities is avoided, being the final solution of high or even analytical purity (typically Ru < 0.5%). Their methodology was effectively applied to the high-throughput synthesis of two different libraries of ROM/RCM products. Other groups reported methodologies involving metathesis in their combinatorial syntheses. Nicolaou et al. used cross-metathesis within the solid phase synthesis of (DL)-muscone and a library of its analogues [32]. Blechert and Schürer reported an efficient solid phase ene-yne cross-metathesis followed by Diels-Alder cycloaddition, achieving, after further elaborations, diversely substituted octahydrobenzazepinones. Lastly, the solid phase synthesis of a library of 3-(aryl)alkenyl-β-lactams **20** as cholesterol absorption inhibitor analogues has been accomplished by Mata and Testero, achieving the desired products with total *E*-selectivity for the double bond configuration, starting from Wang-resin supported **19** (Scheme 2.8) [33].

2.2.3 Nickel-mediated cross-couplings: recent advances

Despite their success in all fields of organic chemistry, the landscape of palladium-mediated cross-couplings still lack efficient methodologies to induce enantioselectivity to the products. This is mainly due to the same issues encountered with

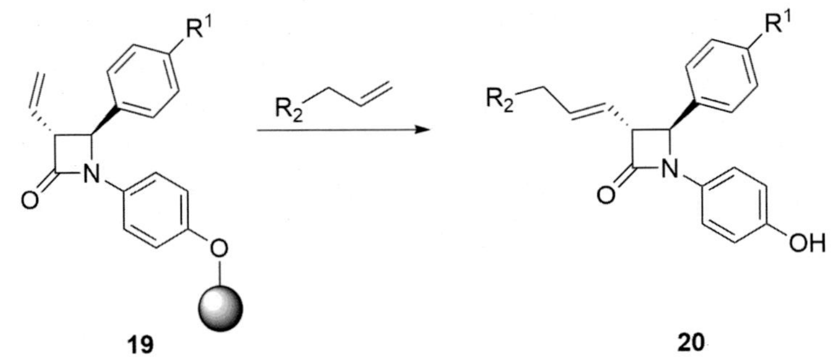

SCHEME 2.8

Combinatorial solid-phase synthesis of a library of potentially active β-lactams.
Conditions: (1) second generation Grubb Ru catalyst, DCM, 40°C; (2) TFA, DCM.

Pd-catalyzed alkyl cross-coupling reactions. Secondary or even primary alkyl halides and alkylmetallic derivatives usually undergo oxidative addition and transmetallation to palladium, respectively, much slower than their $C(sp^2)$- or $C(sp)$-hybridized counterparts. Furthermore, they can suffer from rapid, parasitic β-hydride elimination processes that are unproductive. Despite the powerful tool that ligand engineering proved to be in modulating the efficiency of these elementary steps, the main research interest in the area is now focused on nickel-mediated processes. Several Ni-mediated versions of famous cross-couplings gained popularity. Moreover, it should be recognized the importance of nickel for its ability to react with unconventional electrophiles [34].

Owing to the nature of their coupling partners, enantioselective Negishi and Corriu-Kumada couplings have been developed. Fu et al. achieved excellent yields and enantiomeric excesses for the Corriu-Kumada coupling of α-haloketones [35]. Fu's team published several works proving the efficiency of Ni as a catalyst in enantioselective Negishi cross-coupling, finding excellent yields and enantiomeric excesses. His group successfully managed to cross-couple alkylzinc halides with α-bromoamides [36], benzyl bromides [37,38], α-halonitriles [39], and α-bromosilanes [40]. Similar results were achieved with arylzinc reagents, that have been coupled effectively with propargyl halides [41], benzylic mesylates (prepared in situ with no further purification from the relative alcohols) [42], α-bromo [43] and α-bromo-α-fluoro ketones [44], and even α-halosulfonamides [45].

Another excellent result has been achieved by Vanerikov and Gandelman, that recently reported an enantioconvergent version of Hiyama-Denmark coupling to achieve enantiopure α-trifluoromethyl alcohols, extremely versatile building-blocks in medicinal chemistry [46]. $C(sp^2)$-hybridized trialkyloxysilanes were employed as partners; moreover, *ortho*-, *meta*-, and *para*-substituted arylsilanes are equally amenable to the coupling. The process also proved to be scalable up to 2 g [47].

Fu's group also worked on Ni-catalyzed Suzuki-Miyaura $C(sp^3)$-$C(sp^3)$ cross-coupling. It is worth noticing that also this version of a notorious cross-coupling gave excellent results in terms of yields and enantiomeric/diastereoisomeric excesses. Directing group are required to achieve good selectivity (along with a chiral catalyst), but the scope of the reaction proved to be wide [48−50]. Perhaps the most intriguing possibility is that of establishing two new C−C bonds while configuring a chiral center, in a cascade asymmetric cyclization/cross-coupling fashion. In this approach, the capability of the Ni catalyst to configure two different C−C bonds is exploited [51]. Boron derivatives have been employed in efficient hydroarylations and hydroalkenylations, with excellent enantioselectivity. As a validation of their pathway, Mei and coworkers successfully achieved Ibuprofen **22** in two steps with overall excellent yield and enantiocontrol for **21** (Scheme 2.9) [52].

In contrast with the redox-neutrality of most processes, nickel-mediated ones often involve reductive or oxidative steps. This feature can be efficiently paired with the ability of Ni to drive enantioselective coupling reactions: Diao et al. efficiently implemented an asymmetric reductive diarylation of vinylarenes, resulting in compounds **23**, structures found in potentially active molecules, such as PDE 4A inhibitor **24**, α-thrombin inhibitor **25,** or estrogen receptor ligand **2.26** (Scheme 2.10) [53]. This approach, like many reductive others, requires a sacrificial reducing species, usually zinc metal, cheap and easily available; also organic reductants can be employed [54].

This general methodology can also be applied to late-stage functionalization approaches. Wang et al. reported a very efficient combinatorial reductive cross-coupling on monofluoroalkyl halides, using Mn as sacrificial reductant. The approach proved to be robust and reliable, being the reaction conditions extremely mild and the procedure tolerant to a wide spectrum of functional groups [55].

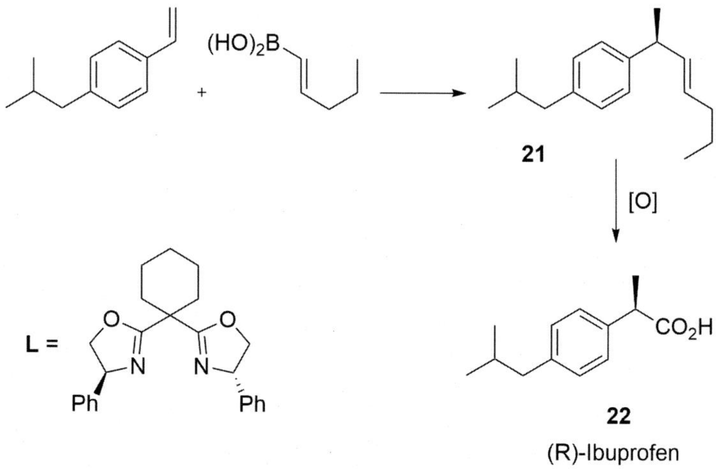

SCHEME 2.9

Enantioselective synthesis of commercial drug (R)-ibuprofen via hydroalkenylation/oxidation. Conditions: Ni(cod)$_2$, ligand **L**, EtOLi, MeOH, 50°C.

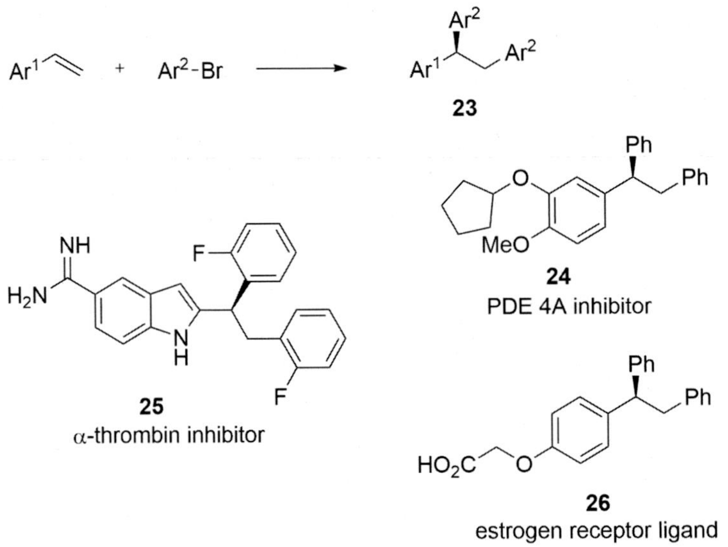

SCHEME 2.10

Reductive arylation of styrenes under Ni catalysis. Conditions: Ni(DME)Br$_2$, (S)-iBu-biOx, Zn, DMPU/THF 1:1, 10°C.

Nickel, along with other metals such as Cu, Fe, and more rarely Pd or organometallic photocatalysts (Ir or Ru), can also be employed in oxidative dehydrogenative cross-couplings (CDC, or *cross dehydrogenative couplings*) [56,57]. Ni-catalyzed double bond migration along an aliphatic chain with total E-selectivity has also been reported, allowing a remote functionalization of important industrial intermediates [58]. The so called C=C chain walk has been reported to occur also under Ru catalysis [59]. Finally, the possibility to undergo SET allows Ni to be a good catalyst even for difficult transformations, such as direct amination of arenes. Leonori et al. recently developed interesting strategies toward N-arylations, exploiting the possibility to generate nitrogen radical under Ni catalysis. Aryl and alkylzinc, arylboronic acids, and aryltriethoxysilanes were all feasible substrates for the process; moreover, in presence of double bonds a remote functionalization through cyclization is possible [60].

2.2.4 Copper, iron, and gold catalyzed reactions: the future in cross-coupling chemistry

It is not unsurprizing that recently, due to the high cost and low availability of palladium, rhodium, iridium, platinum, and ruthenium, the research interest shifted toward other metals as viable catalysts for cross-coupling reactions. The renewed interest in cheaper and more abundant metals as catalysts led to a renaissance of copper-mediated processes [61]; oddly enough, the very first cross-coupling reactions ever described (e.g., Ullmann reaction and Glaser coupling) relied on a Cu salt as catalyst and largely predate any Pd-mediated process [7]. Most Pd-mediated processes have been successfully adapted to be achieved with copper catalysts with some modifications, likewise nickel. The coupling of organic halides with organomagnesium reagents (such as in Corriu-Kumada coupling) has been reported by several groups, and occurs smoothly with alkyl, aryl, vinyl, and allyl Grignard reagents with alkyl chlorides, bromides, iodides, tosylates and mesylates [62,63]. Few examples are reported on alkynyl halides [64] and heteroaryl chlorides [65]. An interesting option that Cu (and also Ni) catalysis offers in this type of coupling is the possibility to exploit as electrophiles even alkyl fluorides, usually unreactive to other catalysts [66,67]. Also iron catalysts allow this particular coupling [68]. Complementarily to the possibility of coupling alkyl electrophiles with a variety of Grignard reagents, copper catalysts have been used to couple effectively alkynyl, aryl, and alkenyl halides with vinyl, (hetero)aryl, and alkynyl tin reagents in a Stille-type coupling [69,70]. Cu-catalyzed versions of the Hiyama-Denmark coupling also offer this possibility; noteworthy results were achieved in coupling vinylsilicon reagents with alkynyl bromides [71] and (hetero)arylsilicon derivatives with (hetero) aryl iodides [72]. Suzuki-Miyaura coupling under Cu catalysis has been developed and allows to couple almost any electrophilic partner with any boron derivative, covering a broad spectrum of synthetic possibilities [73,74]. Additionally, copper can be exploited to achieve enantioselective processes of great interest. Recently, Corey et al. reported an efficient, highly enantioselective methodology to ligate 2 C(sp^3) centers. An enantioselective lithiation of an α-Boc-amine can be achieved using $(-)$-sparteine as a chiral modulator; the lithiated product can be converted into a cuprate by addition of a Cu(I) salt. In presence of a dinitrobenzoate ester and the formation of a Cu(III) species, a reductive elimination occurs, leading to a highly enantiopure diamine with two vicinal chiral centers and C$_2$ symmetry that can be used to develop chiral ligands, catalysts and drugs [75]. The ability of Cu to catalyze enantioselective processes has been exploited also by Liu et al., that reported an innovative method to access quaternary stereodefined chiral centers, forging two C−C bonds and employing unfunctionalized substrates (Scheme 2.11). The combination of copper iodide as precursor and a chiral phosphoric acid **L** as ligand gave excellent results, with excellent yields and enantiomeric ratios for **27**. The strategy works fine on gram scale and affords enantiopure products with a quaternary carbon motif common to several active compounds; moreover, the presence of a fluorinated substituent can be of interest when designing more stable and lipophilic drug-like molecules [76].

Among other metals in which interest grew during time, iron has been exploited to achieve several transformations. Owing to the vast availability of iron and its environmental and economic advantages, Fe-mediated cross-coupling found applications both in research and bulk scale synthesis of pharmaceuticals. Furthermore, iron catalysis occurs through several catalytic cycles, with an orthogonal selectivity with respect to other metals [77]. However, the development of robust coupling reactions has stalled for long, since mechanistic investigations on the catalytic activity

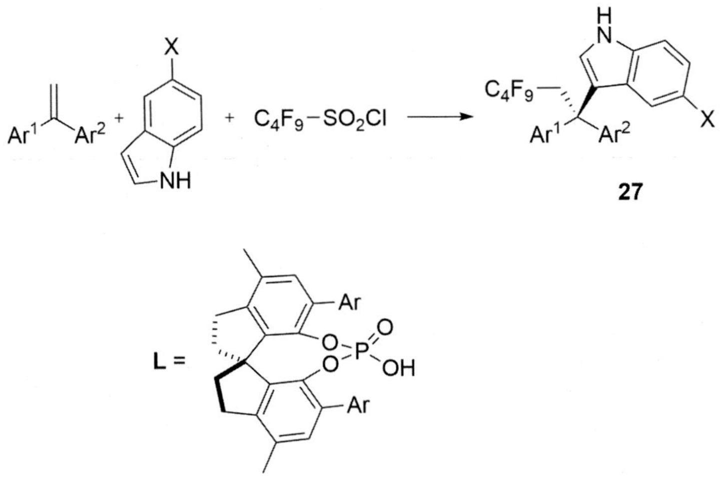

SCHEME 2.11

Asymmetric, Cu-catalyzed 1,2-dicarbofunctionalization of unactivated alkenes.
Conditions: CuI, ligand **L**, Ag$_2$CO$_3$, DCM, 0°C.

of iron lacked. Nevertheless, over time, efficient methods for Corriu-Kumada, Negishi, Suzuki, Sonogashira, and even Heck Fe-mediated cross-couplings were eventually reported [78–80]. Nowadays, iron catalysis proved to be a reliable approach to forge C(sp^2)-C(sp^3) bonds in medicinal chemistry. The most remarkable feature of Fe as catalyst is its strict selectivity: depending on the reaction conditions, chlorides can be preferred to bromides as electrophiles, and heteroaromatic chlorides to aromatic ones. This allows orthogonality to other processes such as Heck reaction (Scheme 2.12), as reported by Mullens et al. in their bulk synthesis of **29** as drug intermediate, where **28** is selectively obtained [81]. Similarly, Lee et al. successfully achieved a cross-coupling on a heteroaryl chloride despite the presence of an aryl one in the same substrate in their synthesis of SGLT2 inhibitors as antidiabetics [82].

At last, it is worth citing the importance of gold catalysis in modern synthesis. As a matter of facts, the last decade saw almost a literal "gold rush" that followed the discovery that gold, likewise most transition metals, has strong catalytic potential despite what common knowledge suggested; unsurprisingly, the discovery of gold catalytic properties has been defined as a "black swan event" in organic synthesis [83]. The once untapped potential of gold (usually in the form of Au(I) complexes) resides in its ability to act as Lewis acid with a unique affinity for π-bonds—along with Pt(II), it has been defined a "π-acid" [84]. This peculiarity is due to relativistic effects, maximum for this element. Gold(I) and gold(III) complexes are usually less air and water sensitive than other metal ones, and they are cheaper than Pd, Ir, Rh, Pt. Moreover, they accept just one ligand, considerably lowering the amount of it required and dwindling its related cost. The biggest difference with respect to Pd

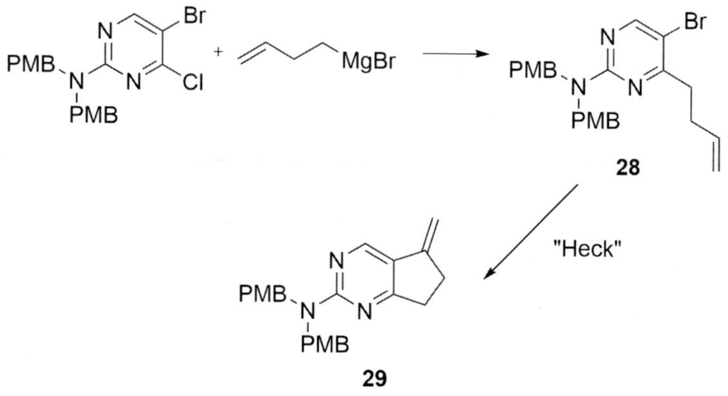

SCHEME 2.12

Selectivity of iron catalysts for chlorides over bromides in the synthesis of pyrimidyl tetrazoles (final products not shown). Conditions for the first coupling: Fe(acac)$_3$, NMP, THF, −20°C, then N,N-DMEDA.

is that protodeauration steps are favored over β-hydride elimination ones, leading to a peculiar reactivity. Gold catalysts allow a variety of transformations [85]. Enynes cycloisomerizations were the first transformations implemented in organic synthesis. Au catalysts also allow to use propargyl esters as α-diazoketones equivalents; moreover, owing to their tolerance toward a broad spectrum of moieties, unconventional nucleophiles can be employed (e.g., not only alcohols or amines, but also epoxides, acetals, ethers, etc.). Last, but not least, hydroaminations and hydroarylations are easily achieved through Au-mediated processes, whereas other catalysts cannot be employed likewise [86]. The potential of gold catalysis in diversity-oriented synthesis has been exploited and proved by Fürstner and Schreiber. Starting from two different diastereoisomers of enyne **30**, through gold catalysis they managed to achieve 10 different naturally occurring sesquiterpenes (Scheme 2.13) [87,88].

Au-mediated processes are extremely stereospecific, and they can be used to introduce naturally occurring cyclopropane motifs with ease. As for SPOS, Testero et al. recently proved that gold chemistry is exploitable also with supported reactants [89]. Despite their linear coordination geometry with one ligand, gold complexes under certain conditions are capable of catalyzing enantioselective processes. This challenging task is usually accomplished using bimetallic gold complexes with atropisomeric bisphosphine ligands; however, several examples of monodentate ligands or chiral counteranions to the gold complex are reported [90]. As mentioned, hydroaminations are difficult to achieve with other metal catalysts, whereas they are operationally simple with gold catalysts; more interestingly, enantioselectivity can be induced through chiral ligands, as reported by Han et al. [91]. Another interesting strategy has been reported by Wang et al. that achieved intermolecular annulation between N-allenamides **31** and enynes **32** in a total enantio-, regio-, and

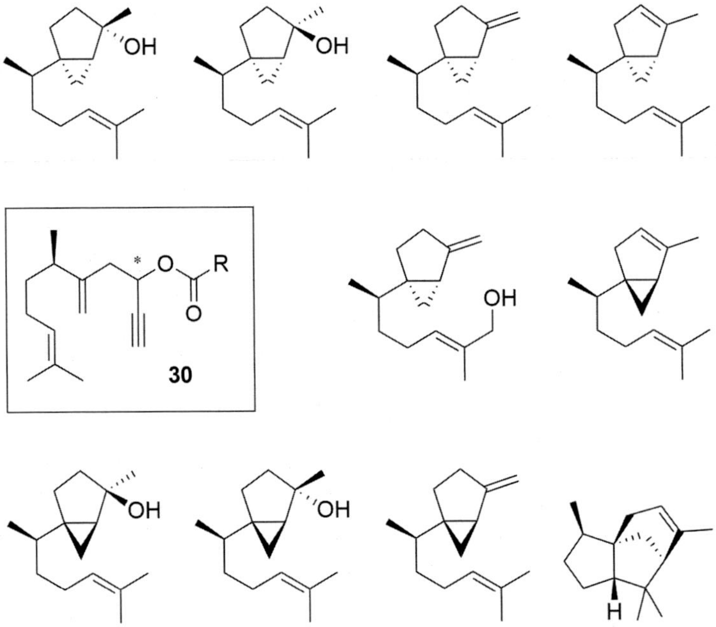

SCHEME 2.13

Enantiospecific synthesis of sesquiterpene analogues via gold catalysis, starting from a common intermediate.

diastereoselective fashion [92]. The reaction proved to be scalable to gram scale and both enantiomers of the product **33** can be selectively prepared. The final product can be further elaborated to introduce diversity (Scheme 2.14).

Concluding, it is important to remind that gold, likewise Pd (see 2.1), Ni, Rh, and other metals, offers the possibility to introduce diversity through ligand-directed divergent approaches [26]. Several remarkable examples are reported in the literature [93–95]; we report here a strategy developed by Jiang et al. that exploited the stereoelectronic effects of different ligands, counterions, and reaction conditions to achieve three different regioselective cycloisomerizations and hydroamidations from a common scaffold [96]. This approach is not only interesting because of the feasibility of introducing scaffold diversity with ease, but also because such reactions involve unactivated $C(sp^2)$-H bonds, highly desirable feature from a step-economical perspective (Scheme 2.15).

2.3 Cycloaddition reactions

Despite their ability to install multiple $C(sp^3)$ centers in a single step, usually in a stereospecific or stereoselective fashion, cycloadditions are quite neglected in drug research [4]. This may be due to the strict electronic and steric requirements

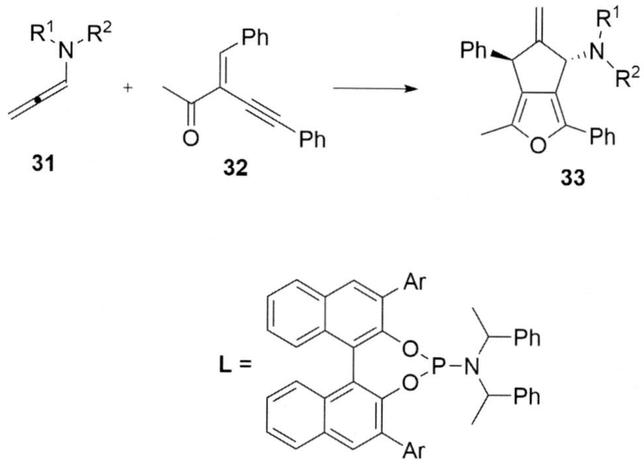

SCHEME 2.14

Regio-, diastereo-, and enantioselective gold catalyzed annulation of *N*-allenamides and enynes. Conditions: AuCl, ligand **L**, AgSbF$_6$, DCM, −15°C.

that any class of cycloadditions features, leading often to a good regioselectivity and stereospecificity, but also to a scarce modularity. Nonetheless, cycloadditions retain an untapped potential to be exploited in drug research; [2 + 2] cycloadditions, Diels-Alder reactions, and 1,3-dipolar cycloadditions are remarkable routes to afford four-,

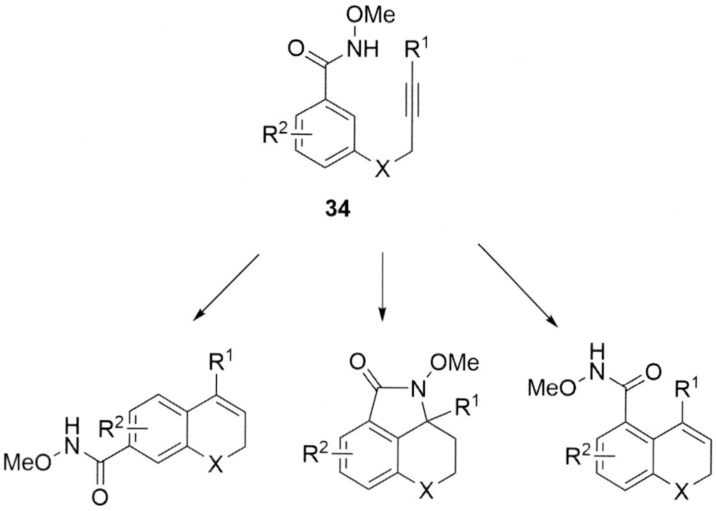

for details on ligands, catalysts and conditions see Ref. [96]

SCHEME 2.15

Ligand-directed divergence under gold catalysis.

six-, and five-membered (hetero)cycles [97]. Complexity and diversifications are usually achieved by varying simultaneously both the cycloaddition components. This can be performed easily through solid phase organic synthesis (SPOS), and several examples are reported and analyzed in literature [97,98]. Cycloadditions can occur by thermal or photochemical activation and can be catalyzed by several metals; thus, chiral ligands to a metallic center can induce enantio- and diastereoselectivity to the process. We report herein some selected examples of cycloadditions that, in our opinion, give a comprehensive outlook of the underrated potential in drug research that these classes of reactions feature.

Brown et al. reported an interesting approach toward [2 + 2] cycloadditions [99]. Commercial alkenols were immobilized on a carboxylated PS resin **35**, derived from Merrifield resin, and converted to cyclobutanone iminium salts **37** via a [2 + 2] cycloaddition with the in situ formed keteneiminium salt of an *N,N*-dialkylamine **36** (Scheme 2.16). The products were subsequently hydrolyzed to the corresponding cyclobutanones and elaborated to achieve γ-lactones (via Baeyer-Villiger ring expansion), γ-lactams (via Beckmann rearrangement), and cyclobutylamine, cyclo-butanol, and cyclobutanone derivatives.

Van der Eycken and Kappe reported an interesting "traceless" solid phase approach on supported 2(1H)-pyrazinones **38**. A tandem microwave assisted Diels-Alder addition with dienophile **2.39**, followed by retrocyclization of **40**, affords the desired pyridine **41** directly in solution, by elimination of an isocyanate resin **42** (Scheme 2.17) [100]. It is also possible to exploit the presence of the chloride moiety of **41** in microwave-assisted cross-coupling reactions to further decorate the scaffold, in a scaffold-divergent fashion [101].

Perhaps the most popular cycloaddition in medicinal chemistry is the 1,3-dipolar cycloaddition between alkynes and azides, also known as the Huisgen reaction [102]. Alkyne and azide moieties are easily introduced within a scaffold; however, despite the usefulness of the triazole ring formed, the classical reaction lacked selectivity, affording a mixture of 1,4- and 1,5-triazoles. A novel surge of interest around this reaction occurred in 2002, when two different groups reported that the process was accelerated by Cu(I) salts, affording selectively the 1,4-regioisomer. Later, in 2005, another group reported a Ru-catalyzed version of the reaction, affording

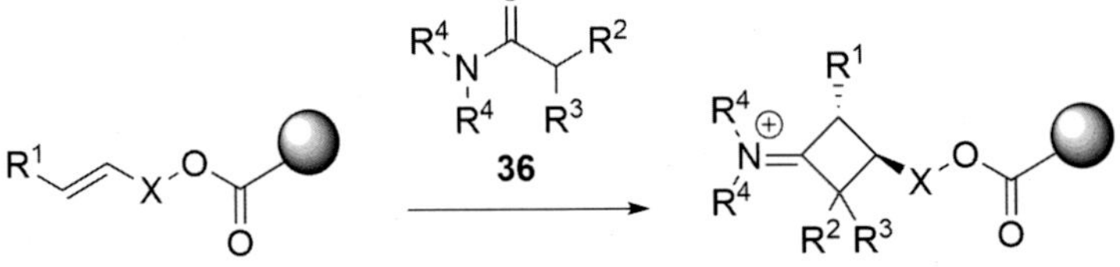

SCHEME 2.16

[2 + 2] cycloaddition of solid-supported alkenols with in situ generated keteniminium salts. Conditions: Tf$_2$O, 2,6-(tBu)$_2$Pyr, DCM.

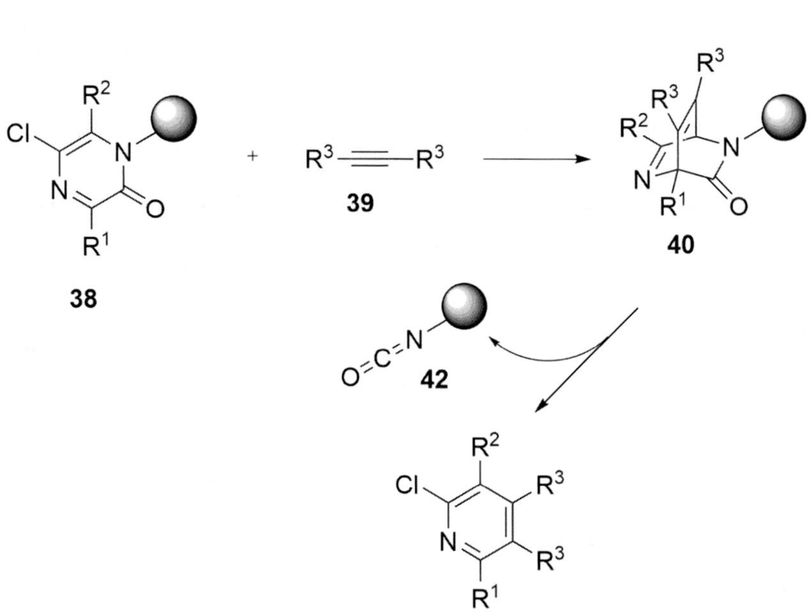

SCHEME 2.17

Microwave-assisted solid-phase synthesis of nitrogen heterocycles via tandem Diels-Alder cyclization/retrocyclization. Conditions: chlorobenzene, MW, 220°C.

selectively the 1,5-regioisomer. Triazoles are extremely stable to reductive, oxidative, basic, and acid conditions, but also to metabolic degradation; their high dipolar moment makes them suitable counterparts in the formation of hydrogen bonds or dipole-dipole interactions. Furthermore, they can be effectively employed as bioisosteres for some labile amide or ester bonds, with 1,4-regioisomers mimicking Z-amide bonds and 1,5-regioisomers mimicking *E*-amide bonds. Last, but not least, they can also act as mimetics for peptide β-strand turns, and acyl phosphate and *E*-olefinic moieties. Among the different approaches that exploited the potential of the CuAAC (*Cu-catalyzed zzide alkyne click*) reaction, the one reported by Zhu et al. exemplifies the utility of this [3 + 2] cycloaddition in the synthesis of macrocycles **43** [103]. A multicomponent approach (for multicomponent reactions, see Par. 4) leads to an alkyne-azide intermediate **42**, that is subsequently cyclized in a one-pot fashion. Their approach allows a facile and rapid synthesis of diverse macrocycles, with up to 16-membered rings (Scheme 2.18).

CuAAC reaction found application also in the so-called *fragment-based drug discovery*. This approach aims to create molecules formed by different moieties linked together, each one interacting with a different binding site of the target. Yao et al. reported a combinatorial, fragment-based approach toward metalloprotease (MMP) inhibitors [104]. Terminal alkynes bearing a succinyl hydroxamate

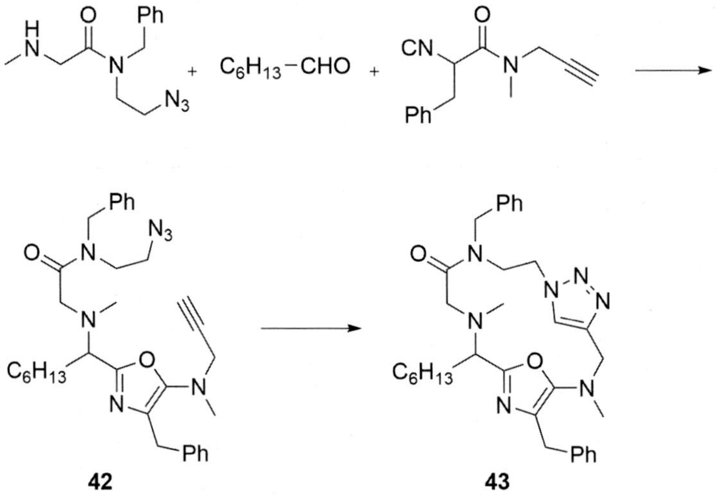

SCHEME 2.18

Synthesis of macrocycles in a one-pot tandem MCR/CuAAC approach. Conditions: toluene, NH$_4$Cl, 80°C, then THF, CuI, DIPEA.

moiety, known to interact with the zinc ion present in the catalytic site of some metalloproteases, were coupled with azides bearing hydrophobic moieties, responsible for the selectivity toward the targeted MMP. The approach allowed the fast generation and screening of a library of 96 compounds, finding two potential drug candidates.

Beside common [2 + 2], [4 + 2], and [3 + 2] cycloaddition strategies, an interesting approach that found application in natural product synthesis is the [5 + 2] cycloaddition [105]. Seven-membered rings are widely diffused among natural and drug products, and cycloadditions allow to generate complex and diverse scaffolds with ease with respect to a linear synthesis followed by ring closure. Several examples of metal-catalyzed or metal-free [5 + 2] cycloadditions are reported; also, enantioselective versions have been developed. Yudin and Wender reported several methods to achieve an effective divergent synthesis of seven-membered carbocycles starting from vinyl cyclopropanes **44** via a Rh-catalyzed [5 + 2] cycloaddition (Scheme 2.19). Compound **44** reacted smoothly with terminal and internal alkynes giving 4,5-disubstituted cyclohept-4-enones **45** [106]. Propargyltrimethylsilanes were employed as allenes equivalents giving the corresponding cycloheptanones **46** [107]; ynol ethers as ketene equivalents are also feasible substrate for the reaction, leading to substituted 1,4-cycloheptandiones **47** [108].

Using chiral catalysts or ligands, cycloadditions can be performed in an enantio- and/or diastereoselective fashion. Configuring quaternary chiral centers is an utmost challenging task; nevertheless, Zhang et al. recently reported a regio- and enantioselective Cu-catalyzed 1,3-dipolar cycloaddition affording in high yields and enantiomeric excesses valuable pyrrolidines [109]. Recently, a similar approach toward

for conditions see Refs. [106-108]

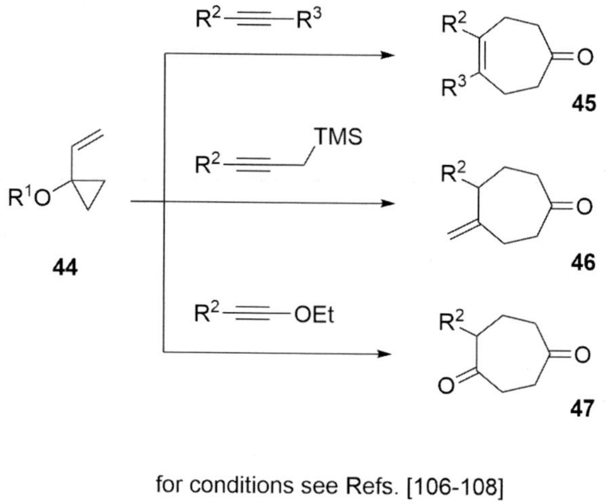

SCHEME 2.19

Scaffold-divergent synthesis via Rh-catalyzed [5 + 2] cycloadditions.

the diversity-oriented synthesis of complex chromeno-pyrrolidines has also been re-ported [110].

Finally, Baran et al. proposed an interesting approach toward enantioselective diversity-oriented synthesis of drug-like molecules [111]. On one hand, cycloaddi-tions feature an unmatched capability of building complex ring systems with great stereocontrol, forging several bonds simultaneously; however, they lack modularity. On the other hand, cross-couplings afford new C−C bond with excellent reliability and modularity. Baran developed a concise synthetic path exploiting the advantages of these two classes of transformations. They established a protocol consisting of cy-cloadditions and cross-coupling reactions, using commercial, cheap maleic anhy-dride as surrogate for a chiral di(pseudo)halide. Maleic anhydride undergoes a cycloaddition in the first step, giving a meso adduct that is subsequently desymme-trized. At last, a cross-coupling/deprotection/cross-coupling sequence on the chiral precursor leads to the final product with the desired stereochemistry. Different cross-coupling reactions (e.g., Kumada, Suzuki, Negishi) can be employed effectively, as well as several cycloaddition reactions ([4 + 2], [3 + 2], [2 + 2], [2 + 1]). As proof of the potential of their approach, they applied it simplifying the formal syntheses of known substances of interest, such as Asenapine **48** (Scheme 2.20).

2.4 Multicomponent reactions

Multicomponent reactions (MCRs) are a class of reactions where three or more start-ing molecules react to afford a product sharing most of the atoms of the original re-actants [112]. As such, they are an extremely powerful tool in combinatorial

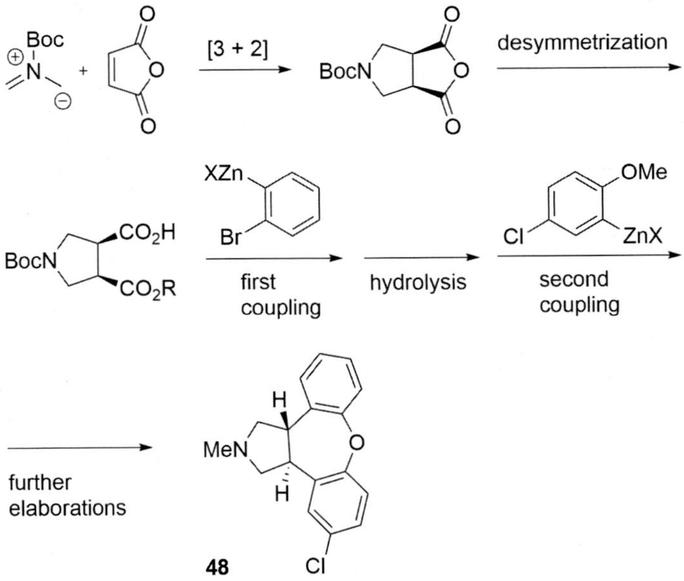

SCHEME 2.20

Formal synthesis of asenapine through a cycloaddition/cross-coupling sequence.

chemistry and drug research. MCRs usually feature a remarkable atom and step economy, they have multiple inputs to be exploited in generating molecular diversity and complexity and they are convergent, giving higher yields with respect to the traditional linear, stepwise approach. Furthermore, they usually form multiple bonds at a time. Many MCRs are name reactions (e.g., Ugi, Passerini, Biginelli, Petasis, Hantzsch, Strecker) [112], and they are usually tolerant to most functional groups, that can be subsequently exploited to achieve further diversification and molecular complexity in a *build/couple/pair* (B/C/P) strategy [3,113]. A great number of reviews and books have been published over the years, remarking their synthetic utility, particularly in biologically oriented synthesis (BIOS) and diversity-oriented synthesis (DOS) [112–117]. An important aspect that should be highlighted when discussing the importance of multicomponent reactions concerns the MCR-accessible chemical space. As we pointed out earlier in this Chapter (Part 1), the most common reactions that are being employed in drug research led to an overproduction of linear/planar shaped molecules, owing to the vast application of cross-coupling and amide bond forming reactions. This consideration should be of utmost importance, if we take into account that the success of a small molecule drug discovery project strictly depends on the chemical space area that has been chosen for investigation. The chemical space accessible through MCRs is way more different from that accessible through traditional methods; this may be the reason behind the fact that MCRs libraries shows higher screening hit rates toward specific target classes (e.g., protein-protein interactions, PPIs). Privileged structures such as cyclopropane rings or spiranes can be easily introduced through MCRs chemistry, whereas other approaches would require long manipulations [112]. Another remarkable aspect of MCRs chemistry is their modularity. Most MCRs can be easily modified through rational design to broaden the spectrum of achievable structures from common substrates. In fact, the notorious Ugi reaction (U-4CR, i.e., Ugi 4-component reaction) is a modification of its elder sibling, the Passerini reaction. Ugi and Passerini reactions feature a variety of modifications, as well as many other MCRs do. Generally, new multicomponent reactions can be designed via *single reactant replacement* (SRR), *modular reaction sequences* (MRS), *condition-based divergence* (CBD), or *combining multicomponent reactions* (MCRs). All these reasons make multicomponent reactions an essential tool for every organic chemist. Trying to enumerate and examine all MCRs would result in an uncomprehensive treatise on a topic that has already been reviewed exhaustively; nonetheless, we will try to give the reader a concise but effective outlook on multicomponent reactions reporting selected examples.

Among the oldest multicomponent reactions known, Passerini reaction (P-3CR) [118,119] is one of those that has been widely exploited and investigated in drug discovery project and library syntheses. P-3CR offers a plethora of possible approaches in BIOS; one of the most elegant and effective strategies hitherto developed is the so-called PADAM (Passerini reaction/amine deprotection/acyl migration) strategy that has been reported by two different groups independently [120,121]. This strategy allows to synthetize libraries of structurally complex β-ketoamides and β-hydroxymethyl amides to be screened as protease inhibitors. The reaction sequence involves a Passerini reaction on a protected (chiral) aldehyde **49**, usually derived from an α-amino acid. Then, the protecting group of adduct **50** is cleaved allowing an O → N acyl migration to afford **51**. Eventually, the β-hydroxymethyl amide **51** can be oxidized to β-ketoamide (Scheme 2.21).

This strategy has been exploited, for example, in the combinatorial optimization of the C-terminal chain (residing near the active site) of HCV NS3 protease inhibitors [122,123]. The strategy proved to be robust and reliable and has been scaled up to kilograms [124] and has also been translated effectively to the solid phase synthesis of peptidomimetic compounds [125]. In addition to the PADAM strategy, Passerini reaction can be effectively employed to create combinatorial libraries of other relevant drug candidates. Dömling et al. effectively applied a combinatorial approach toward the identification of HIV-1 protease inhibitors based on the Passerini reaction and a subsequent cyclization [126].

Ugi reaction, first published by Ivar Ugi in 1962 [127] as a modification of the Passerini reaction, had a strong impact on modern combinatorial synthesis; uncountable applications were found over the years. A noteworthy approach has been reported by Habashita et al. toward the design, synthesis, and screening of low-molecular-weight antagonists of CCR5/CCR2, two classes of G-protein coupled receptors (GPCRs), for HIV infection treatment [128]. The group used the so-called UDC (*Ugi/Deboc/Cyclization*) approach [129] on resin-supported isocyanides **52** to afford libraries of spirodiketopiperazines **2.53**, privileged structures

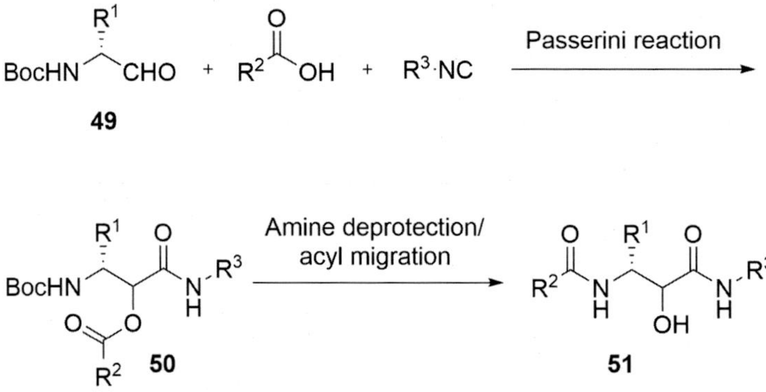

SCHEME 2.21

Classical PADAM strategy (Passerini reaction/amine deprotection/acyl migration).

for GPCRs interaction. After the initial screening, some potential candidates were found, and the subsequent optimization via combinatorial approaches led to the development of a potent lead active on CCR5 and CCR2 (Scheme 2.22).

Basso et al. applied a modification of the classical Ugi reaction, the Ugi 5-centers-4-component reaction (U-5C-4CR) [130,131], to the synthesis of pluripotent substrates (PSs) in the virtual synthesis of compound libraries that were studied as potential antagonists of B-cell lymphoma protein Bcl-X$_L$ [132]. PSs can be differentiated into several libraries with different backbones, achieved by orthogonally elaborating different functional groups on the pluripotent precursor. Starting from the optically pure oxabicyclic amino acid **54**, a stereoselective U-5C-4CR leads to a library of pluripotent peptidomimetic precursors **55** (Scheme 2.23). These substrates

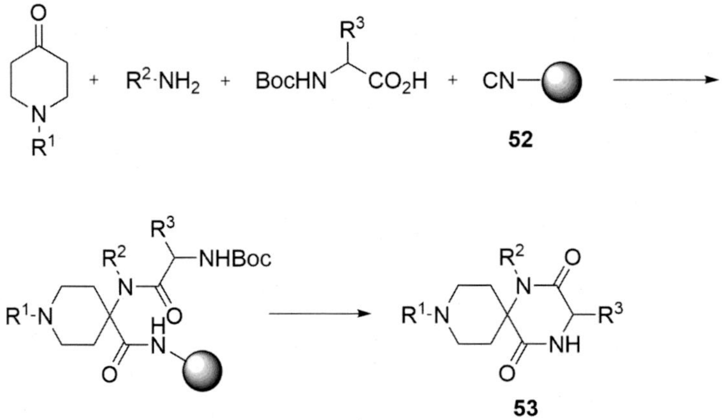

SCHEME 2.22

Solid-phase synthesis of a library of spirodiketopiperazines via Ugi-4CR/deprotection/cyclization protocol.

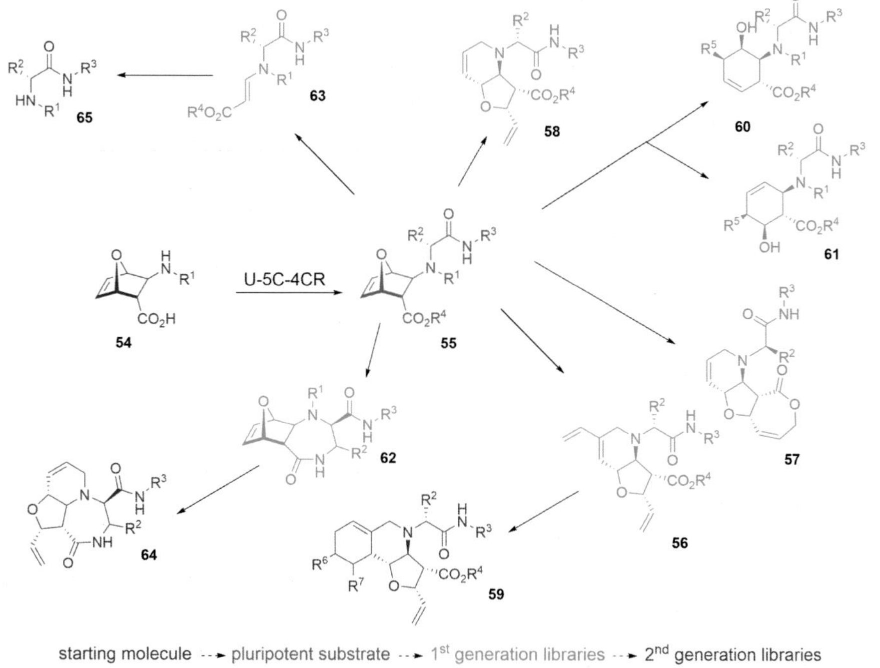

SCHEME 2.23

Combinatorial diastereoselective synthesis of oxabicyclic substrates and derived libraries in the identification of lead compounds as antagonist of protein Bcl-X$_L$ (B-cell lymphoma).

can be further elaborated into libraries **56–58** via tandem ROM/RCM; library **2.59** can be achieved through a [4 + 2] cycloaddition from library **56**. Library **63** can be achieved by *retro*-Diels-Alder with loss of furan and further transformed into chiral amino acids **65**. Libraries **60** and **61** of regioisomeric cyclohexenols were prepared by metal-mediated oxygen bridgehead ring opening of **55**. When an additional amino group is present, a library of bicyclic lactams **62** can be elaborated and then further processed toward library **64** via tandem ROM/RCM. This approach is a remarkable example of the high degree of diversification that can be achieved coupling different, established methodologies (cycloadditions, MCRs, metathesis, cross-couplings).

Oddly enough, even though many multicomponent reactions have been developed independently over time, most of them share at least one component, more commonly two. A glimpse of this intricated scenario is depicted in Schemes 2.24A–C, where Passerini and Ugi modifications are reported. These were achieved by rational SRRs. The Ugi reaction itself, as mentioned earlier, is in fact a modification of the Passerini reaction, employing an imine component in place of a carbonylic one. Both U-4CR and P-3CR can be further modified, exploiting different acidic and/or electrophilic components and achieving accordingly a wide variety of

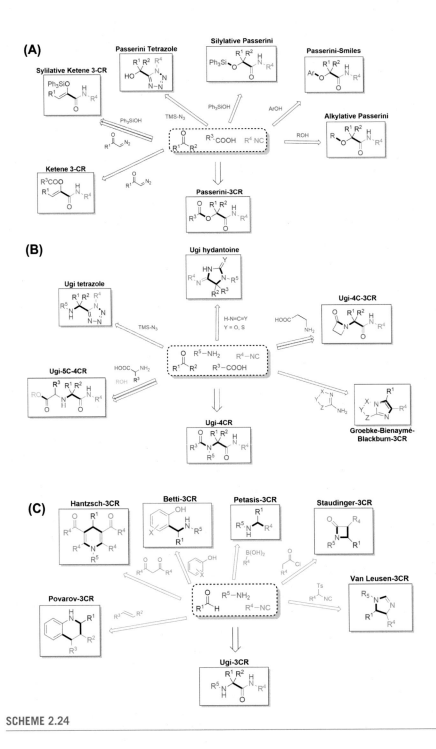

SCHEME 2.24

structures with different applications. Ugi tetrazole, for example, proved to be a valid entry point for combinatorial libraries, even on gram-scale [133]; modified Passerini reactions employing an in situ generated ketene as electrophile (ketene 3-CR [134] and silylative K-3CR [135]) easily generate scaffolds that can be readily converted into drug-like derivatives. Also, with the Groebke-Bienaymé-Blackburn-3CR [136−139] C(sp^2)-rich heterocyclic skeletons can be obtained, in contrast with its siblings P-3CR and U-4CR that usually afford C(sp^3)-rich systems. Similarly, we reported how historically unrelated MCRs can be modified into one another by simple SRR. In this case, totally different mechanistic pathways are followed; U-3CR and Van Leusen-3CR [140], for instance, rely on the peculiar reactivity of isocyanides, while all the other examples reported exploit a completely different reactivity. Nonetheless, a plethora of different structures can be achieved, introducing a remarkable diversification with ease.

As stated earlier, other approaches are possible toward complexity and diversity. From a common precursor synthetized via MCR, Zhu, Bienaymé et al. forged different scaffolds with several substituents in a one-pot fashion simply varying the extra component added [141,142]. Another way to introduce scaffold diversity is condition-based divergence (CBD): Kappe et al. exploited different reaction conditions to induce different multicomponent sequences toward different scaffolds, starting from the very same reagents [143]. Last, but not least, it should be remarked how combining different approaches and different MCRs can lead to complex scaffolds with several entry point for diversity in a small number of steps. A good example has been reported by Al-Tel et al. that, by combination of a GBB-3CR and a U-4CR, achieved five- and six-component one-pot transformations affording drug-like structures, forming up to 8 new bonds with up to 10 points of diversity [144].

However, what MCRs boast in terms of versatility and modularity usually lack in terms of stereocontrol; this is particularly evident with isocyanide-based multicomponent reactions (IMCRs) [113]. The wide-spread Ugi and Passerini reactions showed limited stereoselectivity and tend to afford all the possible stereoisomers even when a chiral input is present among the components. This leads to tedious and time-consuming separations. Nonetheless, in some cases IMCRs on chiral substrates gave good results. Orru et al. effectively coupled biocatalysis with MCR chemistry using monoamine oxidase (MAO-N) from *Aspergillus niger* to obtain chiral imines from pyrrolidines. The chiral substrates gave good induction in Ugi-Joullié-3CR and the products were further elaborated to achieve organocatalysts [145], synthetic alkaloids [146], and drug telaprevir [147]. In any case, the low stereocontrol of IMCRs is usually ascribed to the fact that they are uncatalyzed

(A–C). Modularity of MCRs: it is possible to achieve a variety of scaffolds through a single or double reactant replacement (SRR/DRR). The color of the arrow(s) indicates the component that is replaced. New bonds that are formed are marked in bold.

processes, whereas other MCRs such as the Hantzsch [148], Biginelli [149], Povarov [150], and Mannich [151–153] reactions benefited from the development and advances of chiral organocatalysis and Brønsted acids catalysis. Nevertheless, most recently Houk, Tan et al. reported an asymmetric, chiral phosphoric acid (CPA) catalyzed version of the Ugi-4CR [154]. The presence of the CPA catalysts enhanced both the nucleophilicity of the carboxylic acid and the electrophilicity of the imine, allowing the catalyzed process to be favored over the typical noncatalyzed one. Aside the bifunctional activation of the phosphate moiety, the presence of an apolar pocket on the catalyst allows noncovalent interactions between the substrates and the catalyst itself to occur, determining a strong induction to the overall sequence. With this groundbreaking work being published, a surge of interest in developing asymmetric versions of known IMCRs can be expected, hopefully leading to the possibility of introducing a desired stereochemical diversification via multicomponent processes.

2.5 Photochemical processes

In the last decade, the synthetic toolbox available to organic chemists has been enlarged with the development of novel methods and strategies exploiting light as a source of energy. Recently, there has been a surge of interest in photochemistry and photoredox catalysis, owing to the possibility to easily access and exploit radical chemistry, which allows a reactivity hitherto little investigated. Photochemical processes are orthogonal to most approaches and usually complementary to traditional polar chemistry (for example, photolabile resin linkers have been developed [155], allowing cleavage conditions that do not interfere with the most common protecting strategies). Even though some chemical reactions can occur via direct absorption of light by a functional group (e.g., Paternò-Büchi reaction [156,157], Wolff rearrangement [158], pericyclic reactions [159]), the introduction of photocatalysts allowed to extend the scope of light-driven processes to virtually any molecule. Generally, a catalytic cycle, where the first step occurring is the absorption of light by the photocatalyst, is established. Hence, the excited state catalyst can act either as a reductant or an oxidant, either donating to or accepting an electron e^- from the substrate or another species, respectively. At last, an advanced intermediate or a sacrificial reactant interacts with the photocatalyst, converting it back to the ground state and closing the cycle. With respect to polar reactivity, photocatalyzed processes usually occur via single electron transfer (SET) and/or hydrogen atom transfer (HAT), leading to unprecedented reactivity in mild conditions. Both oxidative and reductive catalytic cycles can lead to oxidative, reductive, or redox-neutral processes, with a plethora of resulting possibilities. Also, some processes can occur by energy transfer (ET) from the excited catalyst to a substrate: in this case, the photocatalyst is usually referred to as photosensitizer. Both metallic complexes and organic dyes can efficiently be employed as catalysts, each one with its unique features. Several reviews have been published over the years, covering the topic exhaustively [160–164].

Herein, we report selected examples to give the reader an effective outlook on the synthetic capabilities of photochemistry in modern synthesis.

Photoredox catalysis allows to exploit a different reactivity of already known substrates. Silyl enol ethers, for example, are renown for acting as nucleophiles in aldol reactions [165] and can be easily achieved by deprotonation of an α-acidic carbonyl compound and subsequent trapping of the enolate with a silyl chloride. Pandey et al. established a practical method to involve silyl enol ethers in difficult oxidative aryl annulations [166]. Using 1,4-dicyanonaphthalene (DCN) as a photoredox catalyst under aerobic conditions ($\lambda > 280$ nm, UV-B), electron-rich aromatics smoothly reacted with silyl enol ethers affording complex spiro systems or five- to eight-membered rings in good yields (Scheme 2.25). Pandey further applied this strategy as a key step in the total synthesis of (+)-2,7-dideoxypancratistatin, starting from (−)-quinic acid [167].

Admittedly, Padley's approach is a virtuous example of how photoredox catalysis can affect the reactivity of a functional group; however, it still relies on UV light that is less abundant, more dangerous and more energetic with respect to visible light. Owing to the potential of exploiting sunlight in a near, greener future, a plethora of processes exploiting blue and green light have been developed. Nicewicz et al. developed several anti-Markownikov hydrofunctionalizations of alkenes under mild conditions upon blue LEDs irradiation ($\lambda = 450$ nm). As stated earlier (Par 2.2.3), hydrofunctionalizations are difficult to achieve, and particularly, hydroaminations require specific metal catalysts, as most of them are sensitive to nitrogen-containing compounds. Via their generic catalytic protocol, Nicewicz group achieved simple and effective hydroetherifications [168], hydrotrifluoromethylations [169], and hydroaminations [170] of alkenes, as well as the direct addition of carboxylic acids to C=C double bonds (Scheme 2.26) [171]. The photocatalyst is an acridinium salt that acts as an oxidant in its excited state, accepting an electron from the olefin that subsequently forms a radical cation. A cocatalyst is required to

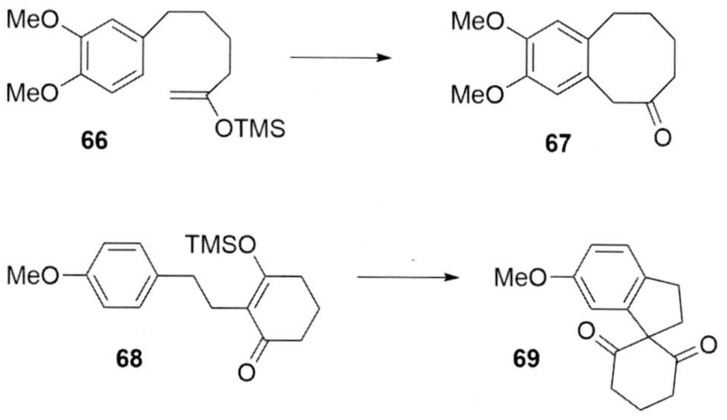

SCHEME 2.25

Photoredox catalyzed annulation with silyl enol ethers. Conditions: DCN, MeCN/H$_2$O 4:1, O$_2$, light (280 nm).

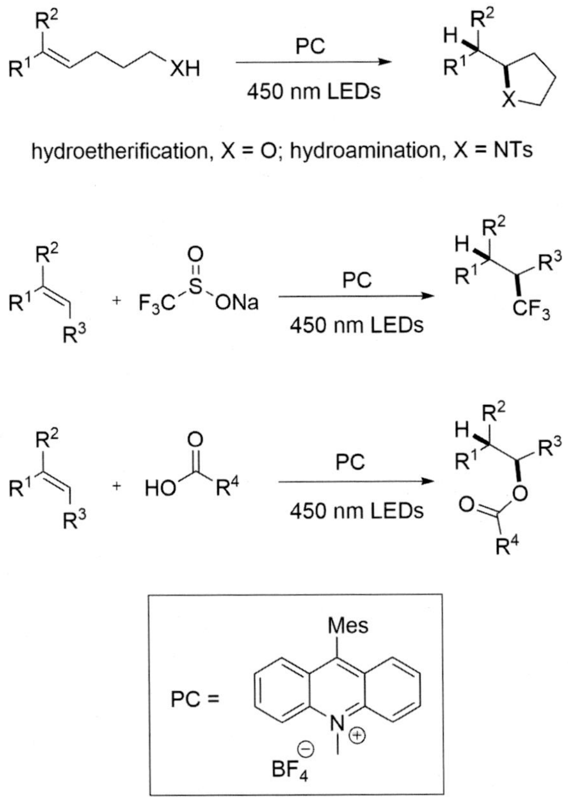

hydroetherification, X = O; hydroamination, X = NTs

SCHEME 2.26

Photoredox catalyzed hydrofunctionalizations of alkenes by Nicewicz group. For conditions see Refs. [168–171].

close the cycle and maintain the net redox neutrality of the process. These methods can be applied as metal-free alternatives to other hydrofunctionalizations.

Other important advances in visible light-driven processes involve the use of singlet oxygen ($^1O_2$) [172]. Singlet oxygen can participate in several transformations ([4 + 2] and [2 + 2] cycloadditions, ene-type reactions, heteroatom oxidations) and can be conveniently produced via *photosensitization*, minimizing the formation of radical or ionic byproducts. Methylene blue (MB), tetraphenylporphirine (TPP), and rose bengal (RB) are the most diffused photosensitizers, exciting to a triplet state upon irradiation. Ground state triplet oxygen ($^3O_2$) is then converted to singlet oxygen upon interaction with the excited photosensitizer that turns back to ground state closing the cycle. Among all the examples reported, we herein report a convenient stereodivergent synthesis of carbasugars, useful synthetic products, and/or intermediates for synthesis and drug discovery [173]. Using TPP as a photosensitizer, Balci et al. developed a convenient, stereodivergent synthesis of *DL-proto-* and *DL-gala-*

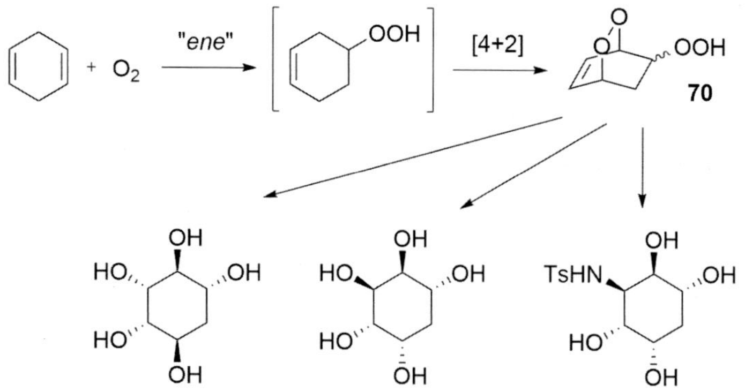

SCHEME 2.27

Stereodivergent carbasugars synthesis via tandem ene-type reaction/[4 + 2] cycloaddition of singlet oxygen with 1,4-cyclohexadiene. For conditions see Refs. [174–176].

quercitol **71** and **72** form **70**, obtained via a singlet oxygen ene-type reaction and subsequent [4 + 2] cycloaddition starting from 1,4-cyclohexadiene [174]. This oxidative approach has been further exploited by the authors to achieve also *DL-gala*-aminoquercitol **73** [175] and branched carbasugars [176] (Scheme 2.27).

Singlet oxygen, as proved by Vassilikogiannakis group and others, can also be a powerful tool to achieve scaffold diversification [177]. Furans can readily react with singlet oxygen and an amine to afford a 2-pyrrolidinone intermediate that can be further elaborated in a tandem fashion toward a plethora of alkaloidal, drug-like molecules.

It appears evident that the importance of photochemical processing in medicinal chemistry and BIOS/DOS is slowly but steadily growing. Recently, Shi et al. tested the scope of some $C(sp^3)$-$C(sp^2)$ photoredox cross-couplings to prove their utility in medicinal chemistry [178]. Using an integrated photoreactor, they successfully demonstrated that both decarboxylative and cross-electrophile couplings are reactions feasible of being applied in drug discovery programs. These reactions were performed on a library of 18 halides with high structural complexity and bearing several functional groups; cross-electrophile couplings proved to be more efficient, and the integrated photoreactor enhanced reaction and success rates. These results are most promising as screening the scope of a reaction on an informer library, i.e., collections of diverse, complex drug-like molecules, is currently one of the highest evaluation methods to advance and establish a novel synthetic sequence [179].

Kuznetsov and Kutateladze developed a novel photoassisted reaction cascade that forges complex molecular drug-like architectures in two to three steps [180]. Oxalyl anilide chromophore **74** can excite upon UV irradiation, and evolves via an excited state intramolecular proton transfer (ESIPT) and a subsequent intersystem

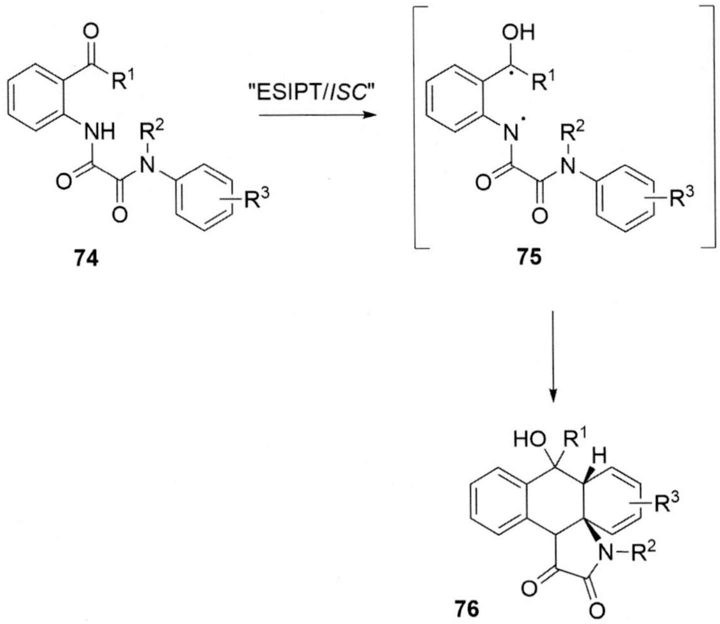

SCHEME 2.28

Photoassisted reaction cascade on oxalyl anilides for DOS applications.

crossing (ISC) to a triplet state diradical **75**, that cyclizes to a primary photoadduct **76** with dearomatization of the anilide moiety (Scheme 2.28). The complex scaffolds achieved can be further elaborated with Pictet-Spengler reactions, intramolecular cyclizations and double bonds functionalizations to access more rigid, drug-like skeletons with various decorations, increasing dramatically the complexity of the final product after each step. The authors reported remarkably good values of drug-likeness, calculated via DataWarrior [181] metrics.

Also, C−S bonds are difficult to establish via traditional routes: in spite of their success, click thiol-ene reactions lack of chemocontrol and overall group tolerance. Photosensitization, instead, allows to achieve hydrothiolations of unactivated multiple bonds with broad functional group tolerance and total chemo- and regioselectivity [182]. As reported by Guldi, Glorius et al., selective hydrothiolations can be performed using a traditional iridium metal catalyst or an organic dye for metal-free conditions. The disulfide-ene reaction occurs smoothly upon blue LEDs irradiation, without interfering with biomolecules and being tolerant to biological conditions. Methylthioether moiety especially is present in a plethora of biologically relevant molecules, and thioalkyl- or thioarylethers moieties can be found in drugs or bioconjugates. The aforementioned features, along with the high yields and the remarkable atom economy of the process, make it highly suitable for medicinal or biology-oriented syntheses.

To conclude, photochemical approaches offer a synthetic potential that is of great interest in assembling small molecules. Light-driven processes allow to achieve high molecular complexity and difficult transformations in a totally orthogonal fashion with respect to traditional chemistry. Moreover, they usually require milder and more environmentally sustainable conditions. Furthermore, they can be easily carried out in flow systems, with potentially increased yields and a more precise control of irradiation and conditions [183]. Accordingly, photoredox catalyzed and photosensitized processes should always be taken in consideration when planning a synthesis, and a wider diffusion of these new methodologies in drug research or medicinal-oriented programs is highly desirable.

2.6 Late-stage functionalizations

As seen throughout the rest of the Chapter, in some cases modern methodologies offer the possibility to use directly unfunctionalized C—H bonds as points of diversification. Due to the intrinsic stability of C—H bonds to most conditions, these transformations can be performed after most of the other ones have been accomplished. This emerging synthetic strategy is usually referred to as *late-stage functionalization* (LSF): C—H bonds are regarded as functional groups and, as such, valid entry points exploitable for introducing diversity into a scaffold, without the need of rethinking or modifying an established synthetic protocol [184]. LSFs hold the potential to rise among the most powerful tools available in drug discovery: they can facilitate the elucidation of SARs with improved efficacy, the optimization of the selectivity of a drug candidate, and the fine-tuning of its absorption-distribution-metabolism-excretion (ADME) properties. In some cases, subtle changes in a scaffold, such as the replacement of a C—H bond with a C—F one, are enough to discriminate whether a drug candidate will be a successful medicinal or a failure. However, some considerations spontaneously arise when dealing with this strategy. Whereas other reactions exploit the presence of certain functional groups and therefore their regioselectivity can be easily predicted, drug-like molecules contain manifold C—H bonds. Understanding the selectivity principles behind LSF reactions is mandatory for their successful incorporation in multistep syntheses. Generally, we can distinguish between *guided reactions*, where C—H discrimination is achieved exploiting the presence of directing groups, either by steric requirements or by molecular recognition, and *innate reactions*, where selectivity is determined by the inherent electronic properties of the C—H moiety. Several reviews described effectively and exhaustively the topic, covering both the mechanistic insights and the fruitful applications in total syntheses of LSF reactions [184–188]. Herein, we report selected examples of methodologies that offer, in our opinion, a representative glimpse of the current scenario and highlight the untapped potential of these emerging approaches in drug discovery.

A virtuous example of an LSF approach has been reported by Hartwig et al. [189]. Using a chiral ligand under iron catalysis in very mild conditions, they exploited the inherent electronic properties of tertiary C—H bonds to achieve an efficient and regioselective azidation. This approach proved to be robust and tolerant to a variety of functional groups and resulted selective for the most electron-rich and remote tertiary C—H bond present on the scaffold. Common directing groups influenced the regioselectivity with their electronic properties rather than with coordination of the catalyst. This approach allows to convert C—H bonds in C—N bonds, a difficult task that even enzymes cannot accomplish directly.

LSFs can be extremely helpful in elucidating SARs and SPRs. Buran et al. demonstrated how the physical properties of a drug candidate can be dramatically changed via C—H oxidations through an LSF approach [190]. Many bioactive compounds of natural occurrence or derivation (such as terpene derivatives) often show unsatisfactory pharmacokinetic profiles. As proved by the authors on betulin and betulinic acid (triterpenic derivatives found in birch tree bark), late-stage chemical or enzymatic oxidations markedly improved the solubility and bioavailability of the substrates, which were divergently decorated with hydroxyl moieties in different position exploiting diverse strategies.

LSFs are emerging also as effective methods to label designated positions on a molecule with specific atoms or (radio)isotopes. A noteworthy example has been reported by Groves et al., that successfully labeled a variety of molecules of biological relevance with ^{18}F using an LSF strategy [191]. ^{18}F is a radioisotope with a half-life of ~110 min, allowing its practical use in positron emission tomography (PET) imaging. PET is widely used in drug discovery as it allows to establish in vivo target engagement and occupancy [192]; Groves and coworkers developed a practical method to incorporate ^{18}F in drugs with no need of preactivation and in short time (ca. 10 min). Using a Mn(salen)OTs catalyst **77** and a source of ^{18}F, a series of COX, ACE, and MAO inhibitors and other bioactive molecules could be easily labeled on benzylic positions, as illustrated in Scheme 2.29 for selected examples **78**—**80**. Groves group also reported late-stages azidations [193] and general late-stage fluorinations [194].

Due to their complexity, natural products are difficult substrates and their elaboration challenging. Modern LSFs work with remarkable site selectivity despite the abundance of C—H bonds and potential directing groups. As proof of this concept, Beckwith, Davies et al. performed carbene insertions on selected C—H bonds on brucine, bicuculline, apovincamine, and other alkaloids and drugs under rhodium catalysis [195]. C—H activations can also occur in a cascade fashion. Yu, Stamos et al. recently published a very elegant triple C—H activation, toward a concise synthesis of complex pyrazoles **83** possessing drug-like scaffolds with no need of any prefunctionalization [196]. Interestingly, the first step occurring is a pyrazole-directed C(sp^3)-H activation of **81**, that under Pd catalysis leads to the formation of a C(sp^3)-C(sp^2) cross-coupling intermediate **82**; this transformation is followed by a double C(sp^2)-H activation, establishing a new C—C bond linking the pyrazole moiety to the aryl one (Scheme 2.30).

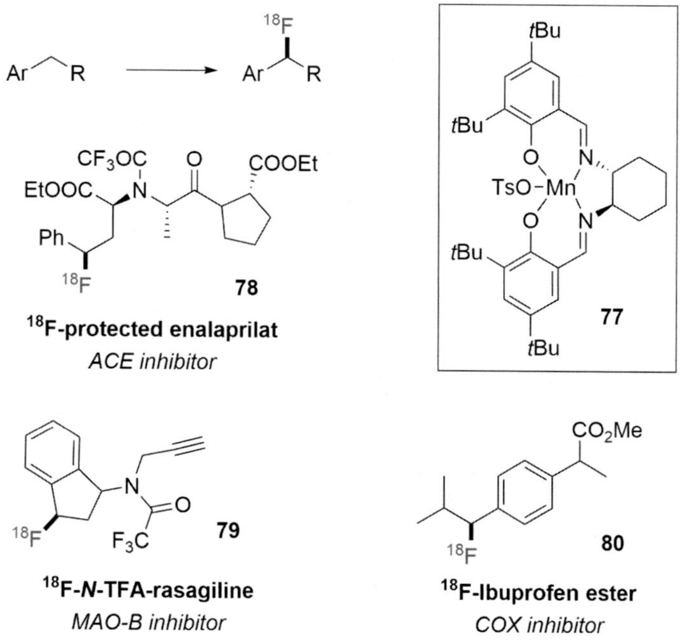

SCHEME 2.29

Radioisotope labeling for PET imaging studies via late-stage benzylic fluorination. Conditions: catalyst **77**, PhIO, [^{18}F]F-, K_2CO_3, acetone, air, 50°C.

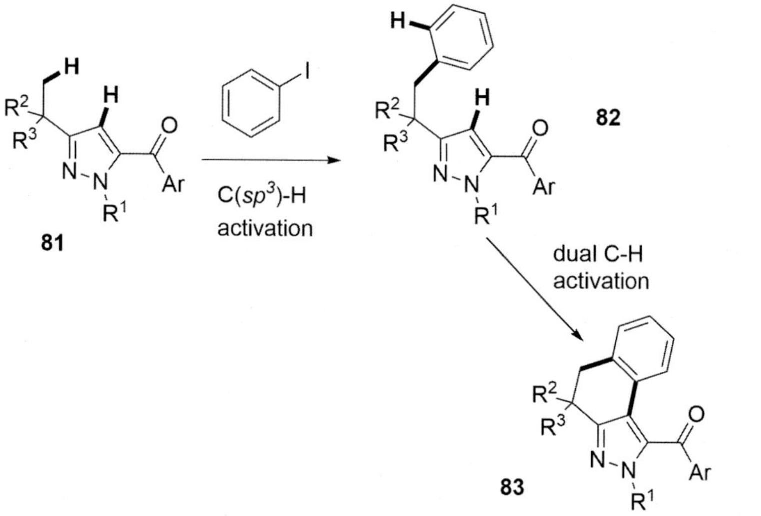

SCHEME 2.30

Triple C−H functionalization exploiting pyrazole directing properties with sequential C(sp^3)-H and dual C(sp^2)-H activation. Conditions: Pd(OTf)$_2$(MeCN)$_4$, Ag$_2$O, AcOH, air, 120°C.

It is worth noticing that C—H functionalizations and late-stage processes in general follow under three reaction classes: metal-catalyzed processes, biotransformations, and photoredox catalyzed processes. As a result, these approaches are usually orthogonal and complementary to one another. Photoredox catalysis proved its efficiency in C(sp^2)-H activation on (hetero)arenes [197,198]. MacMillan et al. recently established a simple photoredox-catalyzed process to perform Minisci reaction, affording drug-like scaffold from unfunctionalized (hetero)arenes and ethers [199].

In conclusion, we ought to remark that, despite their only recent development, LSFs can already be exploited in large-scale synthesis. Multikilogram sequences where late-stage functionalizations are successfully employed have been reported. Noteworthy example can be found, for example, within the synthesis of CETP (cholesteryl ester transfer protein) inhibitor anacetrapib, with a Ru-mediated direct C—H arylation [200], or within the synthesis of the JAK2 (Janus kinase 2) inhibitor LY2784544, where a radical C—H functionalization occurs under vanadium catalysis [201].

2.7 Conclusions and outlook

In this chapter, we have illustrated how novel organic syntheses have widened the portfolio of reactions applicable to the discovery of small molecules, endowed with complex and original structures. Overall, the current scenario offers a number of methodologies that are strictly complementary to one another and that permit the modern medicinal chemist to access even those portions of chemical space that are still unexplored. All the methodologies we have reviewed herein proved to be robust and reliable, as suggested by their slow uptake in the drug discovery field, and pave the way toward developing novel scaffolds to address a plethora of potential therapeutic targets. We can summarize the most important and recent developments in the following themes:

- discovery of more selective metal catalysts in cross-coupling reactions
- generalization of cycloaddition reactions in a substrate-independent manner
- use of chiral catalysis to obtain multicomponent adducts stereoselectively
- development of green and sustainable photochemical transformations

The same points can also be considered as future challenges, as the potential of these synthetic approaches has been only partially revealed and new results in these emerging fields are being reported very rapidly. In addition to these aspects, it is reasonable to expect that in the near future remarkable improvements will involve also more technological areas such as the implementation of efficient flow systems, the development of machine-assisted synthesis or the exploitation of solar energy.

References

[1] J. Drews, Drug discovery: a historical perspective, Science 287 (5460) (2000) 1960−1964.

[2] S.L. Schreiber, Target-oriented and diversity-oriented organic synthesis in drug discovery, Science 287 (5460) (2000) 1964−1969.

[3] T.E. Nielsen, S.L. Schreiber, Towards the optimal screening collection: a synthesis strategy, Angew. Chem. Int. Ed. 47 (1) (2008) 48−56.

[4] D.G. Brown, J. Boström, Analysis of past and present synthetic methodologies on medicinal chemistry: where have all the new reactions gone? J. Med. Chem. 59 (10) (2016) 4443−4458.

[5] T.W.J. Cooper, I.B. CampbellS, J.F. Macdonald, Factors determining the selection of organic reactions by medicinal chemists and the use of these reactions in arrays (small focused libraries), Angew. Chem. Int. Ed. 49 (44) (2010) 8082−8091.

[6] S.P. Nolan, O. Navarro, 11.01 − C−C bond formation by cross-coupling, in: D.M.P. Mingos, R.H. Crabtree (Eds.), Comprehensive Organometallic Chemistry III, Elsevier, Oxford, 2007, pp. 1−37.

[7] C.C.C. Johansson Seechurn, M.O. Kitching, T.J. Colacot, V. Snieckus, Palladium-catalyzed cross-coupling: a historical contextual perspective to the 2010 Nobel prize, Angew. Chem. Int. Ed. 51 (21) (2012) 5062−5085.

[8] S.A. Testero, E.G. Mata, Prospect of metal-catalyzed C−C forming cross-coupling reactions in modern solid-phase organic synthesis, J. Comb. Chem. 10 (4) (2008) 487−497.

[9] R.E. Sammelson, M.J. Kurth, Carbon−Carbon bond-forming solid-phase reactions. Part II, Chem. Rev. 101 (1) (2001) 137−202.

[10] G.L. Bolton, J.C. Hodges, Solid-phase synthesis of substituted benzazepines via intramolecular Heck cyclization, J. Comb. Chem. 1 (2) (1999) 130−133.

[11] M.J. Plunkett, J.A. Ellman, Solid-phase synthesis of structurally diverse 1,4-benzodiazepine derivatives using the Stille coupling reaction, J. Am. Chem. Soc. 117 (11) (1995) 3306−3307.

[12] S. Bräse, M. Schroen, Efficient cleavage−cross-coupling strategy for solid-phase synthesis—a modular building system for combinatorial chemistry, Angew. Chem. Int. Ed. 38 (8) (1999) 1071−1073.

[13] K.C. Nicolaou, N. Winssinger, J. Pastor, F. Murphy, Solid-phase synthesis of macrocyclic systems by a cyclorelease strategy: application of the Stille coupling to a synthesis of (S)-Zearalenone, Angew. Chem. Int. Ed. 37 (18) (1998) 2534−2537.

[14] R.E. Maleczka, I. Terstiege, Development of a one-pot palladium-catalyzed hydrostannylation/stille coupling protocol with catalytic amounts of tin, J. Org. Chem. 63 (26) (1998) 9622−9623.

[15] D.R. Dragoli, L.A. Thompson, J. O'Brie, J.A. Ellman, Parallel synthesis of prostaglandin E1 analogues, J. Comb. Chem. 1 (6) (1999) 534−539.

[16] L.A. Thompson, F.L. Moore, Y.-C. MoonJ, A. Ellman, Solid-phase synthesis of diverse E- and F-series prostaglandins, J. Org. Chem. 63 (7) (1998) 2066−2067.

[17] T. Hanazawa, T. Wada, T. Masuda, S. Okamoto, F. Sato, Novel synthetic approach to 19-nor-1α,25-Dihydroxyvitamin D3 and its derivatives by Suzuki−Miyaura coupling in solution and on solid support, Org. Lett. 3 (24) (2001) 3975−3977.

[18] I.B. Campbell, J. Guo, E. JonesP, G. Steel, A simple solid phase diversity linker strategy using enol phosphonates, Org. Biomol. Chem. 2 (19) (2004) 2725—2727.

[19] D.S. Tan, M.A. Foley, M.D. ShairS, L. Schreiber, Stereoselective synthesis of over two million compounds having structural features both reminiscent of natural products and compatible with miniaturized cell-based assays, J. Am. Chem. Soc. 120 (33) (1998) 8565—8566.

[20] A. Bergh, H. Leffler, A. Sundin, U.J. Nilsson, N. Kann, Cobalt-mediated solid phase synthesis of 3-O-alkynylbenzyl galactosides and their evaluation as galectin inhibitors, Tetrahedron 62 (35) (2006) 8309—8317.

[21] S. Bräse, J.H. Kirchhoff, J. Köbberling, Palladium-catalysed reactions in solid phase organic synthesis, Tetrahedron 59 (7) (2003) 885—939.

[22] V. Zimmermann, S. Bräse, Hartwig—buchwald amination on solid supports: a novel access to a diverse set of 1H-benzotriazoles, J. Comb. Chem. 9 (6) (2007) 1114—1137.

[23] M. Limbeck, H. Wamhoff, T. Rölle, N. Griebenow, Palladium-catalyzed α-arylation of ketones on solid support: scope and limitations, Tetrahedron Lett. 47 (17) (2006) 2945—2948.

[24] C.I. Traficante, C.M.L. DelpiccoloE, G. Mata, Palladium-catalyzed cross-coupling reactions of arylsiloxanes with aryl halides: application to solid-supported organic synthesis, ACS Comb. Sci. 16 (5) (2014) 211—214.

[25] G.C. Fortman, S.P. Nolan, N-Heterocyclic carbene (NHC) ligands and palladium in homogeneous cross-coupling catalysis: a perfect union, Chem. Soc. Rev. 40 (10) (2011) 5151—5169.

[26] Y.-C. Lee, K. Kumar, H. Waldmann, Ligand-directed divergent synthesis of carbo- and heterocyclic ring systems, Angew. Chem. Int. Ed. 57 (19) (2018) 5212—5226.

[27] D. Tsvelikhovsky, S.L. Buchwald, Synthesis of heterocycles via Pd-ligand controlled cyclization of 2-chloro-N-(2-vinyl)aniline: preparation of carbazoles, indoles, dibenzazepines, and acridines, J. Am. Chem. Soc. 132 (40) (2010) 14048—14051.

[28] M.J. Riveira, E.G. Mata, Cross-metathesis on immobilized substrates — application to the generation of synthetically and biologically relevant structures, Eur. J. Org. Chem. (13) (2017) 1675—1693.

[29] A.D. Piscopio, J.F. Miller, K. Koch, A second generation solid phase approach to Freidinger lactams: application of Fukuyama's amine synthesis and cyclative release via ring closing metathesis, Tetrahedron Lett. 39 (18) (1998) 2667—2670.

[30] K.C. Nicolaou, N. Winssinger, J. Pastor, S. Ninkovic, F. Sarabia, Y. He, D. Vourloumis, Z. Yang, T. Li, P. Giannakakou, E. Hamel, Synthesis of epothilones A and B in solid and solution phase, Nature 387 (6630) (1997) 268—272.

[31] J.S. Kingsbury, S.B. Garber, J.M. Giftos, B.L. Gray, M.M. Okamoto, R.A. Farrer, J.T. Fourkas, A.H. Hoveyda, Immobilization of olefin metathesis catalysts on monolithic sol—gel: practical, efficient, and easily recyclable catalysts for organic and combinatorial synthesis, Angew. Chem. Int. Ed. 40 (22) (2001) 4251—4256.

[32] K.C. Nicolaou, J. Pastor, N. Winssinger, F. Murphy, Solid phase synthesis of macrocycles by an intramolecular ketophosphonate reaction. Synthesis of a (dl)-Muscone library, J. Am. Chem. Soc. 120 (20) (1998) 5132—5133.

[33] S.A. Testero, E.G. Mata, Synthesis of 3-(Aryl)alkenyl-β-lactams by an efficient application of olefin cross-metathesis on solid support, Org. Lett. 8 (21) (2006) 4783—4786.

[34] A.H. Cherney, N.T. Kadunce, S.E. Reisman, Enantioselective and enantiospecific transition-metal-catalyzed cross-coupling reactions of organometallic reagents to construct C—C bonds, Chem. Rev. 115 (17) (2015) 9587—9652.

[35] S. Lou, G.C. Fu, Nickel/Bis(oxazoline)-Catalyzed asymmetric Kumada reactions of alkyl electrophiles: cross-couplings of racemic α-bromoketones, J. Am. Chem. Soc. 132 (4) (2010) 1264–1266.

[36] C. Fischer, G.C. Fu, Asymmetric nickel-catalyzed Negishi cross-couplings of secondary α-bromo amides with organozinc reagents, J. Am. Chem. Soc. 127 (13) (2005) 4594–4595.

[37] J.T. Binder, C.J. Cordier, G.C. Fu, Catalytic enantioselective cross-couplings of secondary alkyl electrophiles with secondary alkylmetal nucleophiles: Negishi reactions of racemic benzylic bromides with achiral alkylzinc reagents, J. Am. Chem. Soc. 134 (41) (2012) 17003–17006.

[38] F.O. Arp, G.C. Fu, Catalytic enantioselective Negishi reactions of racemic secondary benzylic halides, J. Am. Chem. Soc. 127 (30) (2005) 10482–10483.

[39] J. Choi, G.C. Fu, Catalytic asymmetric synthesis of secondary nitriles via stereoconvergent Negishi arylations and alkenylations of racemic α-bromonitriles, J. Am. Chem. Soc. 134 (22) (2012) 9102–9105.

[40] G.M. Schwarzwalder, C.D. Matier, G.C. Fu, Enantioconvergent cross-couplings of alkyl electrophiles: the catalytic asymmetric synthesis of organosilanes, Angew. Chem. Int. Ed. 58 (11) (2019) 3571–3574.

[41] S.W. Smith, G.C. Fu, Nickel-catalyzed asymmetric cross-couplings of racemic propargylic halides with arylzinc reagents, J. Am. Chem. Soc. 130 (38) (2008) 12645–12647.

[42] H.-Q. Do, E.R.R. Chandrashekar, G.C. Fu, Nickel/Bis(oxazoline)-Catalyzed asymmetric Negishi arylations of racemic secondary benzylic electrophiles to generate enantioenriched 1,1-diarylalkanes, J. Am. Chem. Soc. 135 (44) (2013) 16288–16291.

[43] P.M. Lundin, J. Esquivias, G.C. Fu, Catalytic asymmetric cross-couplings of racemic α-bromoketones with arylzinc reagents, Angew. Chem. Int. Ed. 48 (1) (2009) 154–156.

[44] Y. Liang, G.C. Fu, Catalytic asymmetric synthesis of tertiary alkyl fluorides: Negishi cross-couplings of racemic α,α-dihaloketones, J. Am. Chem. Soc. 136 (14) (2014) 5520–5524.

[45] J. Choi, P. Martín-Gago, G.C. Fu, Stereoconvergent arylations and alkenylations of unactivated alkyl electrophiles: catalytic enantioselective synthesis of secondary sulfonamides and sulfones, J. Am. Chem. Soc. 136 (34) (2014) 12161–12165.

[46] S. Purser, P.R. Moore, S. Swallow, V. Gouverneur, Fluorine in medicinal chemistry, Chem. Soc. Rev. 37 (2) (2008) 320–330.

[47] A. Varenikov, M. Gandelman, Synthesis of chiral α-trifluoromethyl alcohols and ethers via enantioselective Hiyama cross-couplings of bisfunctionalized electrophiles, Nat. Commun. 9 (1) (2018) 3566.

[48] S.L. Zultanski, G.C. Fu, Catalytic asymmetric γ-alkylation of carbonyl compounds via stereoconvergent Suzuki cross-couplings, J. Am. Chem. Soc. 133 (39) (2011) 15362–15364.

[49] X. Jiang, S. Sakthivel, K. Kulbitski, G. Nisnevich, M. Gandelman, Efficient synthesis of secondary alkyl fluorides via Suzuki cross-coupling reaction of 1-Halo-1-fluoroalkanes, J. Am. Chem. Soc. 136 (27) (2014) 9548–9551.

[50] A. Wilsily, F. Tramutola, N.A. Owston, G.C. Fu, New directing groups for metal-catalyzed asymmetric carbon–carbon bond-forming processes: stereoconvergent alkyl–alkyl Suzuki cross-couplings of unactivated electrophiles, J. Am. Chem. Soc. 134 (13) (2012) 5794–5797.

[51] H. Cong, G.C. Fu, Catalytic enantioselective cyclization/cross-coupling with alkyl electrophiles, J. Am. Chem. Soc. 136 (10) (2014) 3788−3791.

[52] Y.-G. Chen, B. Shuai, X.-T. Xu, Y.-Q. Li, Q.-L. Yang, H. Qiu, K. Zhang, P. Fang, T.-S. Mei, Nickel-catalyzed enantioselective hydroarylation and hydroalkenylation of styrenes, J. Am. Chem. Soc. 141 (8) (2019) 3395−3399.

[53] D. Anthony, Q. Lin, J. Baudet, T. Diao, Nickel-catalyzed asymmetric reductive diarylation of vinylarenes, Angew. Chem. Int. Ed. 58 (10) (2019) 3198−3202.

[54] L.L. Anka-Lufford, K.M.M. Huihui, N.J. Gower, L.K.G. Ackerman, D.J. Weix, Nickel-catalyzed cross-electrophile coupling with organic reductants in non-amide solvents, Chem. Eur J. 22 (33) (2016) 11564−11567.

[55] J. Sheng, H.-Q. Ni, H.-R. Zhang, K.-F. Zhang, Y.-N. Wang, X.-S. Wang, Nickel-catalyzed reductive cross-coupling of aryl halides with monofluoroalkyl halides for late-stage monofluoroalkylation, Angew. Chem. Int. Ed. 57 (26) (2018) 7634−7639.

[56] S.A. Girard, T. Knauber, C.-J. Li, The cross-dehydrogenative coupling of C-H bonds: a versatile strategy for C-C bond formations, Angew. Chem. Int. Ed. 53 (1) (2014) 74−100.

[57] G. Wang, X. Xin, Z. Wang, G. Lu, Y. Ma, L. Liu, Catalytic enantioselective oxidative coupling of saturated ethers with carboxylic acid derivatives, Nat. Commun. 10 (1) (2019) 559.

[58] A. Kapat, T. Sperger, S. Guven, F. Schoenebeck, E-Olefins through intramolecular radical relocation, Science 363 (6425) (2019) 391.

[59] D.B. Grotjahn, C.R. Larsen, J.L. Gustafson, R. Nair, A. Sharma, Extensive isomerization of alkenes using a bifunctional Catalyst: an alkene zipper, J. Am. Chem. Soc. 129 (31) (2007) 9592−9593.

[60] L. Angelini, J. Davies, M. Simonetti, L. MaletSanz, N.S. Sheikh, D. Leonori, Reaction of nitrogen-radicals with organometallics under Ni-catalysis: N-arylations and amino-functionalization cascades, Angew. Chem. Int. Ed. 58 (15) (2019).

[61] S. Thapa, B. Shrestha, S.K. Gurung, R. Giri, Copper-catalysed cross-coupling: an untapped potential, Org. Biomol. Chem. 13 (17) (2015) 4816−4827.

[62] C.-T. Yang, Z.-Q. Zhang, J. Liang, J.-H. Liu, X.-Y. Lu, H.-H. Chen, L. Liu, Copper-catalyzed cross-coupling of nonactivated secondary alkyl halides and tosylates with secondary alkyl grignard reagents, J. Am. Chem. Soc. 134 (27) (2012) 11124−11127.

[63] D.H. Burns, J.D. Miller, H.-K. Chan, M.O. Delaney, Scope and utility of a new soluble copper catalyst [CuBr−LiSPh−LiBr−THF]: a comparison with other copper catalysts in their ability to couple one equivalent of a grignard reagent with an alkyl sulfonate, J. Am. Chem. Soc. 119 (9) (1997) 2125−2133.

[64] G. Cahiez, O. Gager, J. Buendia, Copper-catalyzed cross-coupling of alkyl and aryl grignard reagents with alkynyl halides, Angew. Chem. Int. Ed. 49 (7) (2010) 1278−1281.

[65] L. Hintermann, L. Xiao, A. Labonne, A general and selective copper-catalyzed cross-coupling of tertiary grignard reagents with azacyclic electrophiles, Angew. Chem. Int. Ed. 47 (43) (2008) 8246−8250.

[66] J. Terao, A. Ikumi, H. Kuniyasu, N. Kambe, Ni- or Cu-catalyzed cross-coupling reaction of alkyl fluorides with grignard reagents, J. Am. Chem. Soc. 125 (19) (2003) 5646−5647.

[67] F. Zhu, Z.-X. Wang, Nickel-catalyzed cross-coupling of aryl fluorides and organozinc reagents, J. Org. Chem. 79 (10) (2014) 4285−4292.

[68] Z. Mo, Q. Zhang, L. Deng, Dinuclear iron complex-catalyzed cross-coupling of primary alkyl fluorides with aryl grignard reagents, Organometallics 31 (18) (2012) 6518–6521.

[69] S.-K. Kang, J.-S. Kim, S.-C. Choi, Copper- and manganese-catalyzed cross-coupling of organostannanes with organic iodides in the presence of sodium chloride, J. Org. Chem. 62 (13) (1997) 4208–4209.

[70] J.-H. Li, B.-X. Tang, L.-M. Tao, Y.-X. Xie, Y. Liang, M.-B. Zhang, Reusable copper-catalyzed cross-coupling reactions of aryl halides with organotins in inexpensive ionic liquids, J. Org. Chem. 71 (19) (2006) 7488–7490.

[71] L. Cornelissen, M. Lefrancq, O. Riant, Copper-catalyzed cross-coupling of vinylsiloxanes with bromoalkynes: synthesis of enynes, Org. Lett. 16 (11) (2014) 3024–3027.

[72] S.K. Gurung, S. Thapa, A.S. Vangala, R. Giri, Copper-catalyzed Hiyama coupling of (Hetero)aryltriethoxysilanes with (Hetero)aryl iodides, Org. Lett. 15 (20) (2013) 5378–5381.

[73] C.-T. Yang, Z.-Q. Zhang, Y.-C. Liu, L. Liu, Copper-catalyzed cross-coupling reaction of organoboron compounds with primary alkyl halides and pseudohalides, Angew. Chem. Int. Ed. 50 (17) (2011) 3904–3907.

[74] J.-H. Li, J.-L. Li, D.-P. Wang, S.-F. Pi, Y.-X. Xie, M.-B. Zhang, X.-C. Hu, CuI-catalyzed Suzuki–Miyaura and Sonogashira cross-coupling reactions using DABCO as ligand, J. Org. Chem. 72 (6) (2007) 2053–2057.

[75] E. Bhimireddy, E.J. Corey, Method for highly enantioselective ligation of two chiral C(sp3) stereocenters, J. Am. Chem. Soc. 139 (32) (2017) 11044–11047.

[76] J.-S. Lin, T.-T. Li, J.-R. Liu, G.-Y. Jiao, Q.-S. Gu, J.-T. Cheng, Y.-L. Guo, X. Hong, X.-Y. Liu, Cu/Chiral phosphoric acid-catalyzed asymmetric three-component radical-initiated 1,2-dicarbofunctionalization of alkenes, J. Am. Chem. Soc. 141 (2) (2019) 1074–1083.

[77] A. Piontek, E. Bisz, M. Szostak, Iron-catalyzed cross-couplings in the synthesis of pharmaceuticals: in pursuit of sustainability, Angew. Chem. Int. Ed. 57 (35) (2018) 11116–11128.

[78] R. Jana, T.P. Pathak, M.S. Sigman, Advances in transition metal (Pd,Ni,Fe)-Catalyzed cross-coupling reactions using alkyl-organometallics as reaction partners, Chem. Rev. 111 (3) (2011) 1417–1492.

[79] C. Bolm, J. Legros, J. Le Paih, L. Zani, Iron-catalyzed reactions in organic synthesis, Chem. Rev. 104 (12) (2004) 6217–6254.

[80] T.L. Mako, J.A. Byers, Recent advances in iron-catalysed cross coupling reactions and their mechanistic underpinning, Inorg. Chem. Front. 3 (6) (2016) 766–790.

[81] P. Mullens, E. Cleator, M. McLaughlin, B. Bishop, J. Edwards, A. Goodyear, T. Andreani, Y. Jin, J. Kong, H. Li, M. Williams, M. Zacuto, Two approaches to the chemical development and large-scale preparation of a pyrimidyl tetrazole intermediate, Org. Process Res. Dev. 20 (6) (2016) 1075–1087.

[82] M.J. Kim, J. Lee, S.Y. Kang, S.-H. Lee, E.-J. Son, M.E. Jung, S.H. Lee, K.-S. Song, M. Lee, H.-K. Han, J. Kim, J. Lee, Novel C-aryl glucoside SGLT2 inhibitors as potential antidiabetic agents: pyridazinylmethylphenyl glucoside congeners, Bioorg. Med. Chem. Lett 20 (11) (2010) 3420–3425.

[83] W.A. Nugent, "Black swan events" in organic synthesis, Angew. Chem. Int. Ed. 51 (36) (2012) 8936–8949.

[84] E. Jiménez-Núñez, A.M. Echavarren, Gold-catalyzed cycloisomerizations of enynes: a mechanistic perspective, Chem. Rev. 108 (8) (2008) 3326–3350.

[85] S.A. Shahzad, M.A. Sajid, Z.A. Khan, D. Canseco-Gonzalez, Gold catalysis in organic transformations: a review, Synth. Commun. 47 (8) (2017) 735–755.

[86] A. Fürstner, Gold and platinum catalysis—a convenient tool for generating molecular complexity, Chem. Soc. Rev. 38 (11) (2009) 3208–3221.

[87] T. Luo, S.L. Schreiber, Gold(I)-Catalyzed coupling reactions for the synthesis of diverse small molecules using the build/couple/pair strategy, J. Am. Chem. Soc. 131 (15) (2009) 5667–5674.

[88] A. Fürstner, A. Schlecker, A gold-catalyzed entry into the sesquisabinene and sesquithujene families of terpenoids and formal total syntheses of cedrene and cedrol, Chem. Eur J. 14 (30) (2008) 9181–9191.

[89] A. La-Venia, N.S. Medran, V. Krchňák, S.A. Testero, Synthesis of a small library of imidazolidin-2-ones using gold catalysis on solid phase, ACS Comb. Sci. 18 (8) (2016) 482–489.

[90] W. Zi, F. Dean Toste, Recent advances in enantioselective gold catalysis, Chem. Soc. Rev. 45 (16) (2016) 4567–4589.

[91] Y.-L. Du, Y. Hu, Y.-F. Zhu, X.-F. Tu, Z.-Y. Han, L.-Z. Gong, Chiral gold phosphate catalyzed tandem hydroamination/asymmetric transfer hydrogenation enables access to chiral tetrahydroquinolines, J. Org. Chem. 80 (9) (2015) 4754–4759.

[92] Y. Wang, P. Zhang, D. Qian, J. Zhang, Highly regio-, diastereo-, and enantioselective gold(I)-Catalyzed intermolecular annulations with N-allenamides at the proximal C-C bond, Angew. Chem. Int. Ed. 54 (49) (2015) 14849–14852.

[93] G.-Q. Chen, W. Fang, Y. Wei, X.-Y. Tang, M. Shi, Divergent reaction pathways in gold-catalyzed cycloisomerization of 1,5-enynes containing a cyclopropane ring: dramatic ortho substituent and temperature effects, Chem. Sci. 7 (7) (2016) 4318–4328.

[94] P. Maulcón, R.M. Zcldin, A.Z. GonzálczF, D. Tostc, Ligand-Controllcd acccss to [4 + 2] and [4 + 3] cycloadditions in gold-catalyzed reactions of allene-dienes, J. Am. Chem. Soc. 131 (18) (2009) 6348–6349.

[95] Y.-C. Lee, S. Patil, C. Golz, C. Strohmann, S. Ziegler, K. Kumar, H. Waldmann, A ligand-directed divergent catalytic approach to establish structural and functional scaffold diversity, Nat. Commun. 8 (2017) 14043.

[96] D. Ding, T. Mou, M. Feng, X. Jiang, Utility of ligand effect in homogenous gold catalysis: enabling regiodivergent π-bond-activated cyclization, J. Am. Chem. Soc. 138 (16) (2016) 5218–5221.

[97] L. Feliu, P. Vera-Luque, F. Albericio, M. Álvarez, Advances in solid-phase cycloadditions for heterocyclic synthesis, J. Comb. Chem. 9 (4) (2007) 521–565.

[98] K. Harju, J. Yli-Kauhaluoma, Recent advances in 1,3-dipolar cycloaddition reactions on solid supports, Mol. Divers. 9 (1) (2005) 187–207.

[99] R.C.D. Brown, J. Keily, R. Karim, Solid-phase synthesis of γ-lactams, γ-lactones and cyclobutane derivatives from common resin-bound intermediates, Tetrahedron Lett. 41 (17) (2000) 3247–3251.

[100] N. Kaval, J. Van der Eycken, J. Caroen, W. Dehaen, G.A. Strohmeier, C.O. Kappe, E. Van der Eycken, An exploratory study on microwave-assisted solid-phase Diels–Alder reactions of 2(1H)-Pyrazinones: the elaboration of a new tailor-made acid-labile linker, J. Comb. Chem. 5 (5) (2003) 560–568.

[101] N. Kaval, W. Dehaen, E. Van der Eycken, Solid-phase synthesis of the 2(1H)-Pyrazinone Scaffold: a new approach toward diversely substituted heterocycles, J. Comb. Chem. 7 (1) (2005) 90–95.

[102]   G.C. Tron, T. Pirali, R.A. Billington, P.L. Canonico, G. Sorba, A.A. Genazzani, Click chemistry reactions in medicinal chemistry: applications of the 1,3-dipolar cycloaddition between azides and alkynes, Med. Res. Rev. 28 (2) (2008) 278−308.

[103]   T. Pirali, G.C. Tron, J. Zhu, One-pot synthesis of macrocycles by a tandem three-component reaction and intramolecular [3+2] cycloaddition, Org. Lett. 8 (18) (2006) 4145−4148.

[104]   J. Wang, M. Uttamchandani, J. Li, M. HuS, Q. Yao, Rapid assembly of matrix metalloprotease inhibitors using click chemistry, Org. Lett. 8 (17) (2006) 3821−3824.

[105]   K. Gao, Y.-G. Zhang, Z. Wang, H. Ding, Recent development on the [5+2] cycloadditions and their application in natural product synthesis, Chem. Commun. 55 (13) (2019) 1859−1878.

[106]   S. Baktharaman, N. Afagh, A. Vandersteen, A.K. Yudin, Unprotected vinyl aziridines: facile synthesis and cascade transformations, Org. Lett. 12 (2) (2010) 240−243.

[107]   P.A. Wender, F. Inagaki, M. Pfaffenbach, M.C. Stevens, Propargyltrimethylsilanes as allene equivalents in transition metal-catalyzed [5 + 2] cycloadditions, Org. Lett. 16 (11) (2014) 2923−2925.

[108]   P.A. Wender, C. Ebner, B.D. Fennell, F. Inagaki, B. Schröder, Ynol ethers as ketene equivalents in rhodium-catalyzed intermolecular [5 + 2] cycloaddition reactions, Org. Lett. 19 (21) (2017) 5810−5813.

[109]   S. Xu, Z.-M. Zhang, B. Xu, B. Liu, Y. Liu, J. Zhang, Enantioselective regiodivergent synthesis of chiral pyrrolidines with two quaternary stereocenters via ligand-controlled copper(I)-Catalyzed asymmetric 1,3-dipolar cycloadditions, J. Am. Chem. Soc. 140 (6) (2018) 2272−2283.

[110]   J.-K. Yu, H.-W. Chien, Y.-J. Lin, P. Karanam, Y.-H. Chen, W. Lin, Diversity-oriented synthesis of chromenopyrrolidines from azomethine ylides and 2-hydroxybenzylidene indandiones via base-controlled regiodivergent (3+2) cycloaddition, Chem. Commun. 54 (71) (2018) 9921−9924.

[111]   T.G. Chen, L.M. Barton, Y. Lin, J. Tsien, D. Kossler, I. Bastida, S. Asai, C. Bi, J.S. Chen, M. Shan, H. Fang, F.G. Fang, H.-w. Choi, L. Hawkins, T. Qin, P.S. Baran, Building C(sp3)-rich complexity by combining cycloaddition and C−C cross-coupling reactions, Nature 560 (7718) (2018) 350−354.

[112]   A. Dömling, W. Wang, K. Wang, Chemistry and biology of multicomponent reactions, Chem. Rev. 112 (6) (2012) 3083−3135.

[113]   E. Ruijter, R. Scheffelaar, R.V.A. Orru, Multicomponent reaction design in the quest for molecular complexity and diversity, Angew. Chem. Int. Ed. 50 (28) (2011) 6234−6246.

[114]   I. Akritopoulou-Zanze, Isocyanide-based multicomponent reactions in drug discovery, Curr. Opin. Chem. Biol. 12 (3) (2008) 324−331.

[115]   E. Ruijter, R.V.A. Orru, Multicomponent reactions − opportunities for the pharmaceutical industry, Drug Discov. Today Technol. 10 (1) (2013) e15−e20.

[116]   B.B. Touré, D.G. Hall, Natural product synthesis using multicomponent reaction strategies, Chem. Rev. 109 (9) (2009) 4439−4486.

[117]   A. Váradi, T.C. Palmer, R.N. Dardashti, S. Majumdar, Isocyanide-based multicomponent reactions for the synthesis of heterocycles, Molecules 21 (19) (2016).

[118]   M. Passerini, L. Simone, Gazz. Chim. Ital. 51 (1921) 126−129.

[119]   M. Passerini, G. Ragni, Gazz. Chim. Ital. 61 (1931) 964−969.

[120]   J.E. Semple, T.D. Owens, K. Nguyen, O.E. Levy, New synthetic technology for efficient construction of α-Hydroxy-β-amino amides via the Passerini Reaction1, Org. Lett. 2 (18) (2000) 2769−2772.

[121] L. Banfi, G. Guanti, R. Riva, Passerini multicomponent reaction of protected α-amino-aldehydes as a tool for combinatorial synthesis of enzyme inhibitors, Chem. Commun. (11) (2000) 985—986.

[122] F. Velázquez, S. Venkatraman, M. Blackman, P. Pinto, S. Bogen, M. Sannigrahi, K. Chen, J. Pichardo, A. Hart, X. Tong, V. Girijavallabhan, F.G. Njoroge, Design, synthesis, and evaluation of oxygen-containing macrocyclic peptidomimetics as inhibitors of HCV NS3 protease, J. Med. Chem. 52 (3) (2009) 700—708.

[123] S. Venkatraman, F. Velazquez, W. Wu, M. Blackman, K.X. Chen, S. Bogen, L. Nair, X. Tong, R. Chase, A. Hart, S. Agrawal, J. Pichardo, A. Prongay, K.-C. Cheng, V. Girijavallabhan, J. Piwinski, N.-Y. Shih, F.G. Njoroge, Discovery and Structure—Activity relationship of P1—P3 ketoamide derived macrocyclic inhibitors of hepatitis C virus NS3 protease, J. Med. Chem. 52 (2) (2009) 336—346.

[124] T.D. Owens, J.E. Semple, Atom-economical synthesis of the N(10)—C(17) fragment of cyclotheonamides via a novel Passerini Reaction—Deprotection—Acyl migration Strategy1, Org. Lett. 3 (21) (2001) 3301—3304.

[125] L. Banfi, A. Basso, G. Guanti, R. Riva, Passerini reaction — amine Deprotection — acyl Migration (PADAM): a convenient strategy for the solid-phase preparation of peptido-mimetic compounds, Mol. Divers. 6 (3) (2003) 227—235.

[126] N.A.M. Yehia, W. Antuch, B. Beck, S. Hess, V. Schauer-Vukašinović, M. Almstetter, P. Furer, E. Herdtweck, A. Dömling, Novel nonpeptidic inhibitors of HIV-1 protease obtained via a new multicomponent chemistry strategy, Bioorg. Med. Chem. Lett 14 (12) (2004) 3121—3125.

[127] I. Ugi, Neuere Methoden der präparativen organischen Chemie IV Mit Sekundär-Reaktionen gekoppelte α-Additionen von Immonium-Ionen und Anionen an Isonitrile, Angew. Chem. 74 (1) (1962) 9—22.

[128] H. Habashita, M. Kokubo, S.-i. Hamano, N. Hamanaka, M. Toda, S. Shibayama, H. Tada, K. Sagawa, D. Fukushima, K. Maeda, H. Mitsuya, Design, synthesis, and biological evaluation of the combinatorial library with a new spirodiketopiperazine scaffold. Discovery of novel potent and selective low-molecular-weight CCR5 antagonists, J. Med. Chem. 49 (14) (2006) 4140—4152.

[129] C. Hulme, M.M. Morrissette, F.A. Volz, C.J. Burns, The solution phase synthesis of diketopiperazine libraries via the Ugi reaction: novel application of Armstrong's convertible isonitrile, Tetrahedron Lett. 39 (10) (1998) 1113—1116.

[130] A. Demharter, W. Hörl, E. Herdtweck, I. Ugi, Synthesis of chiral 1,1′-iminodicarbox-ylic acid derivatives from α-amino acids, aldehydes, isocyanides, and alcohols by the diastereoselective five-center—four-component reaction, Angew Chem. Int. Ed. Engl. 35 (2) (1996) 173—175.

[131] A. Basso, L. Banfi, R. Riva, G. Guanti, U-4C-3CR versus U-5C-4CR and stereochemical outcomes using suitable bicyclic β-amino acid derivatives as bifunctional components in the Ugi reaction, Tetrahedron Lett. 45 (3) (2004) 587—590.

[132] S. Di Micco, R. Vitale, M. Pellecchia, M.F. Rega, R. Riva, A. Basso, G. Bifulco, Identification of lead compounds as antagonists of protein bcl-xl with a diversity-oriented multidisciplinary approach, J. Med. Chem. 52 (23) (2009) 7856—7867.

[133] P. Capurro, L. Moni, A. Galatini, C. Mang, A. Basso, Multi-gram synthesis of enantiopure 1,5-disubstituted tetrazoles via ugi-azide 3-component reaction, Molecules 23 (11) (2018) 2758.

[134] A. Basso, L. Banfi, S. Garbarino, R. Riva, Ketene three-component reaction: a metal-free multicomponent approach to stereodefined captodative olefins, Angew. Chem. Int. Ed. 52 (7) (2013) 2096–2099.

[135] F. Ibba, P. Capurro, S. Garbarino, M. Anselmo, L. Moni, A. Basso, Photoinduced multicomponent synthesis of α-silyloxy acrylamides, an unexplored class of silyl enol ethers, Org. Lett. 20 (4) (2018) 1098–1101.

[136] H. Bienaymé, K. Bouzid, A new heterocyclic multicomponent reaction for the combinatorial synthesis of fused 3-aminoimidazoles, Angew. Chem. Int. Ed. 37 (16) (1998) 2234–2237.

[137] C. Blackburn, B. Guan, P. Fleming, K. Shiosaki, S. Tsai, Parallel synthesis of 3-aminoimidazo[1,2-a]pyridines and pyrazines by a new three-component condensation, Tetrahedron Lett. 39 (22) (1998) 3635–3638.

[138] K. Groebke, L. Weber, F. Mehlin, Synthesis of imidazo[1,2-a] annulated pyridines, pyrazines and pyrimidines by a novel three-component condensation, Synlett (6) (1998) 661–663.

[139] S. Shaaban, B.F. Abdel-Wahab, Groebke–Blackburn–Bienaymé multicomponent reaction: emerging chemistry for drug discovery, Mol. Divers. 20 (1) (2016) 233–254.

[140] A.M. Van Leusen, J. Wildeman, O.H. Oldenziel, Chemistry of sulfonylmethyl isocyanides. 12. Base-induced cycloaddition of sulfonylmethyl isocyanides to carbon,-nitrogen double bonds. Synthesis of 1,5-disubstituted and 1,4,5-trisubstituted imidazoles from aldimines and imidoyl chlorides, J. Org. Chem. 42 (7) (1977) 1153–1159.

[141] P. Janvier, H. Bienaymé, J. Zhu, A five-component synthesis of hexasubstituted benzene, Angew. Chem. Int. Ed. 41 (22) (2002) 4291–4294.

[142] P. Janvier, X. Sun, H. Bienaymé, J. Zhu, Ammonium chloride-promoted four-component synthesis of pyrrolo[3,4-b]pyridin-5-one, J. Am. Chem. Soc. 124 (11) (2002) 2560–2567.

[143] V.A. Chebanov, V.E. Saraev, S.M. Desenko, V.N. Chernenko, I.V. Knyazeva, U. Groth, T.N. Glasnov, C.O. Kappe, Tuning of chemo- and regioselectivities in multicomponent condensations of 5-aminopyrazoles, dimedone, and aldehydes, J. Org. Chem. 73 (13) (2008) 5110–5118.

[144] T.H. Al-Tel, R.A. Al-Qawasmeh, W. Voelter, Rapid assembly of polyfunctional structures using a one-pot five- and six-component sequential groebke–blackburn/ugi/passerini process, Eur. J. Org. Chem. (29) (2010) 5586–5593.

[145] A. Znabet, E. Ruijter, F.J.J. de Kanter, V. Köhler, M. Helliwell, N.J. Turner, R.V.A. Orru, Highly stereoselective synthesis of substituted prolyl peptides using a combination of biocatalytic desymmetrization and multicomponent reactions, Angew. Chem. Int. Ed. 49 (31) (2010) 5289–5292.

[146] A. Znabet, J. Zonneveld, E. Janssen, F.J.J. De Kanter, M. Helliwell, N.J. Turner, E. Ruijter, R.V.A. Orru, Asymmetric synthesis of synthetic alkaloids by a tandem biocatalysis/Ugi/Pictet–Spengler-type cyclization sequence, Chem. Commun. 46 (41) (2010) 7706–7708.

[147] A. Znabet, M.M. Polak, E. Janssen, F.J.J. de Kanter, N.J. Turner, R.V.A. Orru, E. Ruijter, A highly efficient synthesis of telaprevir by strategic use of biocatalysis and multicomponent reactions, Chem. Commun. 46 (42) (2010) 7918–7920.

[148] C.G. Evans, J.E. Gestwicki, Enantioselective organocatalytic Hantzsch synthesis of polyhydroquinolines, Org. Lett. 11 (14) (2009) 2957–2959.

[149] L.-Z. Gong, X.-H. Chen, X.-Y. Xu, Asymmetric organocatalytic Biginelli reactions: a new approach to quickly access optically active 3,4-dihydropyrimidin-2-(1H)-ones, Chem. Eur J. 13 (32) (2007) 8920−8926.

[150] H. Liu, G. Dagousset, G. Masson, P. Retailleau, J. Zhu, Chiral brønsted acid-catalyzed enantioselective three-component Povarov reaction, J. Am. Chem. Soc. 131 (13) (2009) 4598−4599.

[151] A. Córdova, S.-i. Watanabe, F. Tanaka, W. Notz, C.F. Barbas, A highly enantioselective route to either enantiomer of both α- and β-amino acid derivatives, J. Am. Chem. Soc. 124 (9) (2002) 1866−1867.

[152] S. Mitsumori, H. Zhang, P. Ha-Yeon Cheong, K.N. Houk, F. Tanaka, C.F. Barbas, Direct asymmetric anti-mannich-type reactions catalyzed by a designed amino acid, J. Am. Chem. Soc. 128 (4) (2006) 1040−1041.

[153] G. Dagousset, F. Drouet, G. Masson, J. Zhu, Chiral brønsted acid-catalyzed enantioselective multicomponent Mannich reaction: synthesis of anti-1,3-diamines using ene-carbamates as nucleophiles, Org. Lett. 11 (23) (2009) 5546−5549.

[154] J. Zhang, P. Yu, S.-Y. Li, H. Sun, S.-H. Xiang, J. Wang, K.N. Houk, B. Tan, Asymmetric phosphoric acid−catalyzed four-component Ugi reaction, Science 361 (6407) (2018) eaas8707.

[155] R.J.T. Mikkelsen, K.E. Grier, K.T. Mortensen, T.E. Nielsen, K. Qvortrup, , photolabile linkers for solid-phase synthesis, ACS Comb. Sci. 20 (7) (2018) 377−399.

[156] E. Paternò, G. Chieffi, Sintesi in chimica organica per mezzo della luce. Nota II. Composti degli idrocarburi non saturi con aldeidi e chetoni, Gazz. Chim. Ital. (39) (1909) 341−361.

[157] G. Büchi, C.G. Inman, E.S. Lipinsky, Light-catalyzed organic reactions. I. The reaction of carbonyl compounds with 2-Methyl-2-butene in the presence of ultraviolet light, J. Am. Chem. Soc. 76 (17) (1954) 4327−4331.

[158] L. Wolff, Ueber diazoanhydride, Justus Liebigs Ann. Chem. 325 (2) (1902) 129−195.

[159] R.B. Woodward, R. Hoffmann, The conservation of orbital symmetry, Angew Chem. Int. Ed. Engl. 8 (11) (1969) 781−853.

[160] D. Ravelli, M. Fagnoni, A. Albini, Photoorganocatalysis. What for? Chem. Soc. Rev. 42 (1) (2013) 97−113.

[161] S. Fukuzumi, K. Ohkubo, Organic synthetic transformations using organic dyes as photoredox catalysts, Org. Biomol. Chem. 12 (32) (2014) 6059−6071.

[162] D.A. Nicewicz, T.M. Nguyen, Recent applications of organic dyes as photoredox catalysts in organic synthesis, ACS Catal. 4 (1) (2014) 355−360.

[163] N.A. Romero, D.A. Nicewicz, Organic photoredox catalysis, Chem. Rev. 116 (17) (2016) 10075−10166.

[164] I.K. Sideri, E. Voutyritsa, C.G. Kokotos, Photoorganocatalysis, small organic molecules and light in the service of organic synthesis: the awakening of a sleeping giant, Org. Biomol. Chem. 16 (25) (2018) 4596−4614.

[165] P. Richardson, Unlocking the potential benefits of flow chemistry in the drug-discovery process, Future Med. Chem. 6 (8) (2014) 845−847.

[166] G. Pandey, M. Karthikeyan, A. Murugan, New intramolecular α-arylation strategy of ketones by the reaction of silyl enol ethers to photosensitized electron transfer generated arene radical Cations: construction of benzannulated and benzospiroannulated compounds, J. Org. Chem. 63 (9) (1998) 2867−2872.

[167] G. Pandey, A. Murugan, M. Balakrishnan, A new strategy towards the total synthesis of phenanthridone alkaloids: synthesis of (+)-2,7-dideoxypancratistatin as a model study, Chem. Commun. (6) (2002) 624−625.

[168] D.S. Hamilton, D.A. Nicewicz, Direct catalytic anti-markovnikov hydroetherification of alkenols, J. Am. Chem. Soc. 134 (45) (2012) 18577—18580.

[169] D.J. Wilger, N.J. Gesmundo, D.A. Nicewicz, Catalytic hydrotrifluoromethylation of styrenes and unactivated aliphatic alkenes via an organic photoredox system, Chem. Sci. 4 (8) (2013) 3160—3165.

[170] T.M. Nguyen, D.A. Nicewicz, Anti-markovnikov hydroamination of alkenes catalyzed by an organic photoredox system, J. Am. Chem. Soc. 135 (26) (2013) 9588—9591.

[171] A.J. Perkowski, D.A. Nicewicz, Direct catalytic anti-markovnikov addition of carboxylic acids to alkenes, J. Am. Chem. Soc. 135 (28) (2013) 10334—10337.

[172] A.A. Ghogare, A. Greer, Using singlet oxygen to synthesize natural products and drugs, Chem. Rev. 116 (17) (2016) 9994—10034.

[173] Y. Kobayashi, Carbasugars: synthesis and functions, in: B.O. Fraser-Reid, K. Tatsuta, J. Thiem (Eds.), Glycoscience: Chemistry and Chemical Biology, Springer Berlin Heidelberg, Berlin, Heidelberg, 2008, pp. 1913—1997.

[174] E. Salamci, H. Seçen, Y. Sütbeyaz, M. Balci, A concise and convenient synthesis of dl-proto-Quercitol and dl-gala-Quercitol via ene reaction of singlet oxygen combined with [2 + 4] cycloaddition to cyclohexadiene, J. Org. Chem. 62 (8) (1997) 2453—2457.

[175] N.I. Kurbanoğlu, M. Çelik, H. Kilic, C. Alp, E. Şahin, M. Balci, Stereospecific synthesis of a dl-gala-aminoquercitol derivative, Tetrahedron 66 (19) (2010) 3485—3489.

[176] Y. Altun, S.D. Dogan, M. Balci, Synthesis of branched carbasugars via photooxygenation and manganese(III) acetate free radical cyclization, Tetrahedron 70 (33) (2014) 4884—4890.

[177] T. Montagnon, D. Kalaitzakis, M. Sofiadis, G. Vassilikogiannakis, Chemoselective photooxygenations of furans bearing unprotected amines: their use in alkaloid synthesis, Org. Biomol. Chem. 14 (37) (2016) 8636—8640.

[178] R. Zhang, G. Li, M. Wismer, P. Vachal, S.L. Colletti, Z.-C. Shi, Profiling and application of photoredox C(sp3)—C(sp2) cross-coupling in medicinal chemistry, ACS Med. Chem. Lett. 9 (7) (2018) 773—777.

[179] P.S. Kutchukian, J.F. Dropinski, K.D. Dykstra, B. Li, D.A. DiRocco, E.C. Streckfuss, L.-C. Campeau, T. Cernak, P. Vachal, I.W. Davies, S.W. Krska, S.D. Dreher, Chemistry informer libraries: a chemoinformatics enabled approach to evaluate and advance synthetic methods, Chem. Sci. 7 (4) (2016) 2604—2613.

[180] D.M. Kuznetsov, A.G. Kutateladze, Step-economical photoassisted diversity-oriented synthesis: sustaining cascade photoreactions in oxalyl anilides to access complex polyheterocyclic molecular architectures, J. Am. Chem. Soc. 139 (46) (2017) 16584—16590.

[181] A.P. Ltd., DataWarrior 4.6.0, 2017. http://openmolecules.org/datawarrior.

[182] M. Teders, C. Henkel, L. Anhäuser, F. Strieth-Kalthoff, A. Gómez-Suárez, R. Kleinmans, A. Kahnt, A. Rentmeister, D. Guldi, F. Glorius, The energy-transfer-enabled biocompatible disulfide—ene reaction, Nat. Chem. 10 (9) (2018) 981—988.

[183] K. Mizuno, Y. Nishiyama, T. Ogaki, K. Terao, H. Ikeda, K. Kakiuchi, Utilization of microflow reactors to carry out synthetically useful organic photochemical reactions, J. Photochem. Photobiol. C Photochem. Rev. 29 (2016) 107—147.

[184] T. Cernak, K.D. Dykstra, S. Tyagarajan, P. Vachal, S.W. Krska, The medicinal chemist's toolbox for late stage functionalization of drug-like molecules, Chem. Soc. Rev. 45 (3) (2016) 546—576.

[185] L. McMurray, F. O'Hara, M.J. Gaunt, Recent developments in natural product synthesis using metal-catalysed C—H bond functionalisation, Chem. Soc. Rev. 40 (4) (2011) 1885—1898.

[186] J. Yamaguchi, A.D. Yamaguchi, K. Itami, C-H bond functionalization: emerging synthetic tools for natural products and pharmaceuticals, Angew. Chem. Int. Ed. 51 (36) (2012) 8960−9009.

[187] K.M. Engle, J.-Q. Yu, Developing ligands for palladium(II)-Catalyzed C−H functionalization: intimate dialogue between ligand and substrate, J. Org. Chem. 78 (18) (2013) 8927−8955.

[188] O. Baudoin, Transition metal-catalyzed arylation of unactivated C(sp3)−H bonds, Chem. Soc. Rev. 40 (10) (2011) 4902−4911.

[189] A. Sharma, J.F. Hartwig, Metal-catalysed azidation of tertiary C−H bonds suitable for late-stage functionalization, Nature 517 (2015) 600.

[190] Q. Michaudel, G. Journot, A. Regueiro-Ren, A. Goswami, Z. Guo, T.P. Tully, L. Zou, R.O. Ramabhadran, K.N. Houk, P.S. Baran, Improving physical properties via C−H oxidation: chemical and enzymatic approaches, Angew. Chem. Int. Ed. 53 (45) (2014) 12091−12096.

[191] X. Huang, W. Liu, H. Ren, R. Neelamegam, J.M. Hooker, J.T. Groves, Late stage benzylic C−H fluorination with [18F]fluoride for PET imaging, J. Am. Chem. Soc. 136 (19) (2014) 6842−6845.

[192] S.M. Ametamey, M. Honer, P.A. Schubiger, Molecular imaging with PET, Chem. Rev. 108 (5) (2008) 1501−1516.

[193] X. Huang, T.M. Bergsten, J.T. Groves, Manganese-catalyzed late-stage aliphatic C−H azidation, J. Am. Chem. Soc. 137 (16) (2015) 5300−5303.

[194] W. Liu, X. Huang, M.-J. Cheng, R.J. Nielsen, W.A. Goddard, J.T. Groves, Oxidative aliphatic C-H fluorination with fluoride ion catalyzed by a manganese porphyrin, Science 337 (6100) (2012) 1322−1325.

[195] J. He, L.G. Hamann, H.M.L. Davies, R.E.J. Beckwith, Late-stage C−H functionalization of complex alkaloids and drug molecules via intermolecular rhodium carbenoid insertion, Nat. Commun. 6 (2015) 5943.

[196] W. Yang, S. Ye, D. Fanning, T. Coon, Y. Schmidt, P. Krenitsky, D. Stamos, J.-Q. Yu, Orchestrated triple C−H activation reactions using two directing groups: rapid assembly of complex pyrazoles, Angew. Chem. Int. Ed. 54 (8) (2015) 2501−2504.

[197] A.C. Sun, R.C. McAtee, E.J. McClain, C.R.J. Stephenson, Advancements in Visible-light-Enabled radical C(sp)2−H alkylation of (Hetero)arenes, Synthesis 51 (05) (2019) 1063−1072.

[198] D.A. DiRocco, K. Dykstra, S. Krska, P. Vachal, D.V. Conway, M. Tudge, Late-stage functionalization of biologically active heterocycles through photoredox catalysis, Angew. Chem. Int. Ed. 53 (19) (2014) 4802−4806.

[199] J. Jin D, W.C. MacMillan, Direct α-arylation of ethers through the combination of photoredox-mediated C−H functionalization and the Minisci reaction, Angew. Chem. Int. Ed. 54 (5) (2015) 1565−1569.

[200] S.G. Ouellet, A. Roy, C. Molinaro, R. Angelaud, J.-F. Marcoux, P.D. O'Shea, I.W. Davies, Preparative scale synthesis of the biaryl core of Anacetrapib via a ruthenium-catalyzed direct arylation reaction: unexpected effect of solvent impurity on the arylation reaction, J. Org. Chem. 76 (5) (2011) 1436−1439.

[201] D. Mitchell, K.P. Cole, P.M. Pollock, D.M. Coppert, T.P. Burkholder, J.R. Clayton, Development and a practical synthesis of the JAK2 inhibitor LY2784544, Org. Process Res. Dev. 16 (1) (2012) 70−81.

Chemoinformatics approaches to assess chemical diversity and complexity of small molecules

Fernanda I. Saldívar-González, José L. Medina-Franco

*Department of Pharmacy, School of Chemistry, Universidad Nacional Autónoma de México,
Mexico City, Mexico*

3.1 Introduction

Chemoinformatics is a tool that emerges from the combination of computer resources and chemical data and is used in the management, visualization and systematic analysis of chemical information [1,2]. Important contributions of chemoinformatics have been made in various areas such as analytical chemistry, organic chemistry, and, more recently, in food informatics [3]. However, thus far, chemoinformatics has had its major impact in drug discovery and development. In this context, concepts such as chemical diversity and complexity play an important role in modern approaches to drug design [1,2,4].

Chemical diversity provides useful information in research programs that seek to prioritize the selection of libraries or sublibraries for experimental evaluation. Compound sets can be selected from an existing corporate or public database, or they can be the result of a systematic process of combinatorial library design [5]. In particular, diversity analysis helps to evaluate the structural novelty of compound libraries. If the purpose of a hit identification project is identifying new molecules, it is convenient to select collections with chemically diverse structures to increase the probability of identifying new scaffolds that can become leads or prototypes for a specific biological target [6]. Likewise, chemoinformatic tools to evaluate diversity serve as a guide in the design of new small molecules, mainly in approaches such as diversity-oriented synthesis (DOS) [7,8].

Similarly, structural complexity is considered an important feature in small molecule libraries. This concept helps to determine synthetic feasibility, and it has also been argued that molecules that are structurally complex are more likely to interact with biological macromolecules in a selective and specific manner [8,9].

The diversity and complexity of molecules in a chemical library can be evaluated in multiple ways, mainly depending on the data under scrutiny and the goals of the

Small Molecule Drug Discovery. https://doi.org/10.1016/B978-0-12-818349-6.00003-0

83

study. In addition to the metrics used, a key aspect of diversity and complexity analysis is molecular representation [10,11]. The most common ways to represent molecules in chemoinformatic applications are molecular descriptors (including physicochemical properties and molecular fingerprints) and chemical scaffolds [12]. Several molecular representations used are tailored to manage efficiently high number (thousands or even millions) of chemical structures present in compound databases. Depending on the type of descriptor and the level of precision desired, the input structures can be in two- or three-dimensions (2D/3D). The choice of molecular representation and the methods to be used depend on the goals of the study. This raises the question of how different methods can be compared in order to identify those most appropriate for a particular application. The objective of this chapter is to provide an overview of the different approaches in which it can be carried out, illustrating their applications with practical examples. The chapter is organized into four major sections. After this Introduction, Section 3.2 discusses approaches for diversity analysis. The next section presents methods for quantifying molecular complexity. Section 3.4 presents Conclusions and Perspectives.

3.2 Diversity analysis

Diversity analysis is an important criterion to design drugs since it can be applied in programs such as compound acquisition, library design, and compound selection for virtual screening. Due to the high relevance of quantifying chemical diversity several chemoinformatic tools have been developed. These tools can lead to different conclusions depending on the metric and molecular representation and are used mainly in the initial stage of the drug design process. In these cases, and according to the similarity principle formulated by Maggiora and Johnson [12,13], it is recommended to use databases as diverse as possible throughout the property space, while exhibiting properties similar to drugs. This has the purpose of increasing the probability of identifying new compounds or scaffolds that can become leads for a specific biological objective. As more information becomes available, for example, a known compound, a pharmacophore, or a quantitative structure-activity relationship model (QSAR), more focused libraries can be designed. However, it is important to keep in mind that even in the design and synthesis of focused libraries, there must be some degree of diversity between the products since the compounds produced are not identical [14]. Fig. 3.1 shows schematically major design approaches for chemical libraries and the main components of diversity that can be evaluated [9].

Thus far, four major types of chemical diversity have been identified in the literature, namely [15–17]:

a) appendage,
b) functional group,
c) stereochemical, and

d) skeletal (scaffold) diversity.

Table 3.1 summarizes the chemoinformatic approaches that can be used to quantify each type of diversity. The table summarizes the advantages, disadvantages, and main applications of each tool. The properties to be quantified and the selection of the chemoinformatic tool(s) depend on the diversity approach(es) that will be followed for designing and synthesizing the new chemical library.

3.2.1 Methods to evaluate chemical diversity

The evaluation of the degree of molecular diversity within a given set of compounds is not trivial. There are several possible valid methods of which some are visual. In recent years, several research groups have developed approaches that quantitatively describe chemical diversity. Tools such as RDKIT, Platform for Unified Molecular Analysis (PUMA), [18] or the workflows developed in KNIME by Naveja et al. [19] can help in the task of assessing chemical diversity. The following section describes the chemoinformatic methods to evaluate diversity and is exemplified by common case studies in drug design.

3.2.1.1 R-group decomposition (qualitative)

This approach determines the possible R-group decompositions of a given target molecule. The query molecule consists of the scaffold and ligand attachment points represented by R-groups. The query molecule can be input as SMARTS, or it can be searched through the implementation of tools such as the Maximum Common Substructure (MCSS) [20]. The advantages of R-group decomposition is that it can be

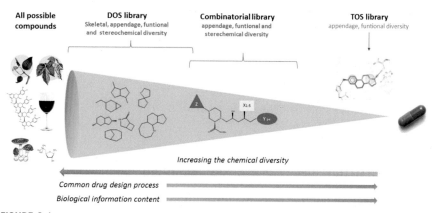

FIGURE 3.1

Different approaches to designing chemical libraries according to their level of diversity. Diversity-oriented synthesis (DOS); combinatorial synthesis, and target-oriented synthesis (TOS).

Adapted from W.R.J.D. Galloway, A. Isidro-Llobet, D.R. Spring, Diversity-oriented synthesis as a tool for the discovery of novel biologically active small molecules, Nat. Commun. 1 (2010) 80.

Table 3.1 Major types of structural diversity and chemoinformatics tools to study each.

Major types of chemical diversity	Information	Chemoinformatics approach	Advantages	Disadvantages
Appendage or building-block	Variation in structural moieties around a common skeleton	• R-group decomposition	It can be used in SAR/SmART analysis	Difficult to interpret when increases the number of molecules with different core scaffolds
Functional group	Variation in the functional groups	• Whole molecule properties (similarity and distance)	Straightforward to interpret	Do not encode information from the structure or substructure
		• SStructural fingerprints	Capture more information of the structure	Difficult to interpret
Stereochemical	Variation in the orientation of potential macromolecule-interacting elements	• Alignment analysis	Provides information about 3D diversity	Increases the time and computational cost
Skeletal (scaffold)	Presence of many distinct molecular skeletons	• Scaffold count analysis • Cyclic system retrieval (CSR) curves • Shannon entropy (SE) • Scaffold tree • Scaffold Hunter	Straightforward to interpret	Do not provide information on the whole molecule

complemented with biological information, which makes possible identifying structural changes that modify biological activity or other properties. However, a shortcoming is that the analysis of the results is increasingly difficult as the number of molecules with different core scaffolds increases. In general, this is the method of choice to analyze combinatorial libraries. To illustrate this point, Fig. 3.2 shows the R-group decomposition of compounds from a combinatorial library of analogues of indinavir [21]. The query molecule is indicated in red.

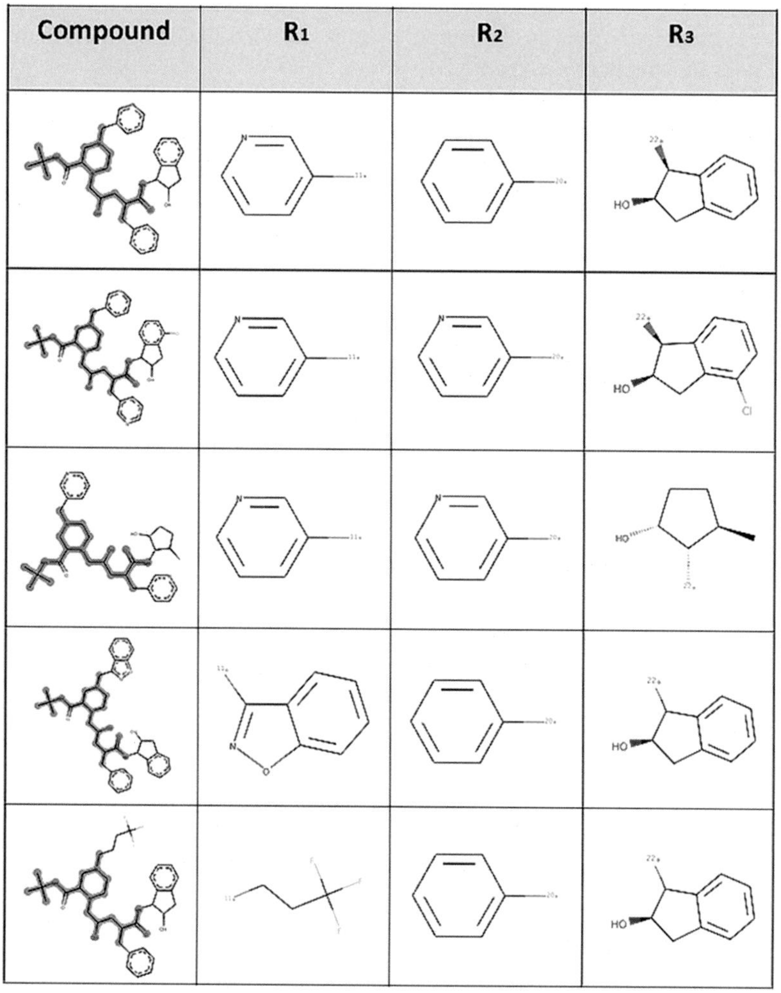

FIGURE 3.2

R-group decomposition of a combinatorial library of analogues of indinavir.

3.2.1.2 Molecular diversity

Molecular diversity is one of the most generalized analyses to describe and compare databases since molecular descriptors capture information of the whole molecule and are usually straightforward to interpret. The molecular descriptors can be evaluated individually by means of dispersion graphs (boxplots, histograms, or density plots), or by means of similarity and distance metrics which quantify diversity as a function of pairwise molecular dissimilarities and that allow the inclusion of more than one descriptor [22]. Molecular descriptors frequently used to describe chemical libraries include molecular weight (MW), number of rotatable bonds (RBs), hydrogen bond acceptors (HBAs), hydrogen bond donors (HBDs), topological polar surface area (TPSA), and the octanol/water partition coefficient (SlogP), which are properties commonly used as descriptors to represent lead-like, drug-like, or medicinally relevant chemical spaces [23,24].

Fig. 3.3 shows the results of an exemplary molecular property analysis of 154,680 natural products from the Universal Natural Product Database (UNPD) [25], 1806 approved drugs (approved drugs), 188 morpholine peptidomimetics from a DOS approach (DOS library) [26], 37 analogues of indinavir retrieved

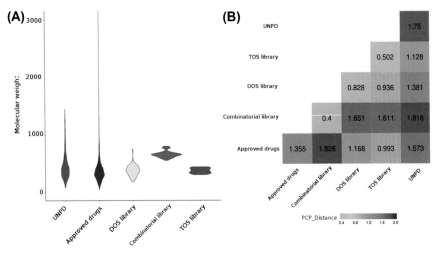

FIGURE 3.3

Molecular property diversity of five different compound databases, namely, 154,680 natural products from the Universal Natural Product Database (UNPD), 1806 approved drugs, 188 morpholine peptidomimetics from a diversity-oriented synthesis (DOS) set, 37 analogues of indinavir from a combinatorial library, and 27 nonnucleoside DNA-methyltransferase inhibitors from lead optimization program e.g., target-oriented synthesis (TOS).

a) Box plot of molecular weight.

b) Distance matrix calculated with the Euclidean distance of six properties of pharmaceutical interest. See text for details.

from a combinatorial library (Combinatorial library) [21], and 27 nonnucleoside DNA-methyltransferase inhibitors from a lead optimization program (TOS library) [27]. The distributions of molecular weight (MW), which was used as a metric to compare the different databases, are summarized in the box plot of Fig. 3.3A. Fig. 3.3B shows the distance matrix calculated with Euclidean distance of six molecular properties (MW, HBD, HBA, TPSA, nRotB, xlogP. The matrix is generated with the median inter- and intradatabase values. Databases in dark red are the most diverse. The distance matrix indicated that, for instance, the combinatorial library is the least diverse (interset distance of 0.4), followed by the TOS library (0.502). In contrast, compounds from UNPD are the most diverse (intraset distance 1.75). It can also be observed that the DOS, TOS, and combinatorial libraries have similar properties to approved drugs and natural products.

3.2.1.3 Structural fingerprints

Although informative and intuitive to interpret, physicochemical descriptors do not capture several structural features contained in the molecules. For instance, it is not uncommon that two compounds with very different chemical structures can have comparable MW and other drug-like properties. Molecular fingerprints are broadly used in a large number of chemoinformatics and computer-aided drug design applications [28]. Fingerprints are especially useful for similarity calculations, such as database searching or clustering, generally measuring similarity as the Tanimoto coefficient [29]. A disadvantage of several fingerprints is that they are difficult to interpret.

Tanimoto similarity is calculated with the expression:

$$T(A, B) = \frac{c}{a + b - c}$$

where T(A, B) is Tanimoto similarity with possible values being any real number between 0 and 1, c is the number of features for which both molecules A and B have a "1" value, a is the number of features for which molecule A has a "1" value, and b is the number of features for which molecule B has a "1" value.

For a database with n compounds, n (n - 1)/2 pairwise comparisons are to be computed and can be stored in a similarity matrix. The matrix can be visualized, for instance, by using a heatmap. However, the visualization becomes more difficult, and in general, the matrix can be quite large as the number of compounds increases. One way to visualize these results and, furthermore, conduct quantitative and statistical analysis with other compound databases is by analyzing the off-values of the diagonal of the similarity matrix, for instance, for the analysis of the cumulative distribution function for each data set.

To exemplify this point, Figs. 3.4A and B shows the cumulative distribution function of the pairwise intraset similarity values of the same five datasets analyzed in Fig. 3.3. The pairwise similarity values were calculated with MACCS keys (166-bit)/Tanimoto and ECFP4/Tanimoto. The table beneath the figure summarizes the statistics of the distribution. In this example, according to the median of the MACCS

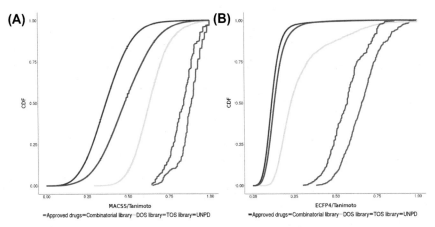

FIGURE 3.4

Cumulative distribution functions of all pairwise similarity comparisons of the five data sets from Fig. 3.3 using the (A) MACCS keys/Tanimoto and (B) ECFP4/Tanimoto coefficient. The table summarizes the statistics of the cumulative distribution functions. Sd: standard deviation; first/third Qu first and third quartiles, respectively.

keys/Tanimoto and ECFP4/Tanimoto, approved drugs is the database with the largest diversity. The least diverse libraries considering the selected molecular fingerprints are combinatorial and TOS libraries.

3.2.1.4 Alignment analysis

Conformational analysis such as the 3D alignment (or superposition) of structures serves as a focal point for the interpretation and understanding of molecular data. This is an important step in the identification of a pharmacophoric pattern for molecules that bind to the same receptor. Similarities can be used in programs of design and optimization of compounds and in the identification of new classes of compounds. If atoms of similar characteristics are placed in the same location, they are expected to produce a similar result [30].

To calculate the chemical similarity of 3D-aligned molecules, programs such as ROCS (rapid overlay of chemical structures) measures the shape and chemical ("color") similarity of two compounds by calculating Tanimoto coefficients from aligned overlap volumes:

$$T = \frac{OAB}{OA + OB - OAB}$$

where OAB is the aligned overlap volume between molecules A and B. Color similarity is calculated from overlaps between dummy atoms marking a predefined set of pharmacophore features defined by the ROCS color force field [31].

At least in principle, it is expected that 3D methods provide a more detailed characterization of a molecule than a simple fingerprint or 2D descriptor [32]. However,

3D methods require a previous conformational analysis, or an energy minimization of the compounds, which in some cases is computationally demanding. There are also reports that 2D methods are comparable or even superior to 3D methods in similarity searching [33]. Alignment analysis can also be used in approaches such as DOS to ensure the 3D diversity of a library of compounds [34–36].

3.2.1.5 Scaffold diversity

A complementary approach to characterize compound databases is through molecular scaffolds or 'chemotypes" i.e., the core structure of a molecule [37]. Similar to physicochemical properties, molecular scaffolds are easy to interpret and facilitate communication with scientists working in different disciplines. Indeed, molecular scaffolds are associated with the concepts "scaffold hopping" [38] and "privileged structures" [39]. This definition has been used to enhance the interpretation of the chemical space and classifying molecules based on their scaffolds, for example, *Scaffold Tree* [40] and *Scaffold Hunter* [41].

Scaffold diversity analysis depends on several factors including the specific approach to describe the scaffolds, the size of the database, and the distribution of the molecules in those scaffold classes. Often, scaffold diversity is measured based on frequency counts. While these measures are correct in the way they are defined they do not provide sufficient information concerning the specific distribution of the molecules across the different scaffolds, particularly the most populated ones. As a complementary metric for the comprehensive scaffold diversity analysis of compound data sets it has been proposed [42].

To illustrate the scaffold content and diversity analysis, Fig. 3.5 shows the scaffold diversity of five compound libraries obtained from different sources (vide supra). The scaffold diversity is measured using cyclic system retrieval (CSR) curves (Fig. 3.5A) where the fraction of cyclic systems is plotted against the cumulative fraction of the database. Quantitatively, the CSR curves can be compared using metrics such as area under the curve (AUC) and the fraction of chemotypes that recover 50% of the molecules in the data set (F_{50}). The graph indicates that the combinatorial library, being the closest to a diagonal (e.g., lower AUC value), is the most diverse. The curves for UNPD and TOS have a rapid increase in its slope, indicating that these datasets have the lowest scaffold diversity. Based on these metrics, the diversity decreases in the relative order: combinatorial library > DOS library > approved drugs > TOS library > UNPD. On the other hand, the distribution of the 10 most frequent scaffolds for the five compound libraries and their respective values of scaled Shannon entropy (SSE) are shown in Fig. 3.5B where values of SSE closer to 0 indicate that all compounds have the same chemotype (minimum diversity) and SSE closer to 1.0 indicate that the compounds are evenly distributed between the n acyclic and/or cyclic systems (maximum diversity). Under this criterion, the combinatorial and DOS libraries show a high relative diversity and the TOS library is the least diverse. The table beneath the figure summarizes the results of the scaffold diversity computed with different metrics.

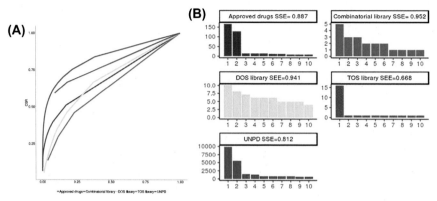

FIGURE 3.5

Scaffold diversity of five different databases: 154,680 natural products from the Universal Natural Product Database (UNPD), 1806 approved drugs, 188 morpholine peptidomimetics from a diversity-oriented synthesis (DOS) set, 37 analogues of indinavir from a combinatorial library, and 27 nonnucleoside DNA-methyltransferase inhibitors from a target-oriented synthesis (TOS) approach. (A) Cyclic system retrieval (CSR) curves for the data sets studied in this work. (B) Distribution and Shannon entropy for the 10 most frequent chemotypes.

3.2.1.6 Consensus diversity analysis: Consensus Diversity Plots

As elaborated above, chemical representation and descriptors are at the core of diversity analysis and basically any chemoinformatic application. Therefore, the perception of the chemical space and assessment of the diversity of a compound collection, in general, is relative to the molecular representation. In order to decrease the dependence of the diversity with molecular representation, it has been proposed to use a consensus approach through the assessment of the global diversity using Consensus Diversity Plots (CDPs) [43]. A CDP is a 2D graph that represents in the same plot up to four measures of diversity. The most common are fingerprint-based, scaffold, whole molecular properties associated with drug-like characteristics, and size of the database. CDPs have been employed to quantify characterize the global diversity of over 7000 compounds tested against 52 epigenetic-related targets [44] and data sets of natural products including NuBBE$_{DB}$ that is a database of natural products from Brazil [44,45] and BIOFACQUIM that is a database of natural products from Mexico [46]. There is a free online server to generate CDPs [43].

Fig. 3.6 illustrates a CDP comparing the global diversity of the same five datasets analyzed in Section 2.1.2: 14,680 natural products from UNPD, 1806 approved drugs, 188 morpholine peptidomimetics from a DOS approach, 37 analogues of indinavir obtained from a combinatorial library, and 27 nonnucleoside DNA-methyltransferase inhibitors from lead optimization program (TOS library). In this example, the thresholds along the X- and Y-axis were defined as the median of the pairwise fingerprint similarity and the median of the AUC for all databases,

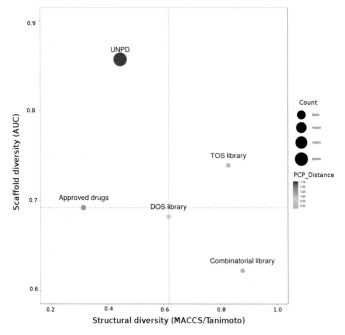

FIGURE 3.6

Exemplary Consensus Diversity Plot. Each data point represents a compound database: 154,680 natural products from the Universal Natural Product Database (UNPD), 1806 approved drugs, 188 morpholine peptidomimetics from a diversity-oriented synthesis (DOS) set, 37 analogues of indinavir from a combinatorial library, and 27 nonnucleoside DNA-methyltransferase inhibitors from a target-oriented synthesis (TOS) approach. Molecular fingerprints diversity is plotted in the X-axis, the scaffold diversity in the Y-axis, the physicochemical properties diversity in a color continuous scale, and the relative number of compounds in the database as the data point size. AUC: area under the curve; PCP: physicochemical properties.

respectively. The CDP plot in Fig. 3.6 indicates that approved drugs and DOS are the data sets with the overall largest fingerprint and scaffold diversity. In contrast, TOS is the data set with the lowest fingerprint and scaffold diversity, relative to the data sets that are compared in the graph. Regarding property diversity, UNPD is the database with the largest diversity.

3.2.2 Visual representation of chemical diversity

The notion of chemical diversity is highly attached to the concept of chemical space. Thus, in addition to the methods discussed above to assess qualitatively or quantitatively chemical diversity, methods to visualize chemical space further complement the integral assessment of diversity.

Two main components are required to generate a visual representation of the chemical space of compound collections: (1) the molecular representation of the molecules to define the multidimensional descriptor space, for example, physicochemical descriptors and molecular fingerprints, and (2) a visualization technique used to reduce the multidimensional space into two or three dimensions. Common visualization methods are principal component analysis (PCA) and self-organizing maps (SOMs) [47]. Other visualizations approaches are multifusion similarity (MFS) maps [48], radar plots, Sammon mapping, activity-seeded structure-based clustering, singular value decomposition, minimal spanning tree, k-means clustering, generative topographic mapping (GTM) [49], hierarchical GTM [50], t-distributed stochastic neighbor embedding (t-SNE) [51], and ChemMaps [52]. Another representation that uses molecular descriptors is the principal moment of inertia (PMI) plot, which represents the shape distribution of the molecules in a library [53]. This plot is commonly used to illustrate the molecular shape diversity in approaches like DOS.

In drug design, it is common to visualize a representation of the chemical space in terms of physicochemical properties of pharmaceutical relevance. It is also informative comparing the chemical space of a compound library of interest with reference libraries of approved drugs to chart the medicinally relevant chemical space [54] and as a guide in the design of libraries focused on bioactivity which is also known as biology-oriented synthesis (BIOS) [55].

To illustrate a visual representation of the property-based chemical space, Fig. 3.7 depicts the comparison of the five libraries analyzed in previous examples (Sections 2.1.2, 2.1.3, 2.1.5, and 2.1.6). The figure shows that the database of UNPD covers most of the chemical space and is also the database with the largest structural diversity and is diverse by physicochemical properties and shapes. In contrast, the combinatorial library and TOS library occupy a more focused region of the property space, which in turn is included within the space of UNPD.

3.3 Molecular complexity

Like similarity, molecular complexity is a subjective concept that can be described differently based on the fields in which one is working. For instance, synthetic and structure complexity could be different. In this regard, Coley et al. developed a metric through a neural network model that defines the synthetic complexity in relation to the expected number of reaction steps needed to produce a target molecule [56]. Ertl and Schuffenhauer also proposed a method to estimate synthetic accessibility of drug-like molecules based on a combination of fragment contributions and molecular complexity. The approach yielded results that were in agreement with an estimation of ease of synthetic accessibility by expert medicinal chemists [57].

In general, chemists define molecular complexity in the same way that many people define art: "they know it when they see it" [58]. However, to facilitate the quantification of complexity, several research groups have developed different

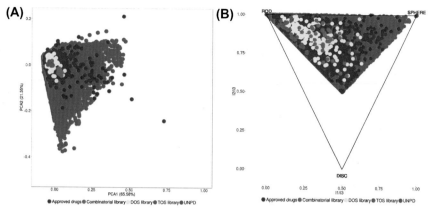

FIGURE 3.7

Molecular diversity and chemical space of five different databases: 154,680 natural products from the Universal Natural Product Database (UNPD), 1806 approved drugs, 188 morpholine peptidomimetics from a diversity-oriented synthesis (DOS) set, 37 analogues of indinavir from a combinatorial library, and 27 nonnucleoside DNA-methyltransferase inhibitors from a target-oriented synthesis (TOS) approach. (A) Principal component analysis of six physicochemical properties of pharmaceutical relevance, (B) principal moment of inertia (PMI) plot.

metrics that facilitate the work in various areas of chemistry. Complexity represents a crucial component in the design of drugs, where it has been associated with selectivity [59,60], safety [59], and the success of compounds in the progress toward clinical development [59]. Likewise, molecular complexity has been associated with the design of libraries for virtual screening [61].

3.3.1 Methods to evaluate chemical complexity

Several descriptors have been published to quantify the molecular complexity, where the quality of the response is judged according to the results that coincide with the intuition of the experiment. Of note, simple complexity metrics are not necessarily metrics with poor performance [62]. The methods described in the literature to assess molecular complexity can be divided into three main categories: (1) graph-theoretical methods, (2) (sub)structure-based approaches, and (3) topological and physicochemical descriptors. Graphic theoretical advances are focused on the connectivity and symmetry of a molecular structure [63]. In contrast, methods based on the substructure analyze molecular structures by counting chemical features and making combinations to get a single score. Perhaps the most intuitive and simple descriptors, yet quite useful, are the topological and physicochemical descriptors such as MW and the number and fraction of chiral centers. Tables 3.2 and 3.3summarize different descriptors and indexes to assess molecular complexity. These and other metrics are reviewed elsewhere [62,63,71].

Table 3.2 Topological and physicochemical descriptors associated with molecular complexity.

Descriptor	Type	Interpretation
Molecular weight	2D	Intuitively associated with larger and possibly more complex structures.
Fraction of carbon atoms with sp^3 hybridization	2D	A higher value of F-sp^3 indicates that the molecule is more likely to have a three-dimensional structure, in other words, that it be a nonplanar structure.
Fraction of chiral carbons (FCC)	2D	A higher value of FCC means larger stereochemical complexity.
Number of rings	2D	Intuitively associated with larger and possibly more complex structures.
Number of heteroatoms	2D	A higher number of heteroatoms tends to increase the compositional complexity of a compound.
Globularity	3D	It is defined as the radius of the molecular surface on the surface of a sphere of the same volume, a value of 1.0 indicates that the molecule has the shape of a sphere.

Table 3.3 Examples of indices proposed to assess molecular complexity.

Index	Description	Reference
Bertz index	Combination of graph theory and information theory to determine the molecular topology (includes terms of symmetry and elemental diversity in a molecule).	[64]
Weiner index	Measurement of molecular branching and cyclicity.	[65]
Randic index	Measurement of molecular branching and cyclicity.	[66]
Hann index	It combines connectivity and substructure descriptors. Useful for assessing complexity in the context of ligand recognition.	[63]
Zagreb index	Graphs are generated from molecules by replacing atoms with vertices and bonds with edges. Zagreb indices increase with size, branching, cyclicity but are not directly affected by symmetry.	[67]
PubChem complexity index	It incorporates terms of symmetry, aromaticity, FCC and Fsp^3. Stereochemistry is not considered explicitly.	[68]
DataWarrior complexity index	No technical information available in the original publication of dataWarrior.	[69]
Scaffold complexity index	Combination of four descriptors: (1) the maximum number of the smallest set of smallest rings, (2) the maximum number of heavy atoms, (3) the maximum number of bonds, where covalent bonds between hydrogen atoms and other atoms are excluded, and (4) the maximum sum of heavy atomic numbers.	[70]

3.4 **Combining diversity and molecular complexity**

So far, a direct relationship between diversity and complexity has not been established; however, in virtual screening approaches, these concepts are closely related and their application can have an impact on the final results.

As described above, in hit identification projects with the aim is identifying new molecules, it is convenient to select collections with chemically diverse structures to increase the probability of identifying new scaffolds that can become leads or prototypes for a specific biological target. In this context, for large compound collections, different filters can be applied to select compounds to be screened. To this end, metrics of molecular complexity have been proposed since according to the experiments of Kuntz [72] and the Hann's model [73], there is a relationship between biological activity and molecular complexity. Therefore, ligands for lead discovery should have a maximum complexity but not higher than necessary to generate a detectable binding affinity, which is the basis for the concept of lead-likeness [74].

Complexity biases introduced by diversity selections can also influence the outcome of the selected subsets [71]. To illustrate this point, Fig. 3.8 shows the

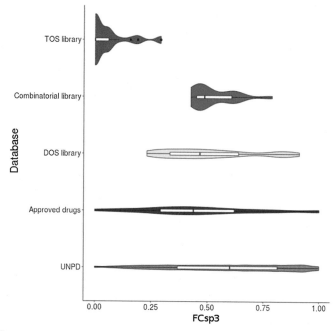

FIGURE 3.8

Distribution of the fraction of sp^3 hybridized carbon of five different compound databases: 154,680 natural products from the Universal Natural Product Database (UNPD), 1806 approved drugs, 188 morpholine peptidomimetics from a diversity-oriented synthesis (DOS) set, 37 analogues of indinavir from a combinatorial library, and 27 nonnucleoside DNA-methyltransferase inhibitors from a target-oriented synthesis (TOS) approach.

distribution of the fraction of sp^3 hybridized carbon of the databases analyzed in the previous sections. As expected and as has already been described in previous studies [60], NPs stand out for the structural complexity higher than compounds from synthesis. In the example of Fig. 3.8, UNPD is the database with the largest distribution of $FCsp^3$ values.

3.5 Conclusions and future directions

Diversity and chemical complexity are two concepts with high relevance in library design and compound selection for experimental evaluation, mainly in HTS. The association between these two concepts has not been defined; however, it is known that a bias in the complexity introduced by the diversity of a chemical library (for example, stereochemical diversity) can have a direct impact on the result of the biological evaluation. So far, different methodologies and metrics have been developed to assess the diversity and chemical complexity in a consistent manner. These approaches have been applied to compare, for instance, the diversity and complexity profiles of drugs approved for clinical use, compounds in clinical development, and screening data sets from different sources such as synthetic or natural products. The development of new tools that describe these concepts in a more global manner has been encouraged to facilitating analysis and decision-making including compound library purchase or compound selection. Diversity and chemical complexity involve many areas of research in the modern design of drugs. Currently, many academic groups are integrating tools for their evaluation and the creation of multidisciplinary works is being encouraged.

Acknowledgments

This work was supported by the National Council of Science and Technology (CONACyT, Mexico) grant number 282785. FIS-G is thankful to CONACyT for the granted scholarship number 629458.

References

[1] J. Gasteiger, Chemoinformatics: achievements and challenges, a personal view, Molecules 21 (2016) 151.

[2] P. Willett, Chemoinformatics: a history, WIREs Comput. Mol. Sci. 1 (2011) 46−56, https://doi.org/10.1002/wcms.1.

[3] Foodinformatics − Applications of Chemical Information to Food Chemistry, Karina Martinez-Mayorga| Springer, n.d. https://www.springer.com/us/book/9783319102252.

[4] A. Puratchikody, S.L. Prabu, A. Umamaheswari, Computer Applications in Drug Discovery and Development, IGI Global, 2018.

[5] M.S. Lajiness, V. Shanmugasundaram, Strategies for the identification and generation of informative compound sets, Methods Mol. Biol. 275 (2004) 111−130.

[6] J.L. Medina-Franco, Chemoinformatic Characterization of the Chemical Space and Molecular Diversity of Compound Libraries, in: Diversity-Oriented Synthesis, 2013, pp. 325−352, https://doi.org/10.1002/9781118618110.ch10.

[7] E. Lenci, A. Guarna, A. Trabocchi, Diversity-oriented synthesis as a tool for chemical genetics, Molecules 19 (2014) 16506−16528.

[8] S.L. Schreiber, Target-oriented and diversity-oriented organic synthesis in drug discovery, Science 287 (2000) 1964−1969.

[9] W.R.J.D. Galloway, A. Isidro-Llobet, D.R. Spring, Diversity-oriented synthesis as a tool for the discovery of novel biologically active small molecules, Nat. Commun. 1 (2010) 80.

[10] J.L. Medina-Franco, G.M. Maggiora, Molecular similarity analysis, in: J. Bajorath (Ed.), Chemoinformatics for Drug Discovery, John Wiley & Sons, Inc, Hoboken, NJ, 2013, pp. 343−399.

[11] R. Sheridan, Why do we need so many chemical similarity search methods? Drug Discov. Today 7 (2002) 903−911.

[12] N. Singh, R. Guha, M.A. Giulianotti, C. Pinilla, R.A. Houghten, J.L. Medina-Franco, Chemoinformatic analysis of combinatorial libraries, drugs, natural products, and molecular libraries small molecule repository, J. Chem. Inf. Model. 49 (2009) 1010−1024.

[13] G. Klopmand, in: M.A. Johnson, G.M. Maggiora (Eds.), Concepts and Applications of Molecular Similarity, vol. 13, John Wiley & Sons, New York, 1990, pp. 539−540, 393 pp. J. Comput. Chem. 1992.

[14] K.M.G. O'Connell, R.J. Warren, D.R. Spring, The Basics of Diversity-Oriented Synthesis, in: Diversity-Oriented Synthesis, 2013, pp. 1−26, https://doi.org/10.1002/9781118618110.ch1.

[15] D.R. Spring, Diversity-oriented synthesis; a challenge for synthetic chemists, Org. Biomol. Chem. 1 (2003) 3867.

[16] M.D. Burke, S.L. Schreiber, A planning strategy for diversity-oriented synthesis, Angew Chem. Int. Ed. Engl. 43 (2004) 46−58.

[17] R.J. Spandl, M. Díaz-Gavilán, K.M.G. O'Connell, G.L. Thomas, D.R. Spring, Diversity-oriented synthesis, Chem. Rec. 8 (2008) 129−142.

[18] M. González-Medina, J.L. Medina-Franco, Platform for unified molecular analysis: PUMA, J. Chem. Inf. Model. 57 (2017) 1735−1740.

[19] J.J. Naveja, F.I. Saldívar-González, N. Sánchez-Cruz, J.L. Medina-Franco, Cheminformatics approaches to study drug polypharmacology, in: K. Roy (Ed.), Multi-Target Drug Design Using Chem-Bioinformatic Approaches, Springer New York, New York, NY, 2019, pp. 3−25.

[20] Y. Cao, T. Jiang, T. Girke, A maximum common substructure-based algorithm for searching and predicting drug-like compounds, Bioinformatics 24 (2008) i366−i374.

[21] Y. Cheng, T.A. Rano, T.T. Huening, F. Zhang, Z. Lu, W.A. Schleif, L. Gabryelski, D.B. Olsen, M. Stahlhut, L.C. Kuo, J.H. Lin, X. Xu, L. Jin, T.V. Olah, D.A. McLoughlin, R.C. King, K.T. Chapman, J.R. Tata, A combinatorial library of indinavir analogues and its in vitro and in vivo studies, Bioorg. Med. Chem. Lett 12 (2002) 529−532.

[22] G.M. Maggiora, Introduction to Molecular Similarity and Chemical Space, Foodinformatics, 2014, pp. 1−81, https://doi.org/10.1007/978-3-319-10226-9_1.

[23] C.A. Lipinski, Lead- and drug-like compounds: the rule-of-five revolution, Drug Discov. Today Technol. 1 (2004) 337−341.

[24] D.F. Veber, S.R. Johnson, H.-Y. Cheng, B.R. Smith, K.W. Ward, K.D. Kopple, Molecular properties that influence the oral bioavailability of drug candidates, J. Med. Chem. 45 (2002) 2615−2623.

[25] J. Gu, Y. Gui, L. Chen, G. Yuan, H.-Z. Lu, X. Xu, Use of natural products as chemical library for drug discovery and network pharmacology, PLoS One 8 (2013) e62839.

[26] E. Lenci, R. Innocenti, G. Menchi, A. Trabocchi, Diversity-oriented synthesis and chemoinformatic analysis of the molecular diversity of sp3-rich morpholine peptidomimetics, Front. Chem. 6 (2018), https://doi.org/10.3389/fchem.2018.00522.

[27] B. Zhong, S. Vatolin, N.D. Idippily, R. Lama, L.A. Alhadad, F.J. Reu, B. Su, Structural optimization of non-nucleoside DNA methyltransferase inhibitor as anti-cancer agent, Bioorg. Med. Chem. Lett 26 (2016) 1272−1275.

[28] A.G. Maldonado, J.P. Doucet, M. Petitjean, B.-T. Fan, Molecular similarity and diversity in chemoinformatics: from theory to applications, Mol. Divers. 10 (2006) 39−79.

[29] N. Nikolova, J. Jaworska, Approaches to measure chemical similarity— a review, QSAR Comb. Sci. 22 (2003) 1006−1026.

[30] M.D. Miller, Molecular Superposition, Encyclopedia of Computational Chemistry, 2002, https://doi.org/10.1002/0470845015.cma029m.

[31] W.-H. Shin, X. Zhu, M.G. Bures, D. Kihara, Three-dimensional compound comparison methods and their application in drug discovery, Molecules 20 (2015) 12841−12862.

[32] M. Thimm, A. Goede, S. Hougardy, R. Preissner, Comparison of 2D similarity and 3D superposition. Application to searching a conformational drug database, J. Chem. Inf. Comput. Sci. 44 (2004) 1816−1822.

[33] R.P. Sheridan, Chemical similarity searches: when is complexity justified? Expert Opin. Drug Discov. 2 (2007) 423−430.

[34] S.K. Ko, H.J. Jang, E. Kim, S.B. Park, Concise and diversity-oriented synthesis of novel scaffolds embedded with privileged benzopyran motif, Chem. Commun. (2006) 2962−2964.

[35] Y. Choi, H. Kim, S.B. Park, A divergent synthetic pathway for pyrimidine-embedded medium-sized azacycles through an N-quaternizing strategy, Chem. Sci. 10 (2019) 569−575.

[36] H. Kim, T.T. Tung, S.B. Park, Privileged substructure-based diversity-oriented synthesis pathway for diverse pyrimidine-embedded polyheterocycles, Org. Lett. 15 (2013) 5814−5817.

[37] A. Schuffenhauer, T. Varin, Rule-based classification of chemical structures by scaffold, Mol. Inform. 30 (2011) 646−664.

[38] G. Schneider, W. Neidhart, T. Giller, G. Schmid, "Scaffold-hopping" by topological pharmacophore search: a contribution to virtual screening, Angew. Chem. Int. Ed. 38 (1999) 2894−2896.

[39] B.E. Evans, K.E. Rittle, M.G. Bock, R.M. DiPardo, R.M. Freidinger, W.L. Whitter, G.F. Lundell, D.F. Veber, P.S. Anderson, R.S. Chang, Methods for drug discovery: development of potent, selective, orally effective cholecystokinin antagonists, J. Med. Chem. 31 (1988) 2235−2246.

[40] A. Schuffenhauer, P. Ertl, S. Roggo, S. Wetzel, M.A. Koch, H. Waldmann, The scaffold tree–visualization of the scaffold universe by hierarchical scaffold classification, J. Chem. Inf. Model. 47 (2007) 47−58.

[41] T. Schäfer, N. Kriege, L. Humbeck, K. Klein, O. Koch, P. Mutzel, Scaffold Hunter: a comprehensive visual analytics framework for drug discovery, J. Cheminf. 9 (2017) 28.

[42] J.L. Medina-Franco, K. Martínez-Mayorga, A. Bender, T. Scior, Scaffold diversity analysis of compound data sets using an entropy-based measure, QSAR Comb. Sci. 28 (2009) 1551−1560.

[43] M. González-Medina, F.D. Prieto-Martínez, J.R. Owen, J.L. Medina-Franco, Consensus Diversity Plots: a global diversity analysis of chemical libraries, J. Cheminf. 8 (2016) 63.

[44] J.J. Naveja, J.L. Medina-Franco, Insights from pharmacological similarity of epigenetic targets in epipolypharmacology, Drug Discov. Today 23 (2018) 141−150.

[45] F.I. Saldívar-González, M. Valli, A.D. Andricopulo, V. da Silva Bolzani, J.L. Medina-Franco, Chemical space and diversity of the Nu BBE database: a chemoinformatic characterization, J. Chem. Inf. Model. 59 (2019) 74−85.

[46] B.A. Pilón-Jiménez, F.I. Saldívar-González, B.I. Díaz-Eufracio, J.L. Medina-Franco, Biofacquim: a Mexican compound database of natural products, Biomolecules 9 (2019), https://doi.org/10.3390/biom9010031.

[47] D.I. Osolodkin, E.V. Radchenko, A.A. Orlov, A.E. Voronkov, V.A. Palyulin, N.S. Zefirov, Progress in visual representations of chemical space, Expert Opin. Drug Discov. 10 (2015) 959−973.

[48] J.L. Medina-Franco, G.M. Maggiora, M.A. Giulianotti, C. Pinilla, R.A. Houghten, A similarity-based data-fusion approach to the visual characterization and comparison of compound databases, Chem. Biol. Drug Des. 70 (2007) 393−412.

[49] M. Reutlinger, G. Schneider, Nonlinear dimensionality reduction and mapping of compound libraries for drug discovery, J. Mol. Graph. Model. 34 (2012) 108−117.

[50] P. Tino, I. Nabney, Hierarchical GTM: constructing localized nonlinear projection manifolds in a principled way, IEEE Trans. Pattern Anal. Mach. Intell. 24 (2002) 639−656.

[51] L. van der Maaten, G. Hinton, Visualizing Data using t-SNE, J. Mach. Learn. Res. 9 (2008) 2579−2605.

[52] J.J. Naveja, J.L. Medina-Franco, ChemMaps: towards an approach for visualizing the chemical space based on adaptive satellite compounds, F1000Res. 6 (2017) 1134.

[53] W.H.B. Sauer, M.K. Schwarz, Molecular shape diversity of combinatorial libraries: a prerequisite for broad bioactivity, J. Chem. Inf. Comput. Sci. 43 (2003) 987−1003.

[54] J.L. Medina-Franco, K. Martinez-Mayorga, N. Meurice, Balancing novelty with confined chemical space in modern drug discovery, Expert Opin. Drug Discov. 9 (2014) 151−165.

[55] H. van Hattum, H. Waldmann, Biology-oriented synthesis: harnessing the power of evolution, J. Am. Chem. Soc. 136 (2014) 11853−11859.

[56] C.W. Coley, L. Rogers, W.H. Green, K.F. Jensen, SCScore, Synthetic complexity learned from a reaction corpus, J. Chem. Inf. Model. 58 (2018) 252−261.

[57] P. Ertl, A. Schuffenhauer, Estimation of synthetic accessibility score of drug-like molecules based on molecular complexity and fragment contributions, J. Cheminf. 1 (2009) 8.

[58] S.H. Bertz, The first general index of molecular complexity, J. Am. Chem. Soc. 103 (1981) 3599−3601, https://doi.org/10.1021/ja00402a071.

[59] F. Lovering, J. Bikker, C. Humblet, Escape from flatland: increasing saturation as an approach to improving clinical success, J. Med. Chem. 52 (2009) 6752−6756.

[60] P.A. Clemons, N.E. Bodycombe, H.A. Carrinski, J.A. Wilson, A.F. Shamji, B.K. Wagner, A.N. Koehler, S.L. Schreiber, Small molecules of different origins

have distinct distributions of structural complexity that correlate with protein-binding profiles, Proc. Natl. Acad. Sci. USA 107 (2010) 18787−18792.

[61] S.H. Nilar, N.L. Ma, T.H. Keller, The importance of molecular complexity in the design of screening libraries, J. Comput. Aided Mol. Des. 27 (2013) 783−792.

[62] O. Méndez-Lucio, J.L. Medina-Franco, The many roles of molecular complexity in drug discovery, Drug Discov. Today 22 (2017) 120−126.

[63] T. Böttcher, An additive definition of molecular complexity, J. Chem. Inf. Model. 56 (2016) 462−470.

[64] R. Barone, M. Chanon, A new and simple approach to chemical complexity. Application to the synthesis of natural products, J. Chem. Inf. Comput. Sci. 41 (2001) 269−272.

[65] D. Bonchev, The overall Wiener index—a new tool for characterization of molecular topology, J. Chem. Inf. Comput. Sci. 41 (2001) 582−592.

[66] M. Randic, Characterization of molecular branching, J. Am. Chem. Soc. 97 (1975) 6609−6615.

[67] S. Nikolić, I.M. Tolić, N. Trinajstić, I. Baučić, On the Zagreb indices as complexity indices, Croat. Chem. Acta 73 (2000) 909−921.

[68] S. Kim, P.A. Thiessen, E.E. Bolton, J. Chen, G. Fu, A. Gindulyte, L. Han, J. He, S. He, B.A. Shoemaker, J. Wang, B. Yu, J. Zhang, S.H. Bryant, PubChem substance and compound databases, Nucleic Acids Res. 44 (2016) D1202−D1213.

[69] T. Sander, J. Freyss, M. von Korff, C. Rufener, DataWarrior: an open-source program for chemistry aware data visualization and analysis, J. Chem. Inf. Model. 55 (2015) 460−473.

[70] J. Xu, A new approach to finding natural chemical structure classes, J. Med. Chem. 45 (2002) 5311−5320.

[71] A. Schuffenhauer, N. Brown, P. Selzer, P. Ertl, E. Jacoby, Relationships between molecular complexity, biological activity, and structural diversity, J. Chem. Inf. Model. 46 (2006) 525−535.

[72] I.D. Kuntz, K. Chen, K.A. Sharp, P.A. Kollman, The maximal affinity of ligands, Proc. Natl. Acad. Sci. USA 96 (1999) 9997−10002.

[73] M.M. Hann, T.I. Oprea, Pursuing the leadlikeness concept in pharmaceutical research, Curr. Opin. Chem. Biol. 8 (2004) 255−263.

[74] M.M. Hann, A.R. Leach, J.N. Burrows, E. Griffen, 4.18 - lead discovery and the concepts of complexity and lead-likeness in the evolution of drug candidates, in: J.B. Taylor, D.J. Triggle (Eds.), Comprehensive Medicinal Chemistry II, Elsevier, Oxford, 2007, pp. 435−458.

Virtual screening of small-molecule libraries

Qingliang Li

National Center for Biotechnology Information, National Library of Medicine, National Institutes of Health, Besthesda, MD 20894, United States

4.1 Introduction

Virtual screening [1—4] is a computational strategy used in drug discovery research to identify active small molecules against a certain biological target from a large chemical library. The term "virtual screening" was coined in the late 1990s, in contrast to the "physical" screening with robots, known as high-throughput screening (HTS). The development of combinatorial chemistry created millions of chemical compounds that can hardly be handled by traditional medicinal chemists; on the other hand, the increasing computational capabilities of computers, including supercomputers, made it possible to screen multiple millions of chemical compounds at an affordable speed and cost.

According the usage of information, virtual screening methods can be divided into two categories: (1) structure-based virtual screening (SBVS) [5] and (2) ligand-based virtual screening (LBVS). LBVS directly search for compounds similar to known active molecules; while, SBVS indirectly search compounds that can fit into the binding site of the biological target. Molecular docking is the core technology of SBVS methods. Target structures are generated by three main sources, X-ray crystallography, nuclear magnetic resonance (NMR), or homology modeling. For LBVS, the similarity measurements between two molecules are dependent on the molecular representation, which can be grouped as two-dimensional (2D) fingerprint similarity searching, pharmacophore matching, and three-dimensional (3D) shape screening.

A typical virtual screening of small-molecule library is a multistep process (Fig. 4.1). It starts with small molecules for screening with a careful preparation of the compounds by removing the unwanted molecules. Depending on the prior knowledge of the target structure and active molecules, LBVS and SBVS can be employed independently or together. Due to the imperfections of the current virtual screening methods, a postscreening process is usually conducted on carefully selecting the hits for further experimental testing.

In this chapter, the content is organized in the following way: (1) structure-based methodologies, (2) ligand-based methodologies, (3) small-molecule libraries and library preparation issues, (4) computational validation, (5) postscreening processing, and (6) perspective.

Small Molecule Drug Discovery. https://doi.org/10.1016/B978-0-12-818349-6.00004-2
2020 Published by Elsevier Inc.

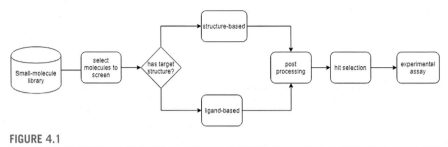

FIGURE 4.1

Typical workflow of virtual screening of a small-molecule library.

4.2 Structure-based virtual screening

SBVS utilizes the structural information of a biological target to screen a chemical library. The biological targets can be RNAs or DNAs, but most of them are proteins. SBVS approaches do not require any compound with known activity in advance; instead, it only needs the biological target structure that has been determined, e.g., X-ray crystal structures. The SBVS approaches comprise three basic pillars: (1) molecular docking, (2) 3D target structure, and (3) chemical compound library. In this section, the first two are discussed, and the small-molecule library will be discussed separately in the next section.

4.2.1 Molecular docking

Molecular docking is the core technique of the SBVS. It is designed to construct the intermolecular complex structure of a chemical compound and the target structure based on geometry and/or physical-chemical principles. A molecular docking tool takes two inputs, i.e., a target structure and a chemical compound, and generates the best interaction mode and an interaction score indicating how well the compound fits in the binding site of the target structure. A typical docking algorithm has two complementary components, namely, (1) a searching strategy and (2) a scoring function. The search strategy is designed to orientate a chemical compound in the binding site of target structure and search the possible interaction mode (positions and poses) toward the target structure. The scoring function is to measure an interaction mode generated by the search strategy and guide the searching process. A docking process usually runs multiple iterations of searching and scoring until reaching the best interaction mode, an energetic minimum state.

A molecular docking is to mimic the molecular recognition between a ligand and its target structure. In this process, both ligand and target structure experience conformational changes. According the treatment of flexibility, molecular docking methods are also divided into flexible-ligand docking and flexible-target docking. In flexible-ligand docking, the ligand is fully flexible that has conformational changes by rotating bonds, while the target structure is static and no conformational change. The conformation of the target structure is usually obtained from complex structure by removing the ligand, so that the binding site is open to accommodate ligands for docking. The searching and sampling space of the possible interaction patterns of a compound could be enormous and time consuming, even only for ligand flexibility, thus an exhaustive search strategy usually can not afford in practice. Docking tools usually incorporate optimization algorithms, such as Genetic Algorithm, Monte Carlo algorithm, to explore the possibilities in a reasonable time [6].

Although most docking tools can handle ligand flexibility well, the treatment of full target flexibility is still an unsolved question, one of the most challenging problems in the field. In general, target flexibility can be local flexibility and global flexibility, which means the flexibility in the binding site and the backbone movement of the entire structure. A variety of approaches have been reported to deal with local flexibility, such as soft docking [7], side chain flexibility [8]. As to global flexibility, ensemble docking [9—12] that cooperates multiple receptor structures is the most widely used method.

Soft docking is the easiest way to accommodate receptor flexibility implicitly by reducing the van der Waals contributions in the total energy score. Thus, it can tolerate a small degree of clashes between the ligand and target to account for local flexibility. The advantage of soft docking approach is the computational efficiency and easy implementation, but its applications are limited to local movement around the binding site. Another approach accounting for local flexibility is to rearrange a few side chains of the residues in binding site by using rotamer library [8,13]. This method allows localized target movement and improves the fit of the ligand at the cost of increasing computational time. Still, only a few selected residues are allowed to move to incorporate local flexibility in this approach. Another method is molecular relaxation [14,15] that first positions ligand into binding site by allowing amount of clashes between the ligand and residues around the binding site, then relaxes or minimizes the complex by moving the protein side chains or backbone. Compared to the approaches mentioned above, this method is able to account for backbone flexibility to some extent.

Ensemble docking [10], also known as multiple target conformation docking [16], is used to dock a ligand to an ensemble of target conformers, then aggregate all the docking results to account receptor flexibility. Ensemble docking is a natural way to extend rigid docking to incorporate receptor flexibility, which becomes the most promising approaches, due to the considerable improvements compared to the single-conformation approaches. In practice, receptor conformations are usually obtained from both experimental approaches, such as X-ray crystal structures, NMR structures, and theoretical approaches, such as MD simulations. As classified by Alonso et al. [17], three types of methods can be used to deal with multiple receptor conformations: (1) combination of the structures into a single docking grid; (2) united description of the receptor with conserved and mobile regions; and (3) individual conformation to be used for docking. The third strategy, also known as the relaxed complex scheme, was first proposed by Lin et al. [18].

4.2.2 Scoring functions

Scoring function is a key component of a docking tool that guides the search process and calculates the final binding score. A scoring function is usually trained by feeding with a collection of compounds with known bioactivities to capture rules that govern the ligand-target interactions, so that it is able to predict the binding affinities of any given new interaction. PDBbind is a valuable source for scoring development, because it provides a comprehensive collection of binding affinities of protein-ligand complex structures obtained from the Protein Data Bank (PDB) database [19]. According to the theories used in scoring function development, the methods can be divided into four categories: (1) force field-based, (2) empirical-based, (3) knowledge-based, and (4) machine learning-based function [20,21].

Force field-based scoring functions predict binding affinities using a sum of energy terms from the classical force field to estimate the strength of protein-ligand interactions. The energy terms are usually noncovalent terms such as van der Waals and electrostatic terms. For example, DOCK [22] and AutoDock [23] are based on AMBER force field [24].

Empirical-based scoring functions calculate the binding score by summing up the contributions and penalties of a group of empirically important factors, such as hydrogen bonding, rotatable bonds, lipophilic contacts, etc, such as Glide score [25], GOLD [26], X-Score [27]. This type of functions is developed using multivariate linear regression or partial least-squares analysis to derive the weights of each term in the function.

Knowledge-based scoring functions estimate the binding scores by summarizing pairwise statistical potentials, which are derived from an inverse Boltzmann analysis between ligand and its target structure. Scoring functions of this type include Drug-Score [28], IT-Score [29], etc.

The fourth type scoring functions, machine learning-based, try to encode the features of the protein-ligand interactions into certain mathematical vectors and train the model by using sophisticated machine learning techniques, such as neural network (NN) [30], support vector machines (SVM) [31], and random forest [32]. Compared to the other three types, machine learning-based scoring functions do not assume any form of the function, instead they learn the function from training data. Ain et al. systematically reviewed the latest improvement of the machine learning-based scoring function, recently [33].

Despite considerable types of scoring functions have been reported, the performance of scoring functions is still the bottleneck of molecular docking tools. None of the existing scoring functions can universally outperform others and accurately calculate the binding affinity of the ligand-target interaction. The limitation of scoring functions directly affects the performance of the structure-based virtual screening, which will likely remain to be the major challenge to the field in the near future. The improvement and development of new scoring functions will continue to be an active direction in research. The latest development of deep learning [34] has demonstrated enormous success in multiple fields, such as computer vision and natural language processing. Compared to traditional machine learning strategies, such as neural network (NN), support vector machines (SVM), and random forest (RF), deep learning can learn features directly from the raw data with minimal or no human intervention. It may be the new direction, at least for now, for scoring function development as well as for computational drug discovery.

4.2.3 Implementations and tools

The pioneering work of molecular docking can be traced back to early 1980s [35]. Since then, a considerable number of docking tools have been developed [6,36]. Some of the most widely used docking tools are listed in Table 4.1, including DOCK [37], AutoDock [23], AutoDock Vina [38], etc.

Due to the many choices of docking tools, performance comparison of these tools are frequently reported. For example, a comprehensive evaluation of 10 docking tools on sampling power and scoring power was reported [40]. It was found that GOLD and LeDock possessed the best capabilities in identifying the correct ligand binding poses and Glide (XP) and GOLD proved to be the two most robust programs on pose predictions. Autodock Vina and GOLD and MOE Dock achieved the best scoring powers. In another comparing study of pose prediction and virtual screening [41], the authors found that ICM, Glide, and Surflex generated ligand pose close to X-ray conformation more often than the other docking tools, and Glide and Surflex outperformed the others when used for virtual screening.

In fact, it is usually observed that comparison studies do not agree with each other on the performance of docking tools. Many reasons may result in such differences. For example, the docking accuracy is specific for a particular protein target family; the target structure preparation and the software parameter settings usually require expert knowledge to conduct a successful docking experiment. According to our experience, there are several considerations for selection docking tools: (1) knowing the docking tool before using it, including theory, parameter settings, and limitations, etc.; (2) well preparing target and ligand structures; (3) creating a validation set of active and inactive ligands if possible to verify the performance and hypothesis; and (4) combining multiple docking tools that are complementary in theory and/or performance.

Table 4.1 Widely used docking tools.

Tool	Availability	Reference	URL
AutoDock	Academic free	[23]	http://autodock.scripps.edu/
AutoDock Vina	Academic free	[38]	http://vina.scripps.edu/
UCSF DOCK	Academic free	[37]	http://dock.compbio.ucsf.edu/
Glide	Commercial	[25]	https://www.schrodinger.com/
GOLD	Commercial	[26]	https://www.ccdc.cam.ac.uk/
MOE Dock	Commercial	[39]	https://www.ccdc.cam.ac.uk/

4.2.4 Sources of 3D target structure

The widely used source of target structures is PDB database [42,43]. PDB is a structural database of large biological molecules, such as proteins and nucleic acids, determined by X-ray crystallography, NMR spectroscopy, and other technologies. Advances in structural biology have generated a great number of 3D structures of biological macromolecules, including therapeutic targets, which offer unparalleled opportunities for structure-based virtual screening and drug discovery. At the time of this work, more than 150,000 biological macromolecular structures are available in the PDB, and it continues to increase every year.

To use these structures in virtual screening, it is important to know the structure quality, especially in the binding site area. Because most of the available structures are from X-ray crystallography, atom positions in these models correspond to the interpretation of the electron density maps by the depositors. Thus, validating the reliability of such coordinates before using them in virtual screening is recommended in practice. In general, the higher quality the better for virtual screening; and complex structures with binding ligand are better than those without binding ligand. Moreover, careful consideration should be taken to prepare for the correct protonation and tautomeric states of amino acids. For instance, in physiological pH, i.e., 7.4, it is generally assumed that the guanidino group of the side chain or arginine (pKa = 12.48) and the ε-amino group of lysine (pKa = 10.54) are protonated (positively charged), whereas the carboxyl groups of the side chains of aspartic acid (pKa = 3.39) and glutamic acid (pKa = 4.07) are not protonated (negatively charged). As histidine can serve as both acid and base over physiological pH range, and both as hydrogen bond donor and acceptor in molecular recognition, it needs to be considered on a case-by-case basis based to its immediate environment [44,45].

In addition, water molecules and metal ions in active-site can significantly contribute to a ligand-target interaction and help to place and orient the ligand into a correct pose in the binding site. The neglection of them would inevitably lead to underestimation of the ligand-target interaction [5].

If no target structure is readily available, computational modeling, such comparative/homology modeling, can be used to build a theoretical model of the target structure [46,47]. Comparative modeling is a methodology to predict the 3D structure based on the observation that proteins with similar sequence usually have similar structure. Generally speaking, it is believed that two protein sequences with an identity of 30% or above share a common 3D structure; a sequence identity over 50% is sufficient for prediction of protein-ligand interaction and drug discovery; a sequence identity between 30% and 50% could facilitate druggability prediction and mutagenesis experiment design [47].

4.3 Ligand-based virtual screening

Ligand-based virtual screening (LBVS) utilizes ligand similarity to retrieve the compounds of interest from a small-molecule database. Compared to the SBVS methods that measure the fitness of a compound candidate against the target structure, LBVS directly search compounds that are similar to the active ones. The basic assumption is that similar molecules may have similar activity or function, though it is not always true if there is an activity cliff [48]. The main differences of LBVS methods are the representations of a molecule and the similarity measurements. Here, we introduce three widely used LBVS methods, namely 2D fingerprint similarity search, 3D shape-based screening, and ligand-based pharmacophore-based matching.

4.3.1 2D fingerprint similarity searching

A 2D molecular fingerprint is devised to encode 2D structural properties of a small molecule into a bit string to allow fast comparison and quantitative similarity assessments of the similarity/difference between two molecules. Each bit (0/1) represents the absence and presence of a certain feature, such as a functional group, a substructure feature, etc. The similarity comparison of the two molecules can be achieved by calculating the similarity between the two fingerprints. In general, there are three types of fingerprints: (1) structural keys-based fingerprints, (2) circular fingerprints, and (3) topological/path-based fingerprints. Some widely used fingerprints and the open-source implementations are listed in Table 4.2.

(1) Structural keys-based fingerprints use an array of bits to represent the presence and absence of certain predefined substructure or features. These fingerprints are restricted by the list of predefined structural keys. The MACCS fingerprints and PubChem fingerprints belong to this category. MACCS fingerprint is a 166 bit-long structural keys that covers most of the interesting chemical features of drug discovery and virtual screening [53]. It was originally developed by MDL, but has also been implemented in other softwares, such as RDKit [51], Open Babel (http://openbabel.org/) [50], and the Chemistry Development Kit (CDK) (https://cdk.github.io/) [49]. As an example of these implementations, the description of

Table 4.2 Popular fingerprints and implementations.

Tool/Library	Implemented fingerprints	Reference	URL
The Chemistry Development Kit (CDK)	Circular fingerprint, MACCS, Pubchem, ShortestPath, etc	[49]	https://cdk.github.io/
Open Babel	MolPrint2D, ECFP, ECFP, etc.	[50]	https://openbabel.org
RDKit	MACCS, circular fingerprint (Morgan fingerprint), topological fingerprint, etc.	[51]	https://www.rdkit.org/
PaDEL-descriptor	Daylight, UNITY, MACCS, PubChem, etc.	[52]	http://www.yapcwsoft.com/dd/padeldescriptor/

the fingerprint in RDKit is reported [54]. Another example is PubChem fingerprint, which has an 881 bit-long vector of different types of substructures and features, including element counts, chemical ring system, atom pairing, atom nearest neighbors, and a group of SMARTS patterns [55]. The fingerprint is used by PubChem [56] for similarity searching and neighboring clustering and is available for download in PubChem system. It has been implemented in the Chemistry Development Kit (CDK) [57] and PaDEL-descriptor [52], an open-source software to calculate molecular descriptors and fingerprints.

(2) Circular fingerprints represent a molecule by using the atom environment of each atom up to a given diameter (number of bond connection). For example, Molprint2D [58] encodes the atom environments of each atom of molecular connectivity table in strings. It is available in Open Babel [50]. The other widely used circular fingerprint is ECFP (Extended-Connectivity Fingerprint) based on Morgan algorithm [59]. It represents molecular structures by circular atom neighborhoods, namely each nonhydrogen atom is encoded into multiple circular layers up to a given diameter. Several diameter variations of these fingerprints, ECFP4 and ECFP6, stand for a diameter of 4 and 6, respectively. This fingerprint is not based on predefined substructural keys. It is one of the most popular fingerprints that are widely used in drug discovery application, and it has been implemented in CDK, RDKit, and PaDEL-descriptor.

(3) Topological or path-based fingerprints represent a molecule by using fragments that follow a path up to a certain number of bonds. Daylight fingerprint is the most prominent of these type of fingerprints [60]. Daylight fingerprint consists of up 2048 bits and encodes all possible connectivity pathways. It has certain variations that can encode linear path and nonlinear connectivity paths.

In addition, there are other types of fingerprints, such as pharmacophore fingerprints [61,62]; hybrid fingerprints that combine both structural keys and connectivity path fragments, e.g., UNITY 2D [63]; other fingerprints encoding protein-ligand interactions, e.g., SIFt (structural interaction fingerprint) [64]. More details about fingerprint similarity search and comparisons in virtual screening were reviewed [65−67].

With fingerprints, the calculation of similarity between two molecules can be obtained by calculating the fingerprint similarity. Many similarity/distance metrics (Table 4.3) have been reported in fingerprint comparison, of which Tanimoto coefficient or Jaccard similarity coefficient is the most widely used.

Table 4.3 Widely used fingerprint similarity metrics.

Name	Equation
Tanimoto coefficient	$\frac{c}{a+b-c}$
Dice coefficient	$\frac{2c}{a+b}$
Cosine similarity	$\frac{c}{\sqrt{ab}}$

Here, a and b represent two fingerprints respectively; c represents the intersection or common part of the two fingerprints of a and b.

Since there are many types of 2D fingerprints, it is of great interest to know which fingerprint performs better and how to select the right fingerprints in virtual screening. An open-source platform has been reported to benchmark fingerprints for ligand-based virtual screening [68]. The study found that the overall performance of all the fingerprints was similar and the intertarget difference was greater than the intratarget difference. Among the 14 tested 2D fingerprints, circular fingerprints were ranked higher and topological torsions fingerprints were always highly ranked regardless of the evaluation methods. Our experience agrees with the observations, as well. In addition, 2D fingerprints have the advantage that they require minimal setup and configuration and less computationally intensive as compared to other virtual screening methods. One of the major disadvantages of 2D fingerprint approaches is the bias toward query molecules [65]. In practice, it is often combined with other virtual screening methods, such as structure-based or shape-based, to avoid such a bias.

4.3.2 3D shape-based screening

The shape of a molecule is the key determinant in molecular recognition in controlling the geometry and energy of molecular interaction. A high degree of shape complementary has been observed in many crystal structures of protein-ligand and protein-protein complexes. Shape-based virtual screening methods use molecular shape, sometimes with other molecular properties like atom types, to find molecules in a database. Putta and Beroza provided an excellent review on shape-based screening methods [69]. In the review, they divided the representations of a molecular shape into four main categories, (1) moment-based, (2) gnomic-based, (3) volume-based, and (4) surface-based. The moment-based methods represent a shape as a set of multipole moments of inertia and the resulting distribution can be represented by a multiple expansion. This method reduces a shape to a set of ordered numbers [70,71]. The gnomic-based approaches map the molecule's shape to a simple shape, e.g., a polyhedron, to encode the complexities of the original shape. Points on the simple shape are encoded as an approximation of the original shape with a set of values [72]. The volume-based representations treat each atom of a molecule as a hard sphere with a radius determined by its van der Waals radius. The hard sphere can be replaced by Gaussian functions [73]. The surface-based methods focus on the interface between the volume interior and exterior and represent a molecular shape as a set of patches on the surface. Among these four types of molecular shape representations, surface-based methods are considered the most accurate in terms of molecular recognition, whereas the volume-based methods are most commonly used. One of the representations of the volume-based methods is ROCS (Rapid Overlay of Chemical Structures), which uses atom-centered Gaussians to accurately represent volumes. In this way, the number of overlaps is much reduced, which enable fast and approximate global maxima to be found. ROCS also extend to ROCS-color by including atom type information in shape-based matching, such

Table 4.4 Widely used shape-based tools.

Tool	Availability	Reference	URL
ROCS	Commercial	[74]	https://www.eyesopen.com/
Phase shape	Commercial	[75]	https://www.schrodinger.com
SHAPEIT	Free	–	http://silicos-it.be.s3-website-eu-west-1.amazonaws.com/index.html

as hydrogen bond donors/acceptors, anion, cation, ring, etc., which greatly improved its performance in virtual screening. Similarly, researchers from Schrodinger implemented Phase Shape in their software. An open-source implementation SHAPEIT also belongs to this catalog (Table 4.4).

Other type types of shape-based methods are those such as ShaEP combined volume-based and electrostatic potential [76]. The USR (ultrafast shape recognition) method calculates the distance distributions of each atomic coordinate from a set of four "reference locations." The first three statistical moments of each of these distributions generate a 12-element vector that captures shape information [77].

Shape-based methods have been widely used in virtual screening, and they are also compared with the structure-based methods. For example, Hawkins et al. compared seven different docking programs to ROCS across 21 different protein systems, and they found that ROCS provided superior performance even when a bioactive conformation of the ligand was not known [78] In fact, there is a general consensus that 3D ligand-based methods perform as well as, if not better than, the structure-based methods in many drug discovery applications including virtual screening and scaffold hopping [79–81].

4.3.3 Pharmacophore-based matching

Pharmacophore is one of the ancient concepts in computer-aided drug design, which was introduced by Monty Kier in a series of papers published in 1967–1971 [82]. The IUPAC definition of pharmacophore, "A pharmacophore is the ensemble of steric and electronic features that is necessary to ensure the optimal supramolecular interactions with a specific biological target structure and to trigger (or to block) its biological response" [83]. It should be emphasized that a pharmacophore does not represent a real molecule or a real interaction between functional groups; instead, it is a purely abstract concept that accounts for the common molecular interaction capacities of a group of compounds toward their targets. The core of the pharmacophore concept is the notion that molecular recognition of a group compounds, and their biological target attributes to a small set of common features, such as hydrogen bond donors, hydrogen bond acceptors, positively/negatively charged groups, and hydrophobic regions [8]. The other key component of pharmacophore is the spatial arrangement of these features.

Table 4.5 Widely used pharmacophore tools.

Tool	Availability	Reference
Pharmer	Free	[85]
Pharao	Free	[86]
Align-it	Free	[86]
Catalyst	Commercial	[87]
MOE	Commercial	[88]
Phase	Commercial	[89]

Traditionally, pharmacophore models are derived from a set of bioactive compounds that interact with the same biological target, known as ligand-based pharmacophore; in contrast, there are also structure-based pharmacophore modeling approaches that use 3D target structure and/or ligand-target complex structure [84] to generate pharmacophore models. In fact, many successful pharmacophore models are generated from protein-ligand complexes nowadays due to the development of crystallographic, NMR, and computational methodologies. The structure-based pharmacophore modeling is beyond the scope of this discussion. Several widely used ligand-based pharmacophore tools are summarized in Table 4.5.

Due to the simplicity and versatility of the pharmacophore concept, pharmacophore modeling has been routinely used in combination with other molecular modeling techniques, though it can be potentially used for virtual screening independently. Caporuscio and Tafi reported a review of synergistic combination of pharmacophore modeling with other molecular modeling approaches such as the hot spot analysis of protein binding sites, molecular dynamics, and docking [90].

4.4 Small-molecule libraries for virtual screening

4.4.1 Sources of small-molecule libraries

In the past decade, many sources of small-molecule libraries become publicly available for free, providing millions of chemical structures for virtual screening and medicinal chemistry. In general, there are four types of small-molecule libraries used in virtual screenings listed in Table 4.6: (1) collections from public repositories, such as PubChem [56], ChEMBL [91], ZINC [92], and ChemSpider [93]; (2) commercial collections from chemical vendors such as Aldrich Market Select, ChemBridge, and BIOVIA Available Chemicals Directory; (3) in-house collections of accumulated chemicals from the past for other purposes; and (4) virtual compound collections that theoretically exist, such as GDB-17 [94].

Although any chemical from the above collections can be used in virtual screening, a conventional virtual screening usually prefers real and existing chemical compounds that are either in-house or available in-stock to purchase, because the

Table 4.6 List of some representative small-molecule libraries for virtual screening.

Database	Availability	Number of compounds	Reference	URL
PubChem	Public	96 million	[56]	https://pubchem.ncbi.nlm.nih.gov
ChEMBL	Public	2 million	[91]	https://www.ebi.ac.uk/chembl/
ChemSpider	Public	74 million	[93]	http://www.chemspider.com/
ZINC	Public	980 million	[92]	http://zinc15.docking.org/
SureChEMBL	Public	17 million	[95]	https://www.surechembl.org/search/
GDB-17	Public	166 billion	[94]	http://gdb.unibe.ch/downloads/
Aldrich Market Select	Commercial	14 million	–	https://www.sigmaaldrich.com/chemistry/chemistry-services/aldrich-market-select.html
ChemBridge	Commercial	1.3 million	–	https://www.chembridge.com/
BIOVIA Available Chemicals Directory (ACD)	Commercial	10 million	–	https://www.3dsbiovia.com/products/collaborative-science/databases/sourcing-databases/biovia-available-chemicals-directory.html
MCULE	Commercial	43 millions	–	https://mcule.com/database/

hits generated from virtual screening eventually need real compounds for experimental assays. Therefore, the availability and price of real compounds are an important consideration in virtual screening. In-house collections are usually the first choice because they are readily available, but the drawback is limited to the existing compounds that may not be designed for the current project. The second choice is the collection of in-stock compounds from chemical vendors considering the cost and time. The other source of small-molecule are public repositories, which may provide additional information, such as bioactivities and their biological targets. Taking PubChem as an example, it is an open chemistry database containing information about chemical structures, chemical and physical properties, biological activities, patents, health, safety, toxicity data, and many others from hundreds of data sources, including government agencies, chemical vendors, and journal publishers. At the time of this work, PubChem has over 235 million of chemical substances and 96 millions of unique compounds from more than 600 depositors including 348 chemical vendors [56]. The last type is the virtual compound libraries that not only contain real existing compounds, but also the ones existing theoretically. One of the applications of the virtual libraries is to facilitate the exploration of the chemical space. Chemical space refers to the space spanned by all possible organic compounds that could be synthesized, which is estimated between 10^{30} and 10^{60} being routinely cited [96]. The most recent release, GDB-17, enumerated 166 billion molecules with up to 17 heavy atoms of C, N, O, S, and halogens forming the chemical universe database. It contains millions of isomers of known drugs, including analogues with high shape similarity to the parent drug, and has much richer scaffold types compared to PubChem [94].

The size of small-molecule libraries has continued to grow in the past years. When performing a virtual screening of a library containing a large number of compounds, it should be taken into account the potential false-positives issue, which is the largest problem in virtual screening [97]. If considering a virtual screen with a false-positive rate of 1%, as an optimistic estimate even for the best method nowadays, a virtual screen on a library of one million molecules would yield 10000 false-positive hits. This number of hits may completely swamp out the signal from the true positive hits in experiments [96].

4.4.2 Preparation of small-molecule libraries for virtual screening

Compounds obtained from any small-molecule libraries need to be cleaned up and prepared for virtual screening. Incorrect preparation and cleanup may cause the eventual failure of a screening. The cleanup is to get rid of unwanted molecules in advance, for example, by removing small fragments of mixture compounds, duplicated compounds, the "frequent hitters" that are interference to biological assays [98], compounds with reactive groups [99], etc. The preparation of a small-molecule library may be specific to the virtual screening tools. In general, it includes structure normalization, adding charges and hydrogens, converting 2D to 3D structures, as well as enumerating tautomeric states. Any mistakes in the preparation process will be propagated to downstream steps and negatively affect the outcomes of a virtual screening. As an example, tautomerism is one of the most underestimated issues in virtual screening campaigns. A molecule exhibits tautomerism if it is representable by two or more structures that are connected by the movement of a hydrogen from one atom to another; for instance, the proton shift from the enol-form to a keto-form of a molecule changes the alcohol group into a carbonyl group. Each tautomer of a single molecule substantially differs from another in electrostatic properties, hydrophobicity, 3D shape, and chemical reactivity. The selection of the wrong tautomer can misguide the assignment of hydrogen bond acceptors and donors, thus leading to false positives and/or false negatives in virtual screening [44].

4.4.3 Filter druglike/leadlike molecules

Filtering druglike or leadlike molecules is another important step in virtual screening, since poor druglike properties are important causes of the costly late-stage failure in drug development [100]. The concept of druglike and, more stringent, leadlike are introduced to determine the characteristics of a drug or a lead to be successful. The seminal paper by Lipinski and colleagues [101] defined the druglike space to restrict the properties of small molecules for orally active drug candidates in drug discovery. It is well-known as the "rule of five," namely, molecular weight (MW) $\leq$ 500, calculated octanol/water partition coefficient (clogP) $\leq$ 5, number of hydrogen bond donors (HBD) $\leq$ 5, number of hydrogen bond acceptors (HBA) $\leq$ 10. In drug design, the molecular weight is often increased in the optimization process in order to improve the affinity and selectivity. Thus, more stringent rules have been proposed, such as the "rule of three" leadlike filter [102,103], namely, molecular weight is $\leq$ 300, the number of hydrogen bond donors is $\leq$ 3, the number of hydrogen bond acceptors is $\leq$3 and ClogP is $\leq$ 3. Additionally, other properties of leadlike are also suggested, such as polar surface area (PSA), number of rotatable bonds, smaller number of rings, and solubility in water [104,105]. Compared to druglike, leadlike is more attractive to achieve leads for optimization. Nevertheless, the actual property cut-offs will depend on the objective of the virtual screening. For example, drug discovery aimed at identifying agents for central nervous system that requires passing the blood-brain barrier or different administration routes may need a different profile of these properties.

To clean up and prepare small-molecule libraries, most commercial softwares have their own module for such purposes. Alternative open-source solutions are also available, for example OpenBabel [50], RDKit [51], CDK [49,57] and PaDEL-descriptor [52] and KNIME [106]. Oprea et al. reported a practical guidance on how to clean up and prepare small-molecule libraries based on their experiences [107]. A KNIME workflow for the preparation of molecules for virtual screening has also been reported [108].

4.5 In silico validation of virtual screening

The basic idea of the in silico validation of virtual screening strategy is to test the ability of the workflow in separating active compounds from the inactive or decoy molecules. In other words, the validation is trying to run a mini retrospective virtual screening to check the performance of the workflow before applying to a large-scale screening.

To conduct such a validation, one needs to create a collection of active and inactive molecules. Active molecules are not always available for a novel target. But, some public databases can be used to find the actives, such PubChem [56], ChEMBL [91], PDBbind [19], etc. Inactive molecules are also available for certain bioassays in PubChem and ChEMBL. In practice, decoy molecules are used instead of inactive molecules in the validation process because of the fact that low amount of inactive results reported in literature. Decoy compounds are generally random druglike molecules, which are much more likely to be inactive than active by chance. In practice, decoy compounds are constructed by searching for compounds that have similar physical descriptors, such as molecular weight, number of rotatable bonds, number of hydrogen donors and acceptors, and octanol-water partition coefficient etc., but are chemically different from the active molecules. Decoy compounds can be obtained either from the existing database such as DUD-E [109], or through using tools, such as DecoyFinder [110].

Once the active and inactive/decoy molecules have been defined in validation set, the workflow of virtual screening is applied to classify the molecules based on their rankings. Each compound can be classified into one of the four classes, i.e., true positive (active compounds that are predicted to be active), false positives (inactive compounds that are predicted to be active), true negatives (inactive compounds that are predicted to be inactive), and false negatives (active compounds that are predicted to be inactive). Then, these parameters can be used to construct confusion matrix and calculate a series of statistical metrics to measure the performance of virtual screening workflow, such as Matthews correlation coefficient (MCC), F1 score, receiver operating characteristics (ROC) curves, Boltzmann enhanced discrimination of ROC (BEDROC), pROC, and enrichment factor (EF). There are also freely available tools reported to calculate and plot the metrics [111].

4.6 Postscreening process

The outcome of a virtual screening, either ligand-based or structure-based, is a long-ordered list of compounds based on their rankings. Because of the imperfection of virtual screening methods, it is still not possible to directly select the best ranking compounds for further experimental testing. Moreover, the typical goal of a screening is to identify a series of molecules with novel structures compared to the known active ones. Therefore, the postscreening process is usually applied to increase the odds of success of a screening. Basically, selecting compounds from a virtual screening for further experimental testing should obtain as much representative as possible and avoid subjective selections. This can be achieved by using unsupervised learning to group hits based on their structure similarities. With clustering, one can select one or more representatives from each group. The structure similarities can use any type of descriptors and methods, but 2D fingerprint similarity is usually employed for such a purpose. Many open-source cheminformatics tools can be used, such as RDKit [51], OpenBabel [50], CDK [49],and PaDEL-Descriptor [52], that have been discussed in the previous section of 2D fingerprint similarity searching. One of the widely used machine learning libraries in python programming language, scikit-learn [112] can be used for structure similarity clustering. The library implements many types of unsupervised learning algorithms, k-means, hierarchical clustering, DBSCAN, etc.

Specifically, for structure-based virtual screening, consensus scoring strategies are also used in postscreening processing. Consensus scoring is using multiple additional scoring functions to rescore the original poses to increase the overall performance because of the fact that each scoring function has its advantages and disadvantages [113–115]. Another approach is to apply a higher level of computation to rescore a subset of the hits with MM/GBSA or MM/PBSA (molecular mechanics with the Poisson-Boltzmann or Generalized Born surface area continuum solvation), which are popular approaches to improve the results of virtual screening and docking, but are too computationally expensive to be directly applied on a large-scale virtual screening [116].

4.7 Perspective

Virtual screening has become a well-established technique that is widely used in academic and industrial medicinal chemistry. Despite their limitations, both SBVS and LBVS approaches have produced considerable success stories of identifying molecules with desired biological activities. Based on literature evidence, SBVS methods still dominate the field, though they have clear disadvantages. For example, the imprecise scoring of the binding poses and binding activities, the treatment of full target flexibility, and the prediction of water-mediated hydrogen bond bridges are still inaccurate and remained to be open questions that challenge the field at least in the near future. The performance of ligand-based approaches is very promising and becomes comparable or superior to structure-based methods in many aspects in virtual screening, such as enrichment analysis.

A considerable number of virtual screening tools have been devised in the past several decades. It could be a benefit and a burden, since all methods and their combinations are far from perfect in terms of finding active molecules in a virtual screening. Moreover, there are a lot of pitfalls that lurk along the way which might render virtual screening less efficient and even useless [117]. One direction of methodology development is to make virtual screening tools smarter to handle various situations automatically and intelligently in each step, such as model setup, parameter settings, data quality filtering, etc.

The strategies of combining the structure-based and ligand-based methods are observed to be quite successful by taking advantage of the strengths of different methodologies. The combination strategies are not new in molecular docking, since consensus scoring approaches [113] have been widely used. However, when medicinal chemistry becomes a Big Data science, how to combine methodologies, as well as data sets from different sources to improve hit selection will be critically important. Methodologies and algorithms need to be developed to handle data fusion and integration of various data types, qualities, and completenesses to support accurate decision making [118].

Deep learning [34], as the latest development of machine learning approaches, has been successfully used in virtual screening, e.g., boosting the docking performance [119], which could outperform traditional methods. But it is too early to tell if deep learning is a way to rescue. It may be wise to fully understand the advantages and limitations of deep learning techniques and never use it blindly or misinterpret the results.

Acknowledgment

This work was carried out by staff of the National Library of Medicine (NLM), National Institutes of Health, with support from NLM.

References

[1] K.A. Carpenter, X. Huang, Machine learning-based virtual screening and its applications to alzheimer's drug discovery: a review, Curr. Pharmaceut. Des. 24 (2018) 3347−3358.

[2] A. Lavecchia, C. Di Giovanni, Virtual screening strategies in drug discovery: a critical review, Curr. Med. Chem. 20 (2013) 2839−2860.

[3] B.K. Shoichet, Virtual screening of chemical libraries, Nature 432 (2004) 862−865.

[4] W.P. Walters, M.T. Stahl, M.A. Murcko, Virtual screening−an overview, Drug Discov. Today 3 (1998) 160−178.

[5] T. Cheng, Q. Li, Z. Zhou, Y. Wang, S.H. Bryant, Structure-based virtual screening for drug discovery: a problem-centric review, AAPS J. 14 (2012) 133−141.

[6] N. Moitessier, P. Englebienne, D. Lee, J. Lawandi, C.R. Corbeil, Towards the development of universal, fast and highly accurate docking/scoring methods: a long way to go, Br. J. Pharmacol. 153 (Suppl. 1) (2008) S7−S26.

[7] F. Jiang, S.-H. Kim, "Soft docking": matching of molecular surface cubes, J. Mol. Biol. 219 (1991) 79−102.

[8] A.R. Leach, Ligand docking to proteins with discrete side-chain flexibility, J. Mol. Biol. 235 (1994) 345−356.

[9] S.-Y. Huang, X. Zou, Ensemble docking of multiple protein structures: considering protein structural variations in molecular docking, Proteins Struct. Funct. Bioinf. 66 (2007) 399−421.

[10] R.M.A. Knegtel, I.D. Kuntz, C.M. Oshiro, Molecular docking to ensembles of protein structures, in: B. Honig (Ed.), J. Mol. Biol. 266 (1997) 424−440.

[11] I.-H. Park, C. Li, Dynamic ligand-induced-fit simulation via enhanced conformational samplings and ensemble dockings: a survivin example, J. Phys. Chem. B 114 (2010) 5144−5153.

[12] S. Rao, P.C. Sanschagrin, J.R. Greenwood, M.P. Repasky, W. Sherman, R. Farid, Improving database enrichment through ensemble docking, J. Comput. Aided Mol. Des. 22 (2008) 621−627.

[13] S.B. Nabuurs, M. Wagener, J. De Vlieg, A flexible approach to induced fit docking, J. Med. Chem. 50 (2007) 6507−6518.

[14] I.W. Davis, D. Baker, RosettaLigand docking with full ligand and receptor flexibility, J. Mol. Biol. 385 (2009) 381–392.

[15] S.-Y. Huang, X. Zou, Advances and challenges in protein-ligand docking, Int. J. Mol. Sci. 11 (2010) 3016–3034.

[16] M. Totrov, R. Abagyan, Flexible ligand docking to multiple receptor conformations: a practical alternative, Curr. Opin. Struct. Biol. 18 (2008) 178–184.

[17] H. Alonso, A.A. Bliznyuk, J.E. Gready, Combining docking and molecular dynamic simulations in drug design, Med. Res. Rev. 26 (2006) 531–568.

[18] J.-H. Lin, A.L. Perryman, J.R. Schames, J.A. McCammon, Computational drug design accommodating receptor flexibility: the relaxed complex scheme, J. Am. Chem. Soc. 124 (2002) 5632–5633.

[19] R. Wang, X. Fang, Y. Lu, C.-Y. Yang, S. Wang, The PDBbind database: methodologies and updates, J. Med. Chem. 48 (2005) 4111–4119.

[20] S.-Y. Huang, S.Z. Grinter, X. Zou, Scoring functions and their evaluation methods for protein–ligand docking: recent advances and future directions, Phys. Chem. Chem. Phys. 12 (2010) 12899–12908.

[21] J. Liu, R. Wang, Classification of current scoring functions, J. Chem. Inf. Model. 55 (2015) 475–482.

[22] T.J.A. Ewing, S. Makino, A.G. Skillman, I.D. Kuntz, Dock 4.0: search strategies for automated molecular docking of flexible molecule databases, J. Comput. Aided Mol. Des. 15 (2001) 411–428.

[23] G.M. Morris, R. Huey, W. Lindstrom, M.F. Sanner, R.K. Belew, D.S. Goodsell, A.J. Olson, AutoDock4 and AutoDockTools4: automated docking with selective receptor flexibility, J. Comput. Chem. 30 (2009) 2785–2791.

[24] S.J. Weiner, P.A. Kollman, D.A. Case, U.C. Singh, C. Ghio, G. Alagona, S. Profeta, P. Weiner, A new force field for molecular mechanical simulation of nucleic acids and proteins, J. Am. Chem. Soc. 106 (1984) 765–784.

[25] R.A. Friesner, J.L. Banks, R.B. Murphy, T.A. Halgren, J.J. Klicic, D.T. Mainz, M.P. Repasky, E.H. Knoll, M. Shelley, J.K. Perry, D.E. Shaw, P. Francis, P.S. Shenkin, Glide: a new approach for rapid, accurate docking and scoring. 1. Method and assessment of docking accuracy, J. Med. Chem. 47 (2004) 1739–1749.

[26] G. Jones, P. Willett, R.C. Glen, A.R. Leach, R. Taylor, Development and validation of a genetic algorithm for flexible docking, in: F.E. Cohen (Ed.), J. Mol. Biol., vol. 267, 1997, pp. 727–748.

[27] R. Wang, L. Lai, S. Wang, Further development and validation of empirical scoring functions for structure-based binding affinity prediction, J. Comput. Aided Mol. Des. 16 (2002) 11–26.

[28] H.F.G. Velec, H. Gohlke, G. Klebe, DrugScore(CSD)-Knowledge-Based scoring function derived from small molecule crystal data with superior recognition rate of near-native ligand poses and better affinity prediction, J. Med. Chem. 48 (2005) 6296–6303.

[29] S.-Y. Huang, X. Zou, An iterative knowledge-based scoring function to predict protein–ligand interactions: I. Derivation of interaction potentials, J. Comput. Chem. 27 (2006) 1866–1875.

[30] J.D. Durrant, J.A. McCammon, NNScore: a neural-network-based scoring function for the characterization of protein-ligand complexes, J. Chem. Inf. Model. 50 (2010) 1865–1871.

[31] S. Das, M.P. Krein, C.M. Breneman, Binding affinity prediction with property-encoded shape distribution signatures, J. Chem. Inf. Model. 50 (2010) 298–308.

[32] P.J. Ballester, J.B.O. Mitchell, A machine learning approach to predicting protein-ligand binding affinity with applications to molecular docking, Bioinformatics 26 (2010) 1169–1175.

[33] Q.U. Ain, A. Aleksandrova, F.D. Roessler, P.J. Ballester, Machine-learning scoring functions to improve structure-based binding affinity prediction and virtual screening, Wiley Interdiscip. Rev. Comput. Mol. Sci. 5 (2015) 405–424.

[34] Y. LeCun, Y. Bengio, G. Hinton, Deep learning, Nature 521 (2015) 436–444.

[35] I.D. Kuntz, J.M. Blaney, S.J. Oatley, R. Langridge, T.E. Ferrin, A geometric approach to macromolecule-ligand interactions, J. Mol. Biol. 161 (2) (1982) 269–288.

[36] S.F. Sousa, P.A. Fernandes, M.J. Ramos, Protein–ligand docking: current status and future challenges, Proteins Struct. Funct. Bioinf. 65 (2006) 15–26.

[37] P.T. Lang, S.R. Brozell, S. Mukherjee, E.F. Pettersen, E.C. Meng, V. Thomas, R.C. Rizzo, D.A. Case, T.L. James, I.D. Kuntz, Dock 6: combining techniques to model RNA–small molecule complexes, RNA 15 (2009) 1219–1230.

[38] O. Trott, A.J. Olson, AutoDock Vina: Improving the speed and accuracy of docking with a new scoring function, efficient optimization, and multithreading, J. Comput. Chem. 31 (2010) 455–461.

[39] C.R. Corbeil, C.I. Williams, P. Labute, Variability in docking success rates due to dataset preparation, J. Comput. Aided Mol. Des. 26 (6) (2012) 775–786.

[40] Z. Wang, H. Sun, X. Yao, D. Li, L. Xu, Y. Li, S. Tian, T. Hou, Comprehensive evaluation of ten docking programs on a diverse set of protein–ligand complexes: the prediction accuracy of sampling power and scoring power, Phys. Chem. Chem. Phys. 18 (2016) 12964–12975.

[41] J.B. Cross, D.C. Thompson, B.K. Rai, J.C. Baber, K.Y. Fan, Y. Hu, C. Humblet, Comparison of several molecular docking programs: pose prediction and virtual screening accuracy, J. Chem. Inf. Model. 49 (2009) 1455–1474.

[42] H.M. Berman, J. Westbrook, Z. Feng, G. Gilliland, T.N. Bhat, H. Weissig, I.N. Shindyalov, P.E. Bourne, The protein Data Bank, Nucleic Acids Res. 28 (1) (2000) 235–242.

[43] https://www.rcsb.org/.

[44] R.C. Braga, V.M. Alves, A.C. Silva, M.N. Nascimento, F.C. Silva, L.M. Liao, C.H. Andrade, Virtual screening strategies in medicinal chemistry: the state of the art and current challenges, Curr. Top. Med. Chem. 14 (16) (2014) 1899–1912.

[45] A.L. Hansen, L.E. Kay, Measurement of histidine pKa values and tautomer populations in invisible protein states, Proc. Natl. Acad. Sci. USA 111 (2014) E1705–E1712.

[46] C.N. Cavasotto, S.S. Phatak, Homology modeling in drug discovery: current trends and applications, Drug Discov. Today 14 (2009) 676–683.

[47] A. Hillisch, L.F. Pineda, R. Hilgenfeld, Utility of homology models in the drug discovery process, Drug Discov. Today 9 (2004) 659–669.

[48] Y. Hu, D. Stumpfe, J. Bajorath, Advancing the activity cliff concept, F1000Research 2 (2013) 199.

[49] E.L. Willighagen, J.W. Mayfield, J. Alvarsson, A. Berg, L. Carlsson, N. Jeliazkova, S. Kuhn, T. Pluskal, M. Rojas-Chertó, O. Spjuth, G. Torrance, C.T. Evelo, R. Guha, C. Steinbeck, The Chemistry Development Kit (CDK) v2.0: atom typing, depiction, molecular formulas, and substructure searching, J. Cheminf. 9 (1) (2017) 33.

[50] N.M. O'Boyle, M. Banck, C.A. James, C. Morley, T. Vandermeersch, G.R. Hutchison, Open Babel: an open chemical toolbox, J. Cheminf. 3 (2011) 33.

[51] G. Landrum, RDKit: Open-Source Cheminformatics, 2006. https://www.rdkit.org/.

[52] C.W. Yap, PaDEL-descriptor: an open source software to calculate molecular descriptors and fingerprints, J. Comput. Chem. 32 (2011) 1466−1474.

[53] J.L. Durant, B.A. Leland, D.R. Henry, J.G. Nourse, Reoptimization of MDL keys for use in drug discovery, J. Chem. Inf. Comput. Sci. 42 (2002) 1273−1280.

[54] https://github.com/rdkit/rdkit-orig/blob/master/rdkit/Chem/MACCSkeys.py.

[55] ftp://ftp.ncbi.nlm.nih.gov/pubchem/specifications/pubchem_fingerprints.txt.

[56] S. Kim, J. Chen, T. Cheng, A. Gindulyte, J. He, S. He, Q. Li, B.A. Shoemaker, P.A. Thiessen, B. Yu, L. Zaslavsky, J. Zhang, E.E. Bolton, PubChem 2019 update: improved access to chemical data, Nucleic Acids Res. 47 (2019) D1102−D1109.

[57] C. Steinbeck, Y. Han, S. Kuhn, O. Horlacher, E. Luttmann, E. Willighagen, The chemistry development Kit (CDK): an open-source java library for chemo- and bioinformatics, J. Chem. Inf. Comput. Sci. 43 (2003) 493−500.

[58] A. Bender, H.Y. Mussa, R.C. Glen, S. Reiling, Similarity searching of chemical databases using atom environment descriptors (MOLPRINT 2D): evaluation of performance, J. Chem. Inf. Comput. Sci. 44 (2004) 1708−1718.

[59] H.L. Morgan, The generation of a unique machine description for chemical structures-A technique developed at chemical abstracts service, J. Chem. Doc. 5 (1965) 107−113.

[60] https://www.daylight.com/meetings/summerschool01/course/basics/fp.html.

[61] M.J. McGregor, S.M. Muskal, Pharmacophore fingerprinting. 1. Application to QSAR and focused library design, J. Chem. Inf. Comput. Sci. 39 (1999) 569−574.

[62] M.J. McGregor, S.M. Muskal, Pharmacophore fingerprinting. 2. Application to primary library design, J. Chem. Inf. Comput. Sci. 40 (2000) 117−125.

[63] https://www.certara.com/.

[64] Z. Deng, C. Chuaqui, J. Singh, Structural interaction fingerprint (SIFt): a novel method for analyzing three-dimensional Protein−Ligand binding interactions, J. Med. Chem. 47 (2004) 337−344.

[65] A. Cereto-Massagué, M.J. Ojeda, C. Valls, M. Mulero, S. Garcia-Vallvé, G. Pujadas, Molecular fingerprint similarity search in virtual screening, Methods 71 (2015) 58−63.

[66] J. Duan, S.L. Dixon, J.F. Lowrie, W. Sherman, Analysis and comparison of 2D fingerprints: insights into database screening performance using eight fingerprint methods, J. Mol. Graph. Model. 29 (2010) 157−170.

[67] I. Muegge, P. Mukherjee, An overview of molecular fingerprint similarity search in virtual screening, Expert Opin. Drug Discov. 11 (2016) 137−148.

[68] S. Riniker, G.A. Landrum, Open-source platform to benchmark fingerprints for ligand-based virtual screening, J. Cheminf. 5 (2013) 26.

[69] S. Putta, P. Beroza, Shapes of things: computer modeling of molecular shape in drug discovery, Curr. Top. Med. Chem. 7 (2007) 1514−1524.

[70] M.L. Mansfield, D.G. Covell, R.L. Jernigan, A new class of molecular shape descriptors. 1. Theory and properties, J. Chem. Inf. Comput. Sci. 42 (2002) 259−273.

[71] Y. Zyrianov, Distribution-based descriptors of the molecular shape, J. Chem. Inf. Model. 45 (2005) 657−672.

[72] A.N. Jain, K. Koile, D. Chapman, Compass: predicting biological activities from molecular surface properties. Performance comparisons on a steroid benchmark, J. Med. Chem. 37 (1994) 2315−2327.

[73] J.A. Grant, B.T. Pickup, A Gaussian description of molecular shape, J. Phys. Chem. 99 (1995) 3503−3510.

[74] T.S. Rush, J.A. Grant, L. Mosyak, A. Nicholls, A shape-based 3-D scaffold hopping method and its application to a bacterial Protein−Protein interaction, J. Med. Chem. 48 (2005) 1489−1495.

[75] G.M. Sastry, S.L. Dixon, W. Sherman, Rapid shape-based ligand alignment and virtual screening method based on atom/feature-pair similarities and volume overlap scoring, J. Chem. Inf. Model. 51 (2011) 2455−2466.

[76] M.J. Vainio, J.S. Puranen, M.S. Johnson, ShaEP: molecular Overlay based on shape and electrostatic potential, J. Chem. Inf. Model. 49 (2009) 492−502.

[77] P.J. Ballester, W.G. Richards, Ultrafast shape recognition to search compound databases for similar molecular shapes, J. Comput. Chem. 28 (2007) 1711−1723.

[78] P.C.D. Hawkins, A.G. Skillman, A. Nicholls, Comparison of shape-matching and docking as virtual screening tools, J. Med. Chem. 50 (2007) 74−82.

[79] P.W. Finn, G.M. Morris, Shape-based similarity searching in chemical databases, Wiley Interdiscip. Rev. Comput. Mol. Sci. 3 (2013) 226−241.

[80] D. Giganti, H. Guillemain, J.-L. Spadoni, M. Nilges, J.-F. Zagury, M. Montes, Comparative evaluation of 3D virtual ligand screening methods: impact of the molecular alignment on enrichment, J. Chem. Inf. Model. 50 (2010) 992−1004.

[81] G.B. McGaughey, R.P. Sheridan, C.I. Bayly, J.C. Culberson, C. Kreatsoulas, S. Lindsley, V. Maiorov, J.-F. Truchon, W.D. Cornell, Comparison of topological, shape, and docking methods in virtual screening, J. Chem. Inf. Model. 47 (2007) 1504−1519.

[82] J.H.V. Drie, Monty kier and the origin of the pharmacophore concept, internet electron, J. Mol. Des. 6 (2007) 271−279.

[83] C.G. Wermuth, C.R. Ganellin, P. Lindberg, L.A. Mitscher, Glossary of terms used in medicinal chemistry, Pure Appl. Chem. 70 (5) (1998) 1129−1143.

[84] S.-Y. Yang, Pharmacophore modeling and applications in drug discovery: challenges and recent advances, Drug Discov. Today 15 (2010) 444−450.

[85] D.R. Koes, C.J. Camacho, Pharmer: efficient and exact pharmacophore search, J. Chem. Inf. Model. 51 (2011) 1307−1314.

[86] J. Taminau, G. Thijs, H. De Winter, Pharao: pharmacophore alignment and optimization, J. Mol. Graph. Model. 27 (2008) 161−169.

[87] https://www.3dsbiovia.com/products/collaborative-science/biovia-discovery-studio/pharmacophore-and-ligand-based-design.html.

[88] https://www.chemcomp.com/.

[89] S.L. Dixon, A.M. Smondyrev, E.H. Knoll, S.N. Rao, D.E. Shaw, R.A. Friesner, PHASE: a new engine for pharmacophore perception, 3D QSAR model development, and 3D database screening: 1. Methodology and preliminary results, J. Comput. Aided Mol. Des. 20 (2006) 647−671.

[90] F. Caporuscio, A. Tafi, Pharmacophore modelling: a forty year old approach and its modern synergies, Curr. Med. Chem. 18 (2011) 2543−2553.

[91] A. Gaulton, L.J. Bellis, A.P. Bento, J. Chambers, M. Davies, A. Hersey, Y. Light, S. McGlinchey, D. Michalovich, B. Al-Lazikani, J.P. Overington, ChEMBL: a large-scale bioactivity database for drug discovery, Nucleic Acids Res. 40 (2012) D1100−D1107.

[92] J.J. Irwin, B.K. Shoichet, ZINC − a free database of commercially available compounds for virtual screening, J. Chem. Inf. Model. 45 (2005) 177−182.

[93] H.E. Pence, A. Williams, ChemSpider: an online chemical information resource, J. Chem. Educ. 87 (2010) 1123–1124.

[94] L. Ruddigkeit, R. van Deursen, L.C. Blum, J.-L. Reymond, Enumeration of 166 billion organic small molecules in the chemical universe database GDB-17, J. Chem. Inf. Model. 52 (2012) 2864–2875.

[95] G. Papadatos, M. Davies, N. Dedman, J. Chambers, A. Gaulton, J. Siddle, R. Koks, S.A. Irvine, J. Pettersson, N. Goncharoff, A. Hersey, J.P. Overington, SureChEMBL: a large-scale, chemically annotated patent document database, Nucleic Acids Res. 44 (2016) D1220–D1228.

[96] W.P. Walters, Virtual chemical libraries: miniperspective, J. Med. Chem. 62 (2019) 1116–1124.

[97] A.N. Jain, Scoring noncovalent protein-ligand interactions: a continuous differentiable function tuned to compute binding affinities, J. Comput. Aided Mol. Des. 10 (1996) 427–440.

[98] G.M. Rishton, Nonleadlikeness and leadlikeness in biochemical screening, Drug Discov. Today 8 (2) (2003) 86–96.

[99] T.I. Oprea, Property distribution of drug-related chemical databases, J. Comput. Aided Mol. Des. 14 (2000) 251–264.

[100] H. van de Waterbeemd, E. Gifford, ADMET in silico modelling: towards prediction paradise? Nat. Rev. Drug Discov. 2 (2003) 192–204.

[101] C.A. Lipinski, F. Lombardo, B.W. Dominy, P.J. Feeney, Experimental and computational approaches to estimate solubility and permeability in drug discovery and development settings, Adv. Drug Deliv. Rev. 23 (1997) 3–25.

[102] M. Congreve, R. Carr, C. Murray, H. Jhoti, A 'Rule of Three' for fragment-based lead discovery? Drug Discov. Today 8 (2003) 876–877.

[103] S.J. Teague, A.M. Davis, P.D. Leeson, T. Oprea, The design of leadlike combinatorial libraries, Angew. Chem. Int. Ed Engl. 38 (1999) 3743–3748.

[104] M.M. Hann, T.I. Oprea, Pursuing the leadlikeness concept in pharmaceutical research, Curr. Opin. Chem. Biol. 8 (2004) 255–263.

[105] T.I. Oprea, A.M. Davis, S.J. Teague, P.D. Leeson, Is there a difference between leads and drugs? A historical perspective, J. Chem. Inf. Comput. Sci. 41 (2001) 1308–1315.

[106] M.R. Berthold, N. Cebron, F. Dill, T.R. Gabriel, T. Kötter, T. Meinl, P. Ohl, K. Thiel, B. Wiswedel, KNIME-the Konstanz information miner: version 2.0 and beyond, ACM SIGKDD Explor. Newsl. 11 (2009) 26–31.

[107] C.G. Bologa, O. Ursu, T.I. Oprea, How to prepare a compound collection prior to virtual screening, in: R.S. Larson, T.I. Oprea (Eds.), Bioinformatics and Drug Discovery, Methods in Molecular Biology, Springer New York, New York, NY, 2019, pp. 119–138.

[108] J.-M. Gally, S. Bourg, Q.-T. Do, S. Aci-Sèche, P. Bonnet, VSPrep: a general KNIME workflow for the preparation of molecules for virtual screening, Mol. Inf. 36 (2017) 1700023.

[109] M.M. Mysinger, M. Carchia, J.J. Irwin, B.K. Shoichet, Directory of useful decoys, enhanced (DUD-E): better ligands and decoys for better benchmarking, J. Med. Chem. 55 (2012) 6582–6594.

[110] A. Cereto-Massagué, L. Guasch, C. Valls, M. Mulero, G. Pujadas, S. Garcia-Vallvé, DecoyFinder: an easy-to-use python GUI application for building target-specific decoy sets, Bioinformatics 28 (2012) 1661–1662.

[111] C. Empereur-Mot, J.-F. Zagury, M. Montes, Screening explorer—an interactive tool for the analysis of screening results, J. Chem. Inf. Model. 56 (2016) 2281–2286.

[112] https://scikit-learn.org.

[113] P.S. Charifson, J.J. Corkery, M.A. Murcko, W.P. Walters, Consensus scoring: a method for obtaining improved hit rates from docking databases of three-dimensional structures into proteins, J. Med. Chem. 42 (1999) 5100–5109.

[114] R.D. Clark, A. Strizhev, J.M. Leonard, J.F. Blake, J.B. Matthew, Consensus scoring for ligand/protein interactions, J. Mol. Graph. Model. 20 (2002) 281–295.

[115] D.R. Houston, M.D. Walkinshaw, Consensus docking: improving the reliability of docking in a virtual screening context, J. Chem. Inf. Model. 53 (2013) 384–390.

[116] S. Genheden, U. Ryde, The MM/PBSA and MM/GBSA methods to estimate ligand-binding affinities, Expert Opin. Drug Discov. 10 (2015) 449–461.

[117] T. Scior, A. Bender, G. Tresadern, J.L. Medina-Franco, K. Martínez-Mayorga, T. Langer, K. Cuanalo-Contreras, D.K. Agrafiotis, Recognizing pitfalls in virtual screening: a critical review, J. Chem. Inf. Model. 52 (2012) 867–881.

[118] L. Richter, G.F. Ecker, Medicinal chemistry in the era of big data, Drug Discov. Today Technol. 14 (2015) 37–41.

[119] J.C. Pereira, E.R. Caffarena, C.N. dos Santos, Boosting docking-based virtual screening with deep learning, J. Chem. Inf. Model. 56 (2016) 2495–2506.

Screening and biophysics in small molecule discovery

Chris G.M. Wilson[1], Michelle R. Arkin[1,2]

Small Molecule Discovery Center and Department of Pharmaceutical Chemistry, University of California San Francisco, San Francisco, CA, United States[1]; Buck Institute for Research on Aging, Novato, CA, United States[2]

5.1 Introduction

Small molecule drugs do not require a defined mechanism of action to be approved for use [1,2]. Nevertheless, a molecular understanding of drug action can be very helpful in defining structure-activity relationships during chemical optimization, for development of biomarkers to select patients and to monitor drug response, for prediction of safety issues in the clinic, and even for supporting patent protection. Biophysical methods directly measure interactions between biomolecules and the interactions of biomolecules with small molecules; use of these methods throughout drug discovery ensures that the project stays on a firm mechanistic footing. Ideally, biophysical methods demonstrate that the potential drugs bind to a single site on a single protein, thereby altering the function of that protein in a defined way. Collectively, we call this "drug-like binding."

This chapter will describe major biophysical technologies used for screening, compound validation, and lead optimization from the perspective of a medicinal chemist engaged in early-stage discovery. This chapter can at best provide only a sense for the range of biophysical screening approaches and technologies available. Comprehensive reviews and informative case studies are referenced for the reader [3,4]. We end with a recent example that illustrates the diversity and integration of methods.

5.2 Background and scope

The equilibrium (Fig. 5.1) between ligand-bound and ligand-free states lies at the core of drug discovery. The essential parameters for evaluating binding are the equilibrium dissociation constant for the small molecule/biomolecule interaction (K_D), the binding kinetics (k_{off}/k_{on}), and the physical and chemical nature of molecular recognition. Each parameter can be probed directly or indirectly (via a reporter), with varying degrees of resolution and descriptive power, and every method comes

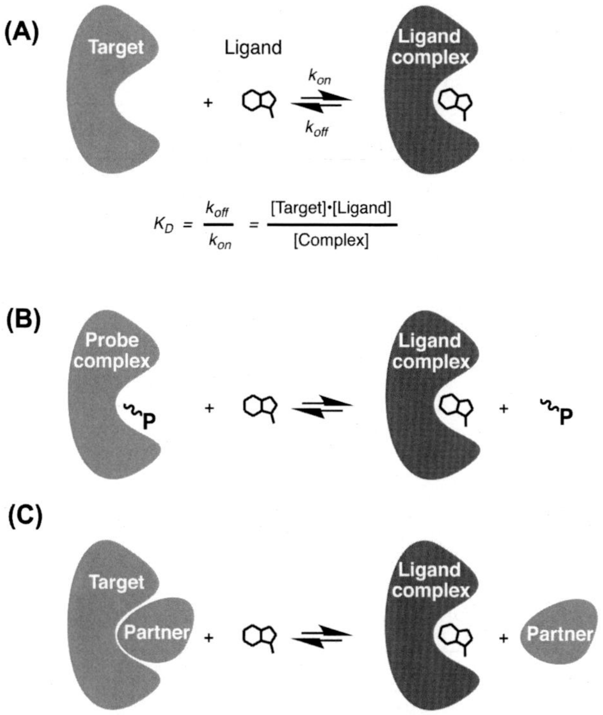

FIGURE 5.1

(A) The target:ligand interaction equilibrium constant (K_D) is defined by the association and dissociation rates (kon and koff, respectively), and by the concentrations of the unbound and bound molecules. (B and C) Binding of a ligand can be indirectly detected using a labeled probe or protein.

with caveats and conditionalities. Logistical constraints, including accessibility and costs, introduce further limitations. Hence, the most appropriate biophysical approach depends on the stage of the drug-discovery program. For instance, new chemical starting points (hits) are generally found through high-throughput screening (HTS), which emphasizes throughput, efficiency and cost-effectiveness over descriptive power; plate-based biophysical measurements are most appropriate at this stage. Conversely, when the project is focused on 10–100s of compounds, a thorough biophysical characterization can be obtained through a combination of complementary, lower throughput strategies. Key features of most commonly used biophysical methods are summarized in Fig. 5.2.

5.2.1 Screening technology

Key enabling factors in small molecule screening include assay standardization in multiwell plate formats (96, 384 and 1536) and the development of compatible

Method	Discovery Phase			Throughput	Power	Difficulty	Accessibility	Cost
	ID >	Validation >	Lead >					
UV-Vis	+	+++	++	++	+	+	+++	+
DLS	+	+++	++	++	+	+	++	+
MSF	+	++	+	+	+	++	+	++
DSF	++	+	+	++	+	+	++	+
FP	+++	++	+	+++	++	+	+++	+
FRET	+++	++	+	+++	++	+	+++	+
TR-FRET	+++	++	+	+++	++	+	++	+
AlphaScreen	+++	++	+	++	++	+	+	++
ELT	+++	+	+	+++	+	++	+	++
SPR	++	+++	+++	++	++	++	+	++
BLI	+	++	+	+++	++	++	+	++
SHG	++	+++	++	+++	++	++	+	++
NMR	++	+++	+++	++	++	++	+	++
AS/MS	++	+	+	++	++	++	+	++
Tethering	+++	++	+	++	++	++	+	++
ITC	+	++	+++	+	++	++	+	++
Xtal	+	++	+++	+	+++	++	+	+++
Cryo-EM	+	++	+++	+	+++	+++	+	+++

FIGURE 5.2

Overview of technologies discussed. More descriptively powerful techniques can often come at the cost of greater technical difficulty and cost and lower throughput and accessibility (UV-Vis, ultraviolet-visible spectroscopy; DLS, dynamic light scattering; MSF, microscale thermophoresis; DSF, differential scanning fluorimetry; FP, fluorescence polarization; FRET, Förster resonance energy transfer; TR-FRET, time-resolved Förster resonance energy transfer; ELT, encoded library technology; SPR, surface plasmon resonance; BLI, biolayer interferometry; SHG, second harmonic generation; NMR, nuclear magnetic resonance; ASMS, affinity selection mass spectrometry; ITC, isothermal titration calorimetry; Xtal, X-ray crystallography; Cryo-EM, cryogenic electron microscopy).

instrumentation [5–7]. Beginning in the late 90s, the Society for Biomolecular Sciences (SBS) proposed standards for multiwell plates and associated labware, leading to the current formats for medium (96), high (384) and ultra-high-throughput (1536) screening. High-throughput assay data is usually gathered using a plate reader, which incorporates one or more detection technology. Dedicated (single mode) readers are common for new technologies, which over time are incorporated into versatile multimode instruments. The highest performance readers feature sensitive and multiple detectors, combinations of filter and monochromator wavelength selection, specialized optics (e.g., laser excitation) and modular multiplate stackers to expedite data collection.

To assemble the assay in SBS-formatted plates, liquid handling instruments range from bulk dispensers (delivery through peristaltic drive mechanisms), to μL−nL tip and pin tool robotics, to acoustic and inkjet nL−pL volume technology. Contact-less transfers (acoustic and inkjet) are increasingly favored because they remove the risk of cross-contamination, strongly favor miniaturization, are extremely flexible (de-livery is not locked to a predetermined pattern) and produce less consumables waste [8]. Acoustic dispensers have also found promising application in low-throughput, high-resolution technologies, including crystallography (nL compound and crystal dispensing, microcrystal suspension) and mass spectrometry [9−13].

Laboratory automation, which consists of robotic plate movers and scheduling software, provides the linkage between plate storage (hotels), dispensers, incubators, and acquisition instruments. Additionally, as the price of automation has fallen, high-resolution instruments (NMR [nuclear magnetic resonance], mass spectrom-etry, ITC, crystallography beamlines) have become increasingly integrated, allowing automated data collection of 100−1000s of samples.

5.2.2 Chemical technology (compounds)

Chemical screening libraries—the raw materials of drug discovery—are generally defined as "diverse" or "focused." Focused libraries are either tailored toward a target class or application, such as scaffolds likely to bind to kinases or molecules suitable for fragment-based drug discovery, or are collections of compounds with known bioactivity, including approved drugs [14]. Diversity libraries contain mole-cules without known function but with "drug-like" properties. Drug-likeness is an evolving and contextually dependent set of parameters, derived from large-scale analysis of successful small molecule therapeutics, and includes molecular weight (<500 Da), measures of lipophilicity, and hydrogen bond donor and acceptor count [15]. Various Rules-of-X (X = 5 for classical Lipinski, 3 for fragments, >5 for "beyond Rule of 5") provide guidelines for likely bioavailability and can aid the researcher in selecting appropriate compound libraries.

The size and chemical composition of a library effects what biophysical ap-proaches can be taken. For instance, isothermal calorimetry is a gold-standard method to measure binding affinity and binding stoichiometry; however, it is a rela-tively insensitive technique that requires solubility of the compounds in the 100 μM range. Fragment-based drug discovery utilizes molecules <300 Da that are expected to bind with K_D values in the 100−1000 μM range; the high solubility of fragments and the small library sizes (100−10,000 members) lend themselves to biophysical methods for screening and characterization.

Purchased compounds are a significant investment and a critical asset for a dis-covery organization, but management of tens to hundreds of thousands of molecules presents a logistics and informatics challenge [16]. Compounds are usually supplied predissolved in DMSO. Realiquoting into limited-use copies (either in plates or in 384-format mini-tube racks) conserves material and preserves its chemical integrity. Steps that limit library deterioration, and particularly aggregation and precipitation due to absorption of water, will ultimately yield higher quality results.

5.2.3 Promiscuous behavior

Extensive analysis of historical screening data has established the reality of promiscuous compounds, so named because they are active in many, unrelated assays. Such compounds inhibit target activity through biologically irrelevant or non-drug-like (unadvanceable) mechanisms, including forming nonselective covalent bonds to proteins or causing protein denaturation. Biophysical methods are commonly used to evaluate and remove these bad actors from further consideration.

The most common phenomenon is **small molecule aggregation**, which can occur spontaneously when drug-like substances are incubated in aqueous media at typical screening concentrations ($\sim 10\,\mu M$). The aggregates are dense, polydisperse ($<100\,nm$), hydrophobic particles that bind, sequester, and denature targets [17−19]. This colloidal property has long been appreciated at the late stages of discovery, where excipients are frequently necessary to stabilize a formulated product [20,21]. Telltale signs that a compound's mechanism of inhibition is due to aggregation are extremely steep dose-response functions (suggesting that monodisperse compounds below the aggregation threshold are inactive), disappearance of activity following centrifugation (aggregates are pelleted) and promiscuous inhibition of unrelated targets. Aggregator risk can be ameliorated during screening through the addition of nonionic detergents (below the detergent's critical micelle concentration, CMC) and carrier proteins (1% purified gelatin or bovine γ-globulin) to assay buffers. Additionally, measurement of the aggregates by dynamic light scattering (DLS) identifies the concentration at which the aggregates form under the screening conditions.

The second source of difficulty are **pan-assay interference compounds (PAINs)** [22,23]. These may contain a range of "bad actor" moieties, including reactive groups that denature the target (e.g., cysteine-reactive Michael acceptors), or react with an assay buffer component and generate a source of denaturant (e.g., redox cycling with thiol reducing agents, forming hydrogen peroxide in situ). There may also be structural motifs responsible for nonspecific target affinity, including highly symmetric flat aromatic ring systems. PAINs structural filters, and the more recently described Bruns-Watson demerit score [24], are filtering and curation metrics that can applied computationally before a purchasing a library and in prioritization of postassay hits. These characteristics are especially hazardous in primary screens, where the tendency to simplify an assay to its minimal components makes it more susceptible to promiscuous mechanisms. Biophysical methods can again be used to measure the binding stoichiometry and reversibility of molecules bound to proteins, effectively weeding out compounds acting through many promiscuous mechanisms.

5.2.4 Scope

It can be reasoned that biophysics encompasses nearly all measurable properties of biological systems. Here, we primarily focus on assays that measure binding between purified proteins and drug-like molecules, and do not describe enzymatic assays or cell-based assays. However, it is worth noting that many of the

methods—such as fluorescence resonance energy transfer (FRET) and bioluminescence resonance energy transfer (BRET)—have an analogous implementation in
cells. To learn about such assays, the reader is referred to excellent recent reviews
[25–27].

5.3 Biophysical methods used in HTS

High-throughput screening emerged in the 1980s as a response by the pharmaceutical industry to increase productivity through standardization, parallelization and
format compression [5,7]. These technologies have evolved to cooptimize relevance,
convenience, cost, and risk. Most common biophysical screens, including fluorescence polarization and energy transfer, monitor competition (or enhancement) of
a reporter probe with the library compounds, and therefore measure the binding
of library compounds indirectly. These probes can be known small molecule ligands,
peptides, or proteins that bind to the enzyme, receptor, protein-protein interface or
protein-nucleic acid interface, etc. Screening using direct detection of protein/
small-molecule interactions would avoid many of the common artifacts associated
with competition experiments, but has been difficult to implement on a large scale
until recently. Two screening methods that measure binding directly are thermal
denaturation and encoded-library technology.

5.3.1 Fluorescence polarization

Fluorescence polarization (FP) and fluorescence anisotropy (FA) measure the binding of fluorescently labeled probes to larger biomolecules. An FP assay uses plane-
polarized light to excite the fluorophore and measures maintenance of polarization in
the emitted light. The reported FP value is the ratio of emission intensities observed
through parallel and perpendicular polarizing filters according to Fig. 5.3 [28,29].

Loss of fluorescence polarization is caused by molecular motion (tumbling) in
solution. The magnitude of polarization is directly correlated with molecular weight;
molecular shape and local flexibility of the fluorophore are additional factors.
Hence, large, slowly tumbling molecules yield high polarization values, and a
fluorophore-labeled probe gains polarization when it binds to the large molecule.
The ideal molecular weight ranges for the probe and "large" biomolecule depend
on the fluorescence lifetime of the fluorophore. Almost all FP assays use organic fluorophores ($t_{1/2} \sim 10$ ns) and ligands in the size range of 500–2000 Da are typical.

$$P = \frac{I_{para} - I_{perp}}{I_{para} + I_{perp}} \qquad r = \frac{I_{para} - I_{perp}}{I_{para} + 2{}^*I_{perp}} \qquad I_{total} = I_{para} + 2{}^*I_{perp}$$

FIGURE 5.3

Relationship between parallel and perpendicular fluorescence intensities, polarization
(P), anisotropy (r), and total intensity.

Larger ligands can be used, but the change in polarization between free and bound ligand will eventually become unmeasurably small.

In HTS formats, FP is usually used as a competition assay, where binding of a small molecule displaces the probe, thereby increasing its tumbling rate and decreasing fluorescence polarization. The assay's dynamic range is governed by the polarization change between free and bound probe and by the K_D of probe/target complex; the lowest molecular weight probe with the tightest affinity ($K_D < 100$ nM) yields the best performing assay. To design a competition FP protocol, the protein is titrated into a constant concentration of labeled probe to identify the protein concentration needed to achieve saturating polarization ($EC_{50} \approx K_D$). The probe concentration is generally set by the fluorescence intensity of the fluorophore and sensitivity of the plate reader; $1-100$ nM is typical. The ideal concentration of protein and probe is determined by the assay goal; conditions with high polarization (EC_{50-80}) are appropriate for an inhibition format, while concentrations at the EC_{10-20} are ideal if the goal is to stabilize the interaction. Unlabeled probe should be titrated into the selected assay condition to demonstrate that the IC_{50} for competition is close to the K_D for the interaction.

FP assays are especially attractive for primary screening and early-stage discovery; they are easily assembled in high-density assay plates, yield a stable signal once equilibrium is reached, and are tolerant to low working volumes. Furthermore, as a ratiometric assay, FP is not strongly sensitive to evaporation. Compound optical interference due to absorbance or autofluorescence can be ameliorated through the selection of red-shifted fluorophores; such interference can also be identified by calculating the total fluorescence intensity in the well. Examples using FP for screening include kinases [30], GPCRs [31], transcription factors [32], and cytokines [33].

5.3.2 Förster (fluorescence) resonance energy transfer

Radiationless energy transfer, caused by dipole-dipole coupling between neighboring excited and ground electronic states, can take place between adjacent fluorophores when their distance is < 10 nm [34]. This proximity dependence makes FRET sensitive to disruption of a ligand/receptor interaction.

In FRET, a "donor" fluorophore is excited and, rather than emit light, the fluorescent energy is transferred to an "acceptor" fluorophore, which then emits light at its characteristic wavelength. The distance at which FRET occurs depends on several features, including the spectral overlap of the fluorophores. Common FRET donor/acceptor pairs include organic dyes such as fluorescein/tetramethylrhodamine and Cy3/Cy5, or fluorescent proteins such as GFP/YFP and mClover/mRuby [35]. The experimental system involves excitation at the donor wavelength and detection at two emission wavelengths (donor and acceptor, see Fig. 5.4). The calculated ratio of acceptor and donor fluorescent intensities gives the transfer efficiency. The assay is designed analogously to FP assays as described above. As with FP, FRET is a ratiometric measurement and therefore relatively insensitive to

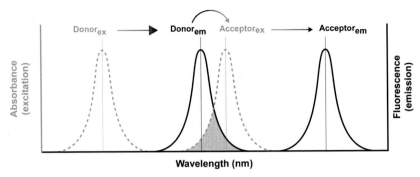

FIGURE 5.4

Principle of FRET between donor and acceptor fluorophores. The efficiency of energy transfer is determined by Donor emission and Acceptor excitation spectral overlap, and the physical proximity of groups.

changes in sample volume. In practice, colorimetric interference from test compounds can be a significant challenge, and should be evaluated by measuring the compounds' absorbance and fluorescence properties and by using secondary assays with orthogonal detection methodologies.

For HTS, FRET is especially valuable in probing protein-protein interactions, where partners are labeled with chemical fluorophores or are genetically tagged with fluorescent proteins [36]. Genetically tagged protein complexes can also be studied within cells by microscopy [37], though in practice this is more challenging than other cell-based formats [25]. FRET is also commonly encountered in enzymatic assays, where a reaction product (e.g., a phosphorylated protein) is detected and quantified through a pair of dye-labeled antibodies (e.g., one recognizing the phosphorylated epitope).

5.3.3 Time-resolved FRET

As noted, compound interference at screening concentrations can dominate FRET detection and render screening data meaningless. Time-resolved (TR)-FRET, known also as homogeneous time-resolved FRET (HTRF), was developed to overcome this issue through lanthanide donor chelates and cryptates [38]. Upon near-UV excitation (~ 340 nm), lanthanide labels emit multiple sharp emission bands that originate from ligand-to-metal charge transfers. These emission energies are accepted by many organic dyes such as fluorescein. While fluorescence from small organic dyes and library compounds will decay on the nanosecond timescale, lanthanides continue to emit for 100s of μs. Thus, after a gating time delay (0.5−1 ms), the majority of interfering fluorescence from small molecules will have ceased, and only lanthanide-to-acceptor FRET will be detectable. Common TR-FRET combinations include Lumi4-Tb/fluorescein or GFP (337/480−488/520), and Europium/Alexa-Fluor 647 (340/615−647/665). One significant limitation of TR-FRET is the low wavelength of excitation (~ 340 nm), where many library compounds will absorb; hence, the light-filtering effect can be significant and must be monitored. TR-FRET has been widely used to measure inhibition of protein-protein interactions, and for kinase screens, where the phosphorylated product is measured [39]. Lanthanide emission lines are sufficiently sharp to allow FRET multiplexing in the same sample, with as many as five interactions probed simultaneously [39,40].

5.3.4 AlphaScreen

FRET and TR-FRET are limited to distances of less <10 nm. Detection of interactions beyond this, e.g., across large receptor complexes, requires a probe with greater range. **Amplified luminescent proximity homogeneous assay (AlphaScreen)** uses the photochemical production of singlet oxygen from a donor bead to stimulate chemiluminescence in an acceptor bead [41]. Singlet oxygen has an aqueous lifetime of ~ 4 µs, reportedly allowing detection over ~ 200 nm. To measure protein-protein interactions, partner-specific antibodies are generally added first, followed by donor and acceptor beads conjugated to secondary antibodies. Assay volumes can be very small ($1-2$ µL per reagent, totaling ~ 6 µL in 1536 well plates), which reduces the high costs of AlphaScreen reagents. Photochemical donor excitation occurs at 680 nm, usually by laser to yield maximal singlet production, and acceptor chemiluminescence is measured across a wide $520-650$ nm window and collected with a comparatively long (500 ms) integration time. Because acceptor beads emit at a shorter wavelength (i.e., appear blue-shifted), interference from fluorescent compounds is also avoided. The cost of Alpha-capable instruments, and especially those equipped with 680 nm lasers that yield the highest signal in the shortest time (~ 3 min per 384 well plate), is higher than for TR-FRET capable fluorimeters and is commonly offered as an upgradeable feature. Alpha technology is therefore suited to highly automated screening workflows against signaling pathway targets [42–44] and is less common at modest scales or during validation and hit follow up.

AlphaScreens are especially notable for originating the concept of pan-assay interference compounds (PAINs), which were first classified through a meta-study of Alpha-identified hit molecules [22,23]. Recent broader analyses, across many different assay types, support the general concept of PAINs while also illustrating that "pan" and "interference" both depend on the assay context [24,45].

5.3.5 Differential scanning fluorimetry

Differential scanning fluorimetry (DSF, Thermofluor) works on the principle that ligand interactions tend to raise the thermostability of proteins; hence, measuring the melting temperature (T_m) of a protein can report on small molecule binding. To measure protein melting temperature in high-throughput, DSF uses an extrinsic fluorescent reporter that fluoresces more brightly when it binds to partially denatured proteins, e.g., the molten globule state(s). In practice, the protein target and test compound are incubated prior to addition of reporter dye such as SYPRO Orange, and

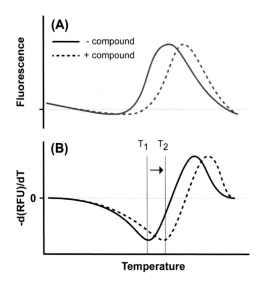

FIGURE 5.5

Differential scanning fluorimetry (DSF) follows the impact of a small molecule analyte on the denaturation temperature (melting midpoint) of the target through extrinsic probe fluorescence (A), measured through the melting midpoint (first derivative, B).

the protein is then denatured in a qPCR instrument while following SYPRO Orange emission intensity (Ex_{470}/Em_{570}). T_m is found from the first derivative (highest rate of change in fluorescence, see Fig. 5.5) of the melting curves, and the change in T_m due to compound is compared to the negative (no compound) control [46–48]. Because extrinsic probes bind to the denatured state, they are themselves a potential source of artifacts. However, given the convenience of assembly, low assay volumes ($\sim 15\ \mu L$), low cost, and modest throughput (~ 90 min per plate), DSF can be an attractive primary HTS, and one of few that directly measures thermodynamic changes in the target protein upon binding. A cell-based implementation of thermal denaturation, known as CETSA, is described in Section 5.4.5.

5.3.6 Encoded library technology

A radical alternative to plate-based assays is combinatorial screening, where targets are challenged with a mixture of small molecules and an affinity selection is used to find the tightest binding molecules. Direct analysis of complex, small molecule mixtures is largely impractical due to the high concentrations needed to detect each member of the library. DNA-encoded libraries (DEL or ELT, for "encoded library technology") solve this problem by attaching library members to highly amplifiable oligonucleotide tags introduced at the end of each split-pool step of ligand synthesis [49,50]. Each ligand molecule is, in principle, identifiable by a unique DNA barcode that records what starting materials were added to make the DNA-tethered compound. Though the synthesis is currently limited to DNA-compatible chemistry, the resulting libraries are numerically large ($\sim 10^9$) and compounds highly water-soluble, making them ideal for affinity-based selections.

ELT screens are conducted with target loaded onto a solid phase support (e.g., biotinylated protein on free or disposable-tip-packed streptavidin beads) and incubated with the ELT library. After washing off unbound compounds, captured library members are eluted, typically by thermal denaturation of target with hot buffer. Selections are repeated 2-3 times, and DNA eluted at each stage is PCR-amplified, quantified, and normalized before analysis by next-generation sequencing. ELT raw data are assembled DNA sequence contigs, which can be treated as alleles in a genetic analysis. DNA enrichment relative to control selections yields small molecule candidates that are synthesized "off-DNA" and validated through traditional methods.

Affinity-selection screening enables flexible experimental design. For instance, running selections in the presence and absence of an active-site ligand can be used to select competitive inhibitors that only bind in the *absence* of the active-site ligand or to select allosteric compounds that bind preferentially in the *presence* of the known ligand. Inherent limitations to the technology include the chemical diversity compatible with DNA, incompatibility with nucleic acid binding targets, and the need to synthesize compounds off-DNA to enable hit validation. These notwithstanding, the ability to rapidly screen extremely large compound collections has led to the expansion of ELT within pharma, contract labs, and academia.

5.4 Biophysical methods used for hit validation

Once compounds are identified through HTS or other discovery methods, these hits should be validated for drug-like mechanisms of action. Biophysical and structural assays demonstrate how and where molecules bind to their targets. With this knowledge in hand, medicinal chemists can confidently develop meaningful structure-activity relationships to optimize binding affinity and selectivity.

A combination of methodologies is usually needed to gain a detailed picture of small molecule/protein interactions. Because these methodologies are time consuming to perform and demand specialized instruments, they have been traditionally used for hit validation rather than screening. However, biophysical and structural methods are often viable screening tools for fragment-based approaches, since libraries are smaller and functional-based assays are less reliable for detecting low-affinity fragments [51−53].

5.4.1 Light scattering

As noted above, of the most common artifacts in small molecule screening arises from aggregation of hydrophobic or amphiphilic molecules in buffer. This aggregation leads to apparent inhibition of enzymes and protein-protein interactions by nonspecific binding and/or denaturation of the proteins. To reduce the impact of aggregation in the first place, biochemical assays should include detergents and inactive proteins such as casein, gelatin, or gamma globulin. However, it is still important to demonstrate that apparent inhibitors do not aggregate under the conditions of the assays. Rayleigh and Mie scattering, which are the parametric (lossless, instantaneous reemission) dipole interactions between light and matter, are key tools in understanding the colloidal properties (aggregation state) of ligands and targets [54].

In **dynamic light scattering (DLS)** the intensity and frequency of scattering are measured at 90 degrees over time, and the behavior deconvoluted by Fourier transform into an average molecular weight and polydispersity. Small molecules in solution should not scatter light efficiently, and the presence of scatter is indicative of a large particle, e.g., an aggregate. Light scattering is especially valuable in establishing a legitimate mechanism of action (hit validation) and in the formulation of compounds for administration [18,55]. DLS is compatible with multiwell plate formats, making it a viable—albeit low throughput—method. Scattering is most commonly applied in a comparative manner, against reference solvent and small molecules or protein standards. Ideally, the aggregation state of compounds is measured in the same buffers and concentrations for both DLS and activity assays; the presence of aggregation at most/all active concentrations strongly suggests that aggregation is the mechanism of activity.

5.4.2 Interaction analysis

A suite of technologies, including surface plasmon resonance, bilayer interferometry, and second harmonic generation, utilize proteins immobilized on a surface to detect changes in an optical property when the protein surface is modified. While SPR and BLI are sometimes marketed as "label free," this understates the potential importance of the surface immobilization on protein stability, conformation, and orientation. We prefer the term label-free to refer only to methodologies such as mass spectrometry and NMR, which rely on intrinsic properties of the ligand and/or biomolecule.

5.4.2.1 Surface plasmon resonance

Surface plasmon resonance (SPR) is a very sensitive method for measuring binding in real time. Biosensors based on SPR contain a glass slide with a gold surface on one face and a biocompatible coating on the other. When a laser is directed at the gold surface, the total internal reflection at the metal/glass/aqueous interface is influenced by the refractive index due to dipole and evanescent (surface plasmon) wave coupling. Proteins immobilized on the coated side alter the refractive index, roughly proportionally to the mass of the immobilized protein. When analytes such as proteins or organic compounds are then flowed over the protein-coated surface, binding to immobilized protein leads to further change in mass/refractive index, which is read as a change in the angle or intensity of the refracted laser. SPR instruments employ disposable sensor chips with microfluidics, coated in a variety of surface

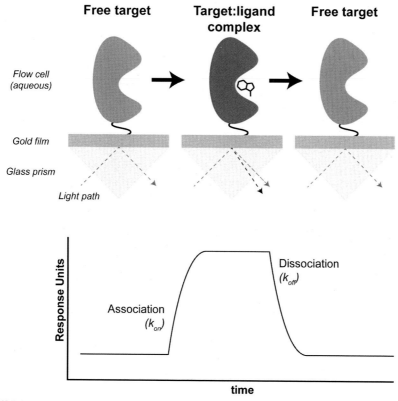

FIGURE 5.6

SPR interaction sensor scheme. Ligand association and dissociation can be monitored through the on- and off-rate kinetics and the response unit (RU) change at equilibrium (the top of the curve). Well-behaved ligands will completely dissociate from target (return to preassociation baseline).

chemistries to enable capture and orientation of targets (see Fig. 5.6). Prediluted ligands are autosampled from multiwell source plates, injected over the sensor surface (association phase), and followed by buffer (dissociation phase). Binding is measured in real time, and the data are read out as response units (RU) versus time (seconds). A typical SPR screen will feature 3–10 min ligand injection cycles, and some instruments measure up to eight samples at a time. Parallel-flow cell instruments can assay up to ~1000 samples per day and include automated plate handling. These factors make SPR suitable for fragment discovery screens and for routine use in hit validation and detailed characterization studies.

SPR has several important features for a hit-validation methodology. First, it is sensitive enough to detect fragment- and drug-sized molecules binding to much larger protein. Modern SPR instruments have reported sensitivities of 0.6 RU, corresponding to 0.6 pg mass change per mm^2 of sensor surface. While this sensitivity allows small molecule detection, it also requires extensive and careful use of matched reference controls, gathered in parallel with ligand samples, to normalize for buffer and DMSO refractive index effects. Second, because RU changes are proportional to mass, one can predict the RU increment for adding a small molecule. For example, starting with 1000 RUs of immobilized 10 kDa target, the theoretical change in RU for a 300 Da ligand, assuming 1:1 binding at 100% occupancy, is 30 RUs (3% change). At each concentration of ligand added, one can determine the binding stoichiometry; plotting stoichiometry versus concentration yields a binding isotherm from which the K_D is determined. Third, association and dissociation kinetics are detected through the real-time measurements. On- and off-rate curve fitting can reveal mechanistic subtleties as well as affinity [56,57], and kinetic measurements can help to optimize drug leads based on slow off-rates.

SPR data are also extremely rich in interpretive detail, and many types of promiscuous activity—including high stoichiometry and nonspecificity—are readily observed. For instance, HTS hits with dissociation constants in the micromolar range should have rapid association and dissociation kinetics; slow kinetics or irreversible binding of a low-affinity compound generally indicates non-drug-like and/or promiscuous binding. Finally, the precise temperature control available in an SPR experiment enables enthalpic and entropic contributions to be determined, through van't Hoff plot analysis [58,59].

5.4.2.2 Biolayer interferometry

The interaction of light at interfaces with variable dielectric properties is also exploited in reflectometric interference spectroscopy (RIfS), implemented commercially as biolayer interferometry (BLI). Instead of manifesting as changes in angle, target binding results in an increased reflection pathlength, and a measurable change in interference pattern [60]. In contrast to SPR flow arrangements, BLI uses a fiber-optic tip. Protein is coated onto the tip using various chemistries and the tip is then "dipped" into the compound solution contained in a 96 or 384 well plate. 8- and 16-probe arrays monitor multiple rows simultaneously. Dissociation is achieved by moving probes from analyte to buffer (dilution). Similar to SPR, the impact of sample on interference signal is monitored continuously, giving kinetic data that reflect on- and off-rates under stopped-flow conditions. Compared to SPR, BLI is technically much simpler to perform, and significantly less susceptible to issues such as precipitation, which plagues microfluidic formats. However, the intrinsic sensitivity of BLI is lower than SPR, and it is more widely used for high-molecular-weight (protein-protein interaction, antibody) screening and analysis. Small molecules may be successfully characterized through indirect measurements, e.g., inhibition of protein-protein interactions, or through carefully controlled direct analysis under well characterized conditions [61].

5.4.2.3 Second harmonic generation

Total internal reflection can also be combined with two-photon excitation of dye-molecule conjugates, leading to second harmonic generation (SHG). The intensity and properties (phase, polarization) of the second harmonic waves are not dependent on changes in mass, but are very sensitive to the orientation of the dye relative to the glass surface. Thus, changes in an SHG signal are interpreted as a change in protein conformation, e.g., upon ligand binding [62–64]. An SHG experiment involves labeling of the protein target with a suitable dye, followed by immobilization of the dye-protein conjugate on an instrument-specific 96/384-well assay plate. The small molecule analyte is then added and SHG intensity is monitored continuously. As implemented in the Biodesy Delta platform (Biodesy Inc), compound addition (50–100 μL) and measurement are rapid, enabling thousands of assays to be performed per day. As with SPR, varying the concentration of compound allows calculation of K_D.

SHG is a relatively new and expensive technique, in terms of both instruments and consumables, with results from first-generation instruments now entering the literature. While the target labeling requirement makes the approach less convenient than SPR, and double modification (SHG probe plus solid-phase capture) raises the potential for artifacts, SHG provides a rapid physical readout of protein conformational changes that may otherwise be complex to measure [65,66]. Current efforts are aimed at triangulating SHG using multiple site-specific dyes to calculate conformational changes upon binding; such information would be very valuable for characterizing compounds' mechanisms of action and explaining allostery.

5.4.3 Nuclear magnetic resonance

Synthetic chemists are most familiar with NMR (nuclear magnetic resonance) spectroscopy for determining chemical structure. In contrast to small molecule NMR, which makes heavy use of J-coupling (spin-spin) experiments, the characterization of noncovalent target-ligand interactions exploits the physical and chemical differences between target and ligand, in order for one to reveal an interaction with the other [67–69]. Methods with relevance to drug discovery fall into "ligand-detected" and "target-detected" categories, depending on the entity ultimately under observation. These also fall conveniently into simple, single dimension homonuclear ^{1}H (proton) measurements, and more involved ^{1}H–^{15}N heteronuclear 2D experiments. The most commonly used methods are outlined here, but there are many additional NMR experiments that can—for example—detect interactions between specific nuclei on the protein and small molecule or between two bound ligands using nuclear Overhauser effects [70]. NMR is compatible with autosampling technology, allow near continuous data acquisition and processing, and making modest scale (100–1000s) compound screens possible [71]. The methods described below are therefore used for screening of small libraries (especially fragment libraries) [72] and for hit validation after HTS.

5.4.3.1 Ligand-detected nuclear magnetic resonance

5.4.3.1.1 ^{1}H saturation transfer difference

Saturation transfer difference (STD) is the most commonly encountered ligand-detected NMR method for small molecule-target studies [73]. The approach relies on a dynamic equilibrium (fast exchange) between bound and free ligand states taking place on the ms timescale. When nuclei on the protein are selectively excited and saturated with tuned RF pulses, magnetization will be transferred from the target to the ligand by dipolar coupling. Hence, irradiation of the protein will yield the 1D NMR spectrum of the compound. In practice, pairs of spectra are collected (one at the target resonance, and one off the target resonance for reference), and a 1D difference spectrum calculated (see Fig. 5.7). Target-to-ligand ratios are kept high (1: 100 is common, e.g., 6 µM protein to 0.6 mM ligand), in part because NMR is intrinsically insensitive and the ligand is under observation, and also to maintain a pool of unexcited free ligand that will suppress back-exchange onto target. Because STD-NMR relies on rapid exchange, it is especially suited low-affinity binders, down to the micromolar range. Structural insights are limited because the target is not under direct observation. However, since the efficiency of magnetization transfer between target and ligand is based on proximity, careful analysis of ligand ^{1}H signal enhancements can reveal clues about the likely binding orientation, and key protons participating in the recognition interface (called "group epitope mapping").

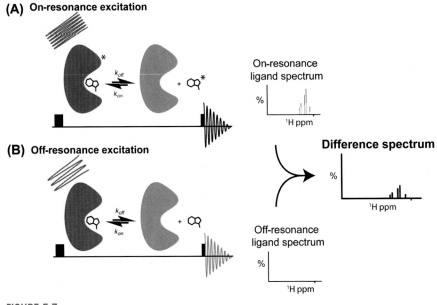

FIGURE 5.7

STD-NMR outline. On-resonance (A) and off-resonance (B) spectra are acquired independently and subtracted, yielding a difference spectrum of ligand protons where magnetization was transferred from target.

¹H differential epitope mapping-STD (DEEP-STD) NMR is a recent development that uses more highly selective RF saturation to excite a limited set of target nuclei, such as aromatic side chain protons [73,74]. If small molecule STD spectra are obtained, it suggests that the molecule is proximal to the targeted amino acid type [75].

5.4.3.1.2 Proton relaxation methods

The size differential between ligand and protein can also be exploited by measuring the increased relaxation rate and decreased diffusion rate of the ligand when it binds to the larger protein [76]. Relaxation results in broader line widths and rapid decrease in peak height when a delay is inserted between excitation and detection. **Target immobilized NMR screening (TINS)** further increases the difference in size between ligand and target by immobilizing the target on a solid surface [77]. When the protein target is captured on an NMR-compatible solid-phase support (CPG glass for nucleic acids, sepharose beads for protein), binding of a ligand strongly inhibits its mobility and tumbling. Ligand peak linewidths are dramatically broadened as a result. Like STD methods, relaxation measurements and TINS use a reference spectrum (ligand in the absence of target, in presence of solid support but no target, etc) to isolate and quantify the ligand-target interaction. In certain cases, immobilization brings the added advantage of stabilizing poorly soluble, difficult targets [78].

5.4.3.2 Protein-detected chemical shift perturbation

The reverse proposition—detecting changes to the NMR of the target in the presence of ligand—generally requires 2D NMR experiments using isotopically labeled protein [67,68]. However, the ready availability of isotopically enriched media (especially ^{15}N) and improvements in expression and purification technologies now make production of labeled protein a common enterprise. ^{1}H$-^{15}$N NMR applications approach a practical limit in the 30 kDa range, while ^{1}H$-^{13}$C NMR methods extend the range to 100s of kDa [79,80]. For protein-detected NMR, large amounts of protein are required due to the high concentrations (20−200 μM) needed for detection.

^{15}N labeled targets are readily examined through ^{1}H$-^{15}$N HSQC (heteronuclear single-quantum correlation/coherence) experiments, which yield 2D spectra reporting on scalar coupling between nitrogen and covalently attached protons. ^{1}H$-^{15}$N cross-peaks arise from the chemical shifts of the N−H proton (typically between 6.5 and 11 ppm) and ^{15}N (between 100 and 130 ppm) chemical shifts. Each amino acid has one backbone amide cross-peak (with the exception of proline, which has no proton), with their precise chemical shift coordinates influenced by the attached side chain (covalent inductive effects) and local environmental factors, especially secondary structure and chemical exchange with solvent. Amide-containing side chains (Asn and Gln) yield characteristic doublets in the ^{15}N dimension. Broad chemical shift dispersion is a consequence of protein structure, which collapses (primarily in the ^{1}H dimension) to random coil values upon denaturation. To obtain the most binding site information, HSQC cross-peaks should be assigned to specific backbone amides in the protein sequence, but this can be a lengthy process, demanding several complementary hetero- and homonuclear experiments [81,82]. In the absence of assignments, the HSQC can still be employed as a target's qualitative fingerprint. $^{1}H-^{13}C$ HSQC or HMQC are analogous experiments that typically use amino acids with selectively labeled ^{13}C-methyl groups (alanine, methionine, valine, leucine, and/or isoleucine).

In **chemical shift perturbation (CSP)** experiments, $^{1}H-^{15}N$ HSQC spectra of the protein, with and without ligand, are acquired and compared (see Fig. 5.8). Provided the samples are otherwise identical, an overlay of ± ligand cross-peak patterns will quickly reveal target residues perturbed by the presence of ligand [83,84]. In general, backbone perturbations reveal the likely location of the ligand binding site. CSPs can also arise from ligand-induced changes in target structure, at locations distant from the binding site. This may be confounding in a naïve (primary) screening setting, but can reveal novel ligandable sites and unanticipated structural connectivity within a target [85]. Provided ligand exchange occurs with an NMR observable time scale, estimates of K_D can be extracted by following CSP in a ligand dose titration.

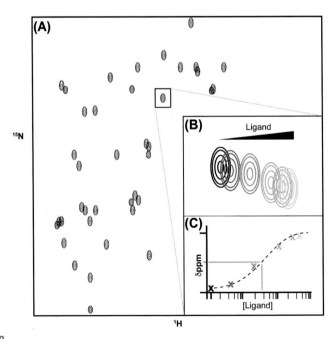

FIGURE 5.8

Principle of $^{1}H-^{15}N$ HSQC chemical shift perturbation experiments. (A) $^{1}H-^{15}N$ HSQC resolves dispersed cross-peaks from target N–H amide groups. Under small-molecule dose challenge (B), increasing ligand induces chemical shift changes in affected residues, which may be quantified (C) for K_D estimate.

5.4.4 Isothermal titration calorimetry

Calorimetry—the measurement of released or absorbed heat associated with a process—is a long-established method that gives access to the key parameters (stoichiometry, entropy, and enthalpy change) that define affinity [86,87]. Mixing a known quantity of small molecule ligand with a known amount of target results in a change in temperature determined by the interaction. ITC instruments measure the electrical power required to maintain the temperature of a sample during a series of ligand injections. Following each injection, the system is left to re-equilibrate to the set temperature, and injections continue until no further temperature changes are detected. Fitting these power spikes describes a binding isotherm, from which one calculates the Gibbs free energy (ΔG), enthalpy (ΔH), entropy (ΔS), stoichiometry, and affinity (K_D) of the interaction. ΔH and ΔS are of special value in conjunction with structure-guided design, where chemical optimization seeks to balance enthalpy (target:ligand interactions) and entropic cost (displacement of ordered water by ligand binding).

A typical ITC instrument (iTC$_{200}$, GE/Malvern) requires significant amounts of ligand ($\sim 40\,\mu L$ at $10\times$ target concentration) and target ($200\,\mu L$). It is also a fairly low-throughput method ($20 +$ injections of $2\,\mu L$ over $2\,h$). ITC was historically limited to single sample instruments, but the advent of automated platforms (MicroCal PEAQ and TA Instruments Auto ITC) reflects a growing demand for fuller thermodynamic insight earlier in the discovery process.

In order to fully capture the titration, the injected ligand concentration must across the K_D, and eventually saturate the target. This may be impossible for weak interactions (such as fragments) where ligand solubility is limiting. In these circumstances, an **enthalpy screen** ranks compounds based on ΔH alone. In enthalpy screening, the usual ITC orientation is reversed, with a known protein concentration injected into a large molar excess of ligand (above the estimated K_D, determined by an orthogonal method) [88]. The recorded temperature change reveals only the enthalpic component of binding, which can drive prioritization of early hits since affinity driven by enthalpy is harder to achieve than by entropy.

5.4.5 Mass spectrometry

Mass spectrometry (MS) includes a diverse set of tools that measure the molecular mass:charge ratio of an ionized material in the gas phase. In principle, protein/ligand complexes can be measured in the gas phase, and there are examples of noncovalent ligands directly resolved through native and extremely soft ionization protocols [89]. However, noncovalent ligand interactions are generally incompatible with reverse-phase chromatography and subsequent ionization, and direct detection is not routinely used. Four applications that do routinely use MS detection for ligand/protein interactions are cellular thermal shift assay (CETSA), affinity selection followed by mass spectrometry (AS/MS), hydrogen/deuterium exchange (HDx), and a screening technology called disulfide trapping (tethering).

5.4.5.1 Cellular thermal shift assay

A noteworthy application of MS is the **cellular thermal shift assay (CETSA)** for small molecule target identification [90]. As described in Section 5.3.6, ligand binding leads to thermal stabilization of the interaction partner. In a CETSA experiment, ligands are added to cells, which are heated to induce protein denaturation. Denatured proteins tend to precipitate, but even thermally depleted cell lysates remain highly complex mixtures. These are now amenable to total proteomic identification and quantification by modern mass spectrometers. Proteins that are enriched in the heated, compound-treated cells compared to heated, untreated cells are "thermally stabilized" by the ligand, and could be the direct molecular target of the small molecule. Candidate protein/ligand interactions from CETSA are further validated for thermal stabilization by Western blot and by other target-validation methods [91].

5.4.6 Affinity selection-mass spectrometry

In affinity selection-mass spectrometry (AS/MS) target-ligand mixtures are incubated, rapidly separated (e.g., by size exclusion chromatography or solid phase extraction), and the fraction containing the eluted target is analyzed for its small molecule content. The correlation of target and ligand(s) in the same fraction is indicative of noncovalent complex formation [92–94]. Identifying highly stabilized, high-affinity (pM) ligand-target complexes is the basis of very high-throughput, very low false-positive (but indirect) **high-affinity mass spectrometry screening (HAMS)** where a pool of compounds are passed over a column containing an immobilized target. Binders are depleted from the pool, and high-resolution quantitative small molecule mass spectrometry quantifies the changes in concentration of pool members [94,95].

5.4.6.1 Hydrogen-deuterium exchange

The chemical exchange of protons at solvent-accessible protein amides discussed previously (Section 5.4.3.1.1) can also be exploited through mass spectrometry. When rapidly transferred to deuterated solvent, the rate of equilibration between ^1H and ^2H nuclei will be impeded in target-ligand complexes. Following quench in acid (low pH suppresses amide exchange) and enzymatic proteolysis to yield fragments of known size and sequence, LC-MS is used to resolve regions of peptide sequence that were protected or perturbed compared to control (no ligand) exchange. With reference to the intact protein structure, these measurements reveal the likely ligand-binding site [96,97].

5.4.6.2 Disulfide trapping

Mass spectrometry is an excellent primary screening modality in cases where covalent modification of a target is the desired outcome. In **disulfide trapping (Tethering)**, cysteine-containing targets are equilibrated under exchange conditions (mild thiolate base) with disulfide-containing small molecule fragments

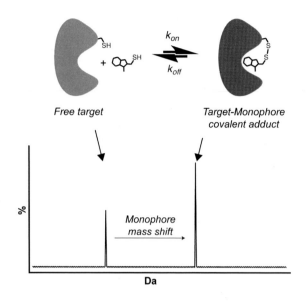

FIGURE 5.9

Sited-directed fragment discovery by disulfide trapping (Tethering). Under appropriate thiol conditions, exchangeable fragments become trapped and are resolved through intact protein LC-MS.

(sometimes called "monophores") [98,99]. Formation of stable (trapped) adducts occurs when there is a thermodynamically favorable interaction between fragment and target and the fragment binds with the proper orientation to make a disulfide bond with the protein (see Fig. 5.9). Cysteines under interrogation can be native or introduced by site-directed mutagenesis; where desired (though usually not necessary), irrelevant or confounding cysteines can also be removed by mutagenesis. A typical Tethering program will challenge 100 nM target with 100 μM monophore, in the presence of 1 mM β-mercaptoethanol (source of thiolate) as single reactions in a 384 well plate, requiring 8 h of reverse-phase LC-MS runtime for 320 fragments. Disulfide fragments can also be pooled. The low pH of the LC-MS solvent system ensures the stability of disulfide adducts, which are resolved and deconvoluted as intact proteins masses. Hit characterization includes monophore and β-mercaptoethanol (competitive) dose response, also assayed through mass spectrometry, to rank potency and identify structural trends among ligands [100–102]. Validated monophores may be optimized in the disulfide-context, and/or converted into noncovalent analogues and used as starting points for traditional fragment development. Alternatively, the disulfide can be replaced with a Michael acceptor moiety and used as a starting point for rational covalent ligand design [103].

5.4.7 Microscale thermophoresis

Microscale thermophoresis (MST) is a relatively new technology for measuring ligand binding. In MST, an infrared laser generates a thermal gradient and the motion of the protein is measured using its intrinsic (tryptophan) fluorescence or the fluorescence of a conjugated fluorophore. The protein's response is sensitive to changes in shape, size, and charge, due to changes in the solvation shell surrounding the target [104−106], and small molecules can alter these properties. Thus, the impact of ligands on the rate of fluorescence recovery (relaxation) is followed, and a K_D for the interaction is determined by dose titration. Measurements are performed in a glass capillary, and therefore use very small amounts of material. However, the capillary format is not automation-friendly, making the technique better suited for ligand characterization than for primary screening.

5.5 Structural methodologies

The preceding techniques offer insight into the strength and/or stoichiometry of binding of a small molecule ligand, but only its approximate location. Atomic level precision about the details of an interaction, meaning discrete distances and angles, is only possible through structural methods such as X-ray crystallography, cryogenic electron microscopy, or—in favorable cases—NMR. It is wise to remember that ligand/protein structures are in fact *structural models* of very complex data that incompletely define the system; hence, structures bring with them limitations and the potential for artifacts. In fundamental research, structural models provide deep insight as to how proteins function; in drug discovery, they are generally used as hypothesis-generating tools, providing direction to iterative cycles of structure-guided design, chemical synthesis, and biological testing.

5.5.1 X-ray crystallography

The detailed principles of X-ray crystallography lie outside this discussion [107−109]. In summary, the first step is generating well-ordered single crystals that contain the protein and bound ligands. Ligands are either "soaked" into existing protein crystals, relying on diffusion and uptake of ligand into an established lattice, or are cocrystallized with the protein. Cocrystallization is often needed when ligand binding causes a significant conformational change in the target, since this new conformation might be incompatible with the known crystal form. The crystals are then subjected to strong X-ray beams, which are diffracted by the electrons in the crystal. The electron density (called "real space"), is related to the X-ray diffraction pattern (called "reciprocal space"). Developing a structural model from the diffraction data requires knowledge of intensities (which can be measured) and phases (which cannot). For protein-sized systems, phases must be derived from isomorphous or anomalous dispersion experiments, or modeled by molecular replacement of an analogous crystal structure, such as the unliganded protein or a close homolog.

Model building from diffraction data is the attempt to satisfy and explain the observed electron density, using a combination of established physical constraints such as bond lengths, dihedral angles, and valencies, and known properties of the target, e.g., the primary sequence of amino acids (or nucleotides, for nucleic acid structures). Early stages of model refinement are automated, followed by manual inspection and adjustments using interactive molecular graphics. Models are used to compute theoretical (calculated) electron density, which is then compared to the experimentally observed density. The R-factor (ratio of the unaccounted for density/observed density) is used as a quality metric for a model, where smaller values indicate a better fit. Modeling starts with the protein (or nucleic acid), followed by modeling the ligand into regions of electron density that are not accounted for by the protein. A visual survey of the observed map will quickly reveal patches of unaccounted for density; while most will be small blobs attributable to bound water molecules, metal ions, or other solutes, true small molecule ligands will ideally have complete density that conforms to a plausible conformation of the ligand.

At its finest, crystallography will provide the highest quality understanding of the target-ligand interaction, as well as insights into structural flexibility of target and ligand (isotropic B-factor analysis). The ability to detect details of ligand/target interactions is determined by the resolution of the diffraction data and the fractional occupancy of the binding site. Provided occupancy is high, unambiguous ligand identification should be possible with 2 Å data, and ligands are often well-modeled in lower resolution structures. Partial occupancy of the site and alternative ligand conformations can weaken parts of the density attributable to the ligand. Above 3 Å, little beyond the target backbone (peptide and a side chain in proteins; helices, purine or pyrimidine bases in nucleic acids) can be reliably assigned. When interpreting protein and protein/ligand structural models, it is important to keep in mind the crystallization conditions (buffer, salt, pH can all alter structure), the acquisition temperature (room temperature or cryogenic) and the effects of the crystal lattice.

Due to the high technical demands of the method, crystallography has traditional been employed at mid-to-late stages of lead development to guide refinement of a compound series. However, recent progress in handling microcrystals has allowed access to structures from previously unusable material (notably GPCRs) and enabled effective use of such material in screening [110]. Standardization of consumables has facilitated automation of crystal growth (96 well plates), preparation (contactless acoustic dispensing of nL ligand droplets), harvesting, mounting and collecting data with robotic platforms [9,10,12,111]. High-throughput crystallography, especially in fragment discovery, can now be considered an accessible resource, albeit one tied to the newest and brightest synchrotron X-ray sources supported at national laboratories.

5.5.2 Cryogenic electron microscopy

While crystallography remains the gold standard for structural detail, not all targets have yielded to the technology. Cryogenic electron microscopy (Cryo-EM) has emerged as an exciting and complementary technique, in which a high-energy beam of electrons is used to "illuminate" the sample [112−114]. In **single particle reconstruction**, target molecules are captured in a noncrystalline (vitreous) state on an electron microscopy sample grid under cryogenic conditions. The low temperature promotes stability and limits damage from high-energy electron exposure, helping to preserve detail. Rapid sequential transmission EM image frames (20−60 Hz) are acquired under variable conditions (e.g., increasing electron intensity to reveal internal features, detector tilt to expand viewed poses), yielding many 2D projections of randomly oriented target particles. Data sets for individual particles are combined and sorted (classified) into sets. Particle orientation and intensities are compared according to the central slice theorem and deconvoluted by 3D Fourier transform into a physical density map. Model building follows crystallographic principles, with computational threading of target sequence to satisfy experimentally observed density.

Cryo-EM has been extremely successful with large and asymmetric proteins/ complexes, because these yield feature-rich images that assist particle sorting and alignment. Technology development is aimed at solving the structures of smaller targets. Target molecules and complexes must be of extremely high quality and purity, to ensure consistency and minimize noise. Sample preparation and data acquisition methods are fairly well established, with the bulk of resources expended on the computational power required to develop 3D maps. While Cryo-EM is an extremely productive method (849 publicly released structures in 2018; more than 300 in the first 4 months of 2019), with automated acquisition and deconvolution workflows capable of producing completed structures in 12 h, the technique is more nascent in terms of small molecule discovery. Ligand costructures solved [115] include ion channels [116], enzymes [117,118], and translational regulators [119]. Comparatively few studies have so far been published, but in some cases, resolution has been well below 3 Å [117,120]. This resolution is sufficient for ligand-based screening and guided docking, but Cryo-EM has not yet shown the rapid turnaround to be useful for lead optimization. One hurdle to democratization of Cryo-EM is the high cost and infrastructure requirements for the most powerful instruments, but there is rapid progress in optimization of the methods, dissemination of skills, and availability of reagents. As the technology continues to mature, it seems likely that Cryo-EM will be increasingly adopted in drug discovery.

5.6 Case study using biophysical methods in concert to discovery small molecule stabilizers of the 14-3-3/ estrogen receptor complex

A recent collaborative program including our group has used a multidisciplinary and multitechnique approach to identify and characterize novel ligands for protein-protein stabilization [121]. Enhancement of ligand engagement is especially

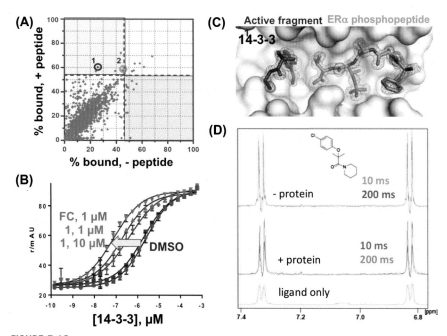

FIGURE 5.10

Biophysical methods were used to discover disulfide-bound fragments that stabilized the 14-3-3/ERα protein-protein interaction. (A) Screening disulfides for binding to 14-3-3 in the absence (X-axis) or presence (Y-axis) of ERα-phosphopeptide using mass spectrometry. The green quadrant represents cooperative hits; hits in the blue quadrant are competitive with ERα. (B) Fluorescence polarization using fluorescein-labeled ERα peptide shows increased binding (increased polarization) in the presence of fragment 1 and fusicoccin (FC). (C) Structural details of disulfide trapped ligands determined by X-ray crystallography. (D) ^{1}H-NMR of nondisulfide fragment provides evidence for binding to the 14-3-3/ERα-peptide complex. Resonances from the fragment relax faster (lose intensity) in the presence of 14-3-3 and ERα-peptide, especially at longer delay times (10 ms vs. 200 ms).

Figures adapted with permission from E. Sijbesma, K.K. Hallenbeck, S. Leysen, P.J. de Vink, L. Skóra, W. Jahnke, L. Brunsveld, M.R. Arkin, C. Ottmann, Site-directed fragment-based screening for the discovery of protein–protein interaction stabilizers, J. Am. Chem. Soc. 141 (2019) 3524–3531. https://doi.org/10.1021/ jacs.8b11658. Copyright (2019) American Chemical Society.

desirable, because it offers the possibility of restoring or augmenting normal function. Few prospective studies have sought to discover stabilizers of protein-protein interactions, but several drugs and natural products have been shown retrospectively to act by this mechanism [122]. The intracellular scaffold protein 14-3-3 binds to phosphoserine and phosphothreonine sites, and alters the stability and subcellular location of bound clients. 14-3-3 binds to phosphorylated estrogen receptor α (ERα), thereby blocking ERα dimerization, nuclear translocation, and transcriptional activation. ERα signaling is a direct driver of ca. 70% of breast cancers. While early disease responds well to antiestrogen therapy, late-stage cancers are still estrogen-responsive but are resistant to current therapies. The fungal natural product fusicoccin A (MW = 680 Da) is a stabilizer of the 14-3-3:ERα interaction in human cells, with derivatives having demonstrated antitumorigenic activity [122–124].

We aimed to explore the wider potential for 14-3-3:ERα stabilizers through new and more drug-like chemical matter using a suite of biophysical methods. First, the team ran a disulfide-trapping screen (Section 5.4.5.4) to identify fragments that bound at the fusicoccin A site and stabilized a peptide derived from ERα. 1600 disulfide-containing fragments were screened for binding to a native or engineered cysteine residues on 14-3-3, in the presence and absence of ERα-peptide. Each cysteine-disulfide screen yielded fragments that were competitive, cooperative or neutral with respect to ERα-peptide (see Fig. 5.10A). Cooperative fragments were characterized by fluorescence polarization using fluorescein-labeled ERα phospho-peptides, confirming that disulfide-bound fragments stabilized ERα binding up to 40-fold (Fig. 5.10B). Selectivity for the 14-3-3/ERα complex was demonstrated by mass spectrometry and FP against alternative clients, including TASK3 (stabilization similar to ERα), ExoS and TAZ (significantly less stabilization). Detailed structural characterization of fragment adducts was performed by growth of 14-3-3:ERα crystals, followed by soaking with the disulfide fragments (see Fig. 5.10C). Highly complementary hydrophobic surfaces were observed in the strongest stabilizers between 14-3-3 and covalent ligand, and between the covalent ligand and ERα peptide, confirming the fragment to be an effective and selective intermolecular bridging ligand. ^{1}H NMR experiments demonstrated that the fragment also bound to the 14-3-3/ERα complex in the absence of the disulfide-tether. Linking multiple biophysical and structural methods was instrumental in discovering, characterizing, and optimizing novel stabilizers of the 14-3-3/ERα complex.

References

[1] E. Gregori-Puigjane, V. Setola, J. Hert, B.A. Crews, J.J. Irwin, E. Lounkine, L. Marnett, B.L. Roth, B.K. Shoichet, Identifying mechanism-of-action targets for drugs and probes, Proc. Natl. Acad. Sci. U.S.A. 109 (2012) 11178–11183, https://doi.org/10.1073/pnas.1204524109.

[2] O. Ursu, J. Holmes, J. Knockel, C.G. Bologa, J.J. Yang, S.L. Mathias, S.J. Nelson, T.I. Oprea, DrugCentral: online drug compendium, Nucleic Acids Res. 45 (2017) D932–D939, https://doi.org/10.1093/nar/gkw993.

[3] X. Du, Y. Li, Y.-L. Xia, S.-M. Ai, J. Liang, P. Sang, X.-L. Ji, S.-Q. Liu, Insights into protein–ligand interactions: mechanisms, models, and methods, Int. J. Mol. Sci. 17 (2016) 144, https://doi.org/10.3390/ijms17020144.

[4] K. Wanat, E. Brzezińska, A.W. Sobańska, Aspects of drug-protein binding and methods of analyzing the phenomenon, Curr. Pharmaceut. Des. 24 (2018) 2974–2985, https://doi.org/10.2174/1381612824666180808145320.

[5] J. Hughes, S. Rees, S. Kalindjian, K. Philpott, Principles of early drug discovery: principles of early drug discovery, Br. J. Pharmacol. 162 (2011) 1239−1249, https://doi.org/10.1111/j.1476-5381.2010.01127.x.

[6] L.M. Mayr, P. Fuerst, The future of high-throughput screening, J. Biomol. Screen. 13 (2008) 443−448, https://doi.org/10.1177/1087057108319644.

[7] D.A. Pereira, J.A. Williams, Origin and evolution of high throughput screening, Br. J. Pharmacol. 152 (2007) 53−61, https://doi.org/10.1038/sj.bjp.0707373.

[8] S. Ekins, J. Olechno, A.J. Williams, Dispensing processes impact apparent biological activity as determined by computational and statistical analyses, PLoS One 8 (2013) e62325, https://doi.org/10.1371/journal.pone.0062325.

[9] C.M. Cuttitta, D.L. Ericson, A. Scalia, C.G. Roessler, E. Teplitsky, K. Joshi, O. Campos, R. Agarwal, M. Allaire, A.M. Orville, R.M. Sweet, A.S. Soares, Acoustic transfer of protein crystals from agarose pedestals to micromeshes for high-throughput screening, Acta Crystallogr. D Biol. Crystallogr. 71 (2015) 94−103, https://doi.org/10.1107/S1399004714013728.

[10] F. Guo, W. Zhou, P. Li, Z. Mao, N.H. Yennawar, J.B. French, T.J. Huang, Precise manipulation and patterning of protein crystals for macromolecular crystallography using surface acoustic waves, Small 11 (2015) 2733−2737, https://doi.org/10.1002/smll.201403262.

[11] E. Teplitsky, K. Joshi, D.L. Ericson, A. Scalia, J.D. Mullen, R.M. Sweet, A.S. Soares, High throughput screening using acoustic droplet ejection to combine protein crystals and chemical libraries on crystallization plates at high density, J. Struct. Biol. 191 (2015) 49−58, https://doi.org/10.1016/j.jsb.2015.05.006.

[12] P. Wu, C. Noland, M. Ultsch, B. Edwards, D. Harris, R. Mayer, S.F. Harris, Developments in the implementation of acoustic droplet ejection for protein crystallography, J. Lab. Autom. 21 (2016) 97−106, https://doi.org/10.1177/2211068215598938.

[13] H. Zhang, Acoustic dispensing-mass spectrometry: the next high throughput bioanalytical platform for early drug discovery, Bioanalysis 9 (2017) 1619−1621, https://doi.org/10.4155/bio-2017-4980.

[14] Z. Gong, G. Hu, Q. Li, Z. Liu, F. Wang, X. Zhang, J. Xiong, P. Li, Y. Xu, R. Ma, S. Chen, J. Li, Compound libraries: recent advances and their applications in drug discovery, Curr. Drug Discov. Technol. 14 (2017), https://doi.org/10.2174/1570163814666170425155154.

[15] S. Mignani, J. Rodrigues, H. Tomas, R. Jalal, P.P. Singh, J.-P. Majoral, R.A. Vishwakarma, Present drug-likeness filters in medicinal chemistry during the hit and lead optimization process: how far can they be simplified? Drug Discov. Today 23 (2018) 605−615, https://doi.org/10.1016/j.drudis.2018.01.010.

[16] J.R. Archer, History, evolution, and trends in compound management for high throughput screening, Assay Drug Dev. Technol. 2 (2004) 675−681, https://doi.org/10.1089/adt.2004.2.675.

[17] S.L. McGovern, B.T. Helfand, B. Feng, B.K. Shoichet, A specific mechanism of nonspecific inhibition, J. Med. Chem. 46 (2003) 4265−4272, https://doi.org/10.1021/jm030266r.

[18] J.J. Irwin, D. Duan, H. Torosyan, A.K. Doak, K.T. Ziebart, T. Sterling, G. Tumanian, B.K. Shoichet, An aggregation advisor for ligand discovery, J. Med. Chem. 58 (2015) 7076−7087, https://doi.org/10.1021/acs.jmedchem.5b01105.

[19] B.K. Shoichet, Screening in a spirit haunted world, Drug Discov. Today 11 (2006) 607−615, https://doi.org/10.1016/j.drudis.2006.05.014.

[20] A.N. Ganesh, J. Logie, C.K. McLaughlin, B.L. Barthel, T.H. Koch, B.K. Shoichet, M.S. Shoichet, Leveraging colloidal aggregation for drug-rich nanoparticle formulations, Mol. Pharm. 14 (2017) 1852–1860, https://doi.org/10.1021/acs.molpharmaceut.6b01015.

[21] R.G. Strickley, Solubilizing excipients in oral and injectable formulations, Pharm. Res. 21 (2004) 201–230.

[22] J.B. Baell, G.A. Holloway, New substructure filters for removal of pan assay interference compounds (PAINS) from screening libraries and for their exclusion in bioassays, J. Med. Chem. 53 (2010) 2719–2740, https://doi.org/10.1021/jm901137j.

[23] J.B. Baell, J.W.M. Nissink, Seven year itch: pan-assay interference compounds (PAINS) in 2017—utility and limitations, ACS Chem. Biol. 13 (2018) 36–44, https://doi.org/10.1021/acschembio.7b00903.

[24] R.F. Bruns, I.A. Watson, Rules for identifying potentially reactive or promiscuous compounds, J. Med. Chem. 55 (2012) 9763–9772, https://doi.org/10.1021/jm301008n.

[25] G.S. Sittampalam, N.P. Coussens, K. Brimacombe, A. Grossman, M. Arkin, D. Auld, C. Austin, J. Baell, B. Bejcek, J.M.M. Caaveiro, T.D.Y. Chung, J.L. Dahlin, V. Devanaryan, T.L. Foley, M. Glicksman, M.D. Hall, J.V. Haas, J. Inglese, P.W. Iversen, S.D. Kahl, S.C. Kales, M. Lal-Nag, Z. Li, J. McGee, O. McManus, T. Riss, O.J. Trask, J.R. Weidner, M.J. Wildey, M. Xia, X. Xu (Eds.), Assay Guidance Manual, Eli Lilly & Company and the National Center for Advancing Translational Sciences, Bethesda, MD, 2004. http://www.ncbi.nlm.nih.gov/books/NBK53196/.

[26] K.H. Robinson, J.R. Yang, J. Zhang, FRET and BRET-based biosensors in live cell compound screens, in: J. Zhang, Q. Ni, R.H. Newman (Eds.), Fluorescent Protein-Based Biosensors, Humana Press, Totowa, NJ, 2014, pp. 217–225, https://doi.org/10.1007/978-1-62703-622-1_17.

[27] Y. Yan, T.-H. Xu, K.G. Harikumar, L.J. Miller, Bioluminescence resonance energy transfer assay (BRET assay), Bio Protoc. 7 (2017), https://doi.org/10.21769/BioProtoc.2904.

[28] M.D. Hall, A. Yasgar, T. Peryea, J.C. Braisted, A. Jadhav, A. Simeonov, N.P. Coussens, Fluorescence polarization assays in high-throughput screening and drug discovery: a review, Methods Appl. Fluoresc. 4 (2016) 022001, https://doi.org/10.1088/2050-6120/4/2/022001.

[29] J.C. Owicki, Fluorescence polarization and anisotropy in high throughput screening: perspectives and primer, J. Biomol. Screen. 5 (2000) 297–306, https://doi.org/10.1177/108705710000500501.

[30] S. Lee, V.S. Hong, Development and application of a high-throughput fluorescence polarization assay to target pim kinases, Assay Drug Dev. Technol. 14 (2016) 50–57, https://doi.org/10.1089/adt.2015.685.

[31] P. Heine, G. Witt, A. Gilardi, P. Gribbon, L. Kummer, A. Plückthun, High-throughput fluorescence polarization assay to identify ligands using purified g protein-coupled receptor, SLAS Discov. 24 (9) (2019) 915–927, https://doi.org/10.1177/2472555219837344.

[32] P.-C. Shih, Y. Yang, G.N. Parkinson, A. Wilderspin, G. Wells, A high-throughput fluorescence polarization assay for discovering inhibitors targeting the DNA-binding domain of signal transducer and activator of transcription 3 (STAT3), Oncotarget 9 (2018), https://doi.org/10.18632/oncotarget.26013.

[33] J.A. Cisneros, M.J. Robertson, M. Valhondo, W.L. Jorgensen, A fluorescence polarization assay for binding to macrophage migration inhibitory factor and crystal structures for complexes of two potent inhibitors, J. Am. Chem. Soc. 138 (2016) 8630–8638, https://doi.org/10.1021/jacs.6b04910.

[34] W.P. Janzen, Screening technologies for small molecule discovery: the state of the art, Chem. Biol. 21 (2014) 1162–1170, https://doi.org/10.1016/j.chembiol.2014.07.015.

[35] B. Bajar, E. Wang, S. Zhang, M. Lin, J. Chu, A guide to fluorescent protein FRET pairs, Sensors 16 (2016) 1488, https://doi.org/10.3390/s16091488.

[36] Z.-H. Wei, H. Chen, C. Zhang, B.-C. Ye, FRET-based system for probing protein-protein interactions between σR and RsrA from streptomyces coelicolor in response to the redox environment, PLoS One 9 (2014) e92330, https://doi.org/10.1371/journal.pone.0092330.

[37] A. Margineanu, J.J. Chan, D.J. Kelly, S.C. Warren, D. Flatters, S. Kumar, M. Katan, C.W. Dunsby, P.M.W. French, Screening for protein-protein interactions using Förster resonance energy transfer (FRET) and fluorescence lifetime imaging microscopy (FLIM), Sci. Rep. 6 (2016) 28186, https://doi.org/10.1038/srep28186.

[38] F. Degorce, HTRF: a technology tailored for drug discovery – a review of theoretical aspects and recent applications, Curr. Chem. Genom. 3 (2009) 22–32, https://doi.org/10.2174/1875397300903010022.

[39] R.A. Horton, K.W. Vogel, Multiplexing terbium- and europium-based TR-FRET readouts to increase kinase assay capacity, J. Biomol. Screen. 15 (2010) 1008–1015, https://doi.org/10.1177/1087057110368993.

[40] D. Geißler, S. Stufler, H.-G. Löhmannsröben, N. Hildebrandt, Six-color time-resolved Förster resonance energy transfer for ultrasensitive multiplexed biosensing, J. Am. Chem. Soc. 135 (2013) 1102–1109, https://doi.org/10.1021/ja310317n.

[41] A. Yasgar, A. Jadhav, A. Simeonov, N.P. Coussens, AlphaScreen-based assays: ultra-high-throughput screening for small-molecule inhibitors of challenging enzymes and protein-protein interactions, in: W.P. Janzen (Ed.), High Throughput Screening, Springer, New York, NY, 2016, pp. 77–98, https://doi.org/10.1007/978-1-4939-3673-1_5.

[42] J. Wang, P. Fang, P. Chase, S. Tshori, E. Razin, T.P. Spicer, L. Scampavia, P. Hodder, M. Guo, Development of an HTS-compatible assay for discovery of melanoma-related microphthalmia transcription factor disruptors using AlphaScreen technology, SLAS Discov. 22 (2017) 58–66, https://doi.org/10.1177/1087057116675274.

[43] E. Weber, I. Rothenaigner, S. Brandner, K. Hadian, K. Schorpp, A high-throughput screening strategy for development of RNF8-Ubc13 protein–protein interaction inhibitors, SLAS Discov. 22 (2017) 316–323, https://doi.org/10.1177/1087057116681408.

[44] S. Taouji, S. Dahan, R. Bosse, E. Chevet, Current screens based on the AlphaScreen™ technology for deciphering cell signalling pathways, Curr. Genom. 10 (2009) 93–101, https://doi.org/10.2174/138920209787847041.

[45] S.J. Capuzzi, E.N. Muratov, A. Tropsha, Phantom PAINS: problems with the utility of alerts for P an- assay interference compound S, J. Chem. Inf. Model. 57 (2017) 417–427, https://doi.org/10.1021/acs.jcim.6b00465.

[46] N. Bai, H. Roder, A. Dickson, J. Karanicolas, Isothermal analysis of ThermoFluor data can readily provide quantitative binding affinities, Sci. Rep. 9 (2019) 2650, https://doi.org/10.1038/s41598-018-37072-x.

[47] S.N. Krishna, C.-H. Luan, R.K. Mishra, L. Xu, K.A. Scheidt, W.F. Anderson, R.C. Bergan, A fluorescence-based thermal shift assay identifies inhibitors of mitogen activated protein kinase kinase 4, PLoS One 8 (2013) e81504, https://doi.org/10.1371/journal.pone.0081504.

[48] G. Senisterra, I. Chau, M. Vedadi, Thermal denaturation assays in chemical biology, Assay Drug Dev. Technol. 10 (2012) 128−136, https://doi.org/10.1089/adt.2011.0390.

[49] V. Kunig, M. Potowski, A. Gohla, A. Brunschweiger, DNA-encoded libraries — an efficient small molecule discovery technology for the biomedical sciences, Biol. Chem. 399 (2018) 691−710, https://doi.org/10.1515/hsz-2018-0119.

[50] D. Neri, R.A. Lerner, DNA-encoded chemical libraries: a selection system based on endowing organic compounds with amplifiable information, Annu. Rev. Biochem. 87 (2018) 479−502, https://doi.org/10.1146/annurev-biochem-062917-012550.

[51] S.G. Patching, Surface plasmon resonance spectroscopy for characterisation of membrane protein−ligand interactions and its potential for drug discovery, Biochim. Biophys. Acta 1838 (2014) 43−55, https://doi.org/10.1016/j.bbamem.2013.04.028.

[52] M. Elinder, M. Geitmann, T. Gossas, P. Källblad, J. Winquist, H. Nordström, M. Hämäläinen, U.H. Danielson, Experimental validation of a fragment library for lead discovery using SPR biosensor technology, J. Biomol. Screen. 16 (2011) 15−25, https://doi.org/10.1177/1087057110389038.

[53] W. Huber, F. Mueller, Biomolecular interaction analysis in drug discovery using surface plasmon resonance technology, Curr. Pharmaceut. Des. 12 (2006) 3999−4021, https://doi.org/10.2174/138161206778743600.

[54] J. Stetefeld, S.A. McKenna, T.R. Patel, Dynamic light scattering: a practical guide and applications in biomedical sciences, Biophys. Rev. 8 (2016) 409−427, https://doi.org/10.1007/s12551-016-0218-6.

[55] L. Peltonen, Practical guidelines for the characterization and quality control of pure drug nanoparticles and nano-cocrystals in the pharmaceutical industry, Adv. Drug Deliv. Rev. 131 (2018) 101−115, https://doi.org/10.1016/j.addr.2018.06.009.

[56] I. Navratilova, A.L. Hopkins, Fragment screening by surface plasmon resonance, ACS Med. Chem. Lett. 1 (2010) 44−48, https://doi.org/10.1021/ml900002k.

[57] J. Hyde, A.C. Braisted, M. Randal, M.R. Arkin, Discovery and characterization of cooperative ligand binding in the adaptive region of interleukin-2, Biochemistry 42 (2003) 6475−6483, https://doi.org/10.1021/bi034138g.

[58] G.A. Papalia, A.M. Giannetti, N. Arora, D.G. Myszka, Thermodynamic characterization of pyrazole and azaindole derivatives binding to p38 mitogen-activated protein kinase using Biacore T100 technology and van't Hoff analysis, Anal. Biochem. 383 (2008) 255−264, https://doi.org/10.1016/j.ab.2008.08.010.

[59] H. Roos, R. Karlsson, H. Nilshans, A. Persson, Thermodynamic analysis of protein interactions with biosensor technology, J. Mol. Recognit. 11 (1998) 204−210, https://doi.org/10.1002/(SICI)1099-1352(199812)11:1/6<204::AID-JMR424>3.0.CO;2-T.

[60] J. Concepcion, K. Witte, C. Wartchow, S. Choo, D. Yao, H. Persson, J. Wei, P. Li, B. Heidecker, W. Ma, R. Varma, L.-S. Zhao, D. Perillat, G. Carricato, M. Recknor, K. Du, H. Ho, T. Ellis, J. Gamez, M. Howes, J. Phi-Wilson, S. Lockard, R. Zuk, H. Tan, Label-free detection of biomolecular interactions using BioLayer interferometry for kinetic characterization, Comb. Chem. High Throughput Screen. 12 (2009) 791−800.

[61] C.A. Wartchow, F. Podlaski, S. Li, K. Rowan, X. Zhang, D. Mark, K.-S. Huang, Biosensor-based small molecule fragment screening with biolayer interferometry,

J. Comput. Aided Mol. Des. 25 (2011) 669−676, https://doi.org/10.1007/s10822-011-9439-8.

[62] T.A. Young, B. Moree, M.T. Butko, B. Clancy, M. Geck Do, T. Gheyi, J. Strelow, J.J. Carrillo, J. Salafsky, Second-harmonic generation (SHG) for conformational measurements: assay development, optimization, and screening, in: Methods in Enzymology, Elsevier, 2018, pp. 167−190, https://doi.org/10.1016/bs.mie.2018.09.017.

[63] B. Moree, K. Connell, R.B. Mortensen, C.T. Liu, S.J. Benkovic, J. Salafsky, Protein conformational changes are detected and resolved site specifically by second-harmonic generation, Biophys. J. 109 (2015) 806−815, https://doi.org/10.1016/j.bpj.2015.07.016.

[64] F. Vanzi, Protein conformation and molecular order probed by second-harmonic-generation microscopy, J. Biomed. Opt. 17 (2012) 060901, https://doi.org/10.1117/1.JBO.17.6.060901.

[65] Y. Hantani, K. Iio, R. Hantani, K. Umetani, T. Sato, T. Young, K. Connell, S. Kintz, J. Salafsky, Identification of inactive conformation-selective interleukin-2-inducible T-cell kinase (ITK) inhibitors based on second-harmonic generation, FEBS Open Bio 8 (2018) 1412−1423, https://doi.org/10.1002/2211-5463.12489.

[66] B. Moree, G. Yin, D.F. Lázaro, F. Munari, T. Strohäker, K. Giller, S. Becker, T.F. Outeiro, M. Zweckstetter, J. Salafsky, Small molecules detected by second-harmonic generation modulate the conformation of monomeric α-synuclein and reduce its aggregation in cells, J. Biol. Chem. 290 (2015) 27582−27593, https://doi.org/10.1074/jbc.M114.636027.

[67] T. Sugiki, K. Furuita, T. Fujiwara, C. Kojima, Current NMR techniques for structure-based drug discovery, Molecules 23 (2018) 148, https://doi.org/10.3390/molecules23010148.

[68] D.M. Dias, A. Ciulli, NMR approaches in structure-based lead discovery: recent developments and new frontiers for targeting multi-protein complexes, Prog. Biophys. Mol. Biol. 116 (2014) 101−112, https://doi.org/10.1016/j.pbiomolbio.2014.08.012.

[69] M. Pellecchia, I. Bertini, D. Cowburn, C. Dalvit, E. Giralt, W. Jahnke, T.L. James, S.W. Homans, H. Kessler, C. Luchinat, B. Meyer, H. Oschkinat, J. Peng, H. Schwalbe, G. Siegal, Perspectives on NMR in drug discovery: a technique comes of age, Nat. Rev. Drug Discov. 7 (2008) 738−745, https://doi.org/10.1038/nrd2606.

[70] M. Pellecchia, D.S. Sem, K. Wüthrich, NMR in drug discovery, Nat. Rev. Drug Discov. 1 (2002) 211−219, https://doi.org/10.1038/nrd748.

[71] C. Peng, S.W. Unger, F.V. Filipp, M. Sattler, S. Szalma, Automated evaluation of chemical shift perturbation spectra: new approaches to quantitative analysis of receptor-ligand interaction NMR spectra, J. Biomol. NMR 29 (2004) 491−504, https://doi.org/10.1023/B:JNMR.0000034351.37982.9e.

[72] T. Oltersdorf, S.W. Elmore, A.R. Shoemaker, R.C. Armstrong, D.J. Augeri, B.A. Belli, M. Bruncko, T.L. Deckwerth, J. Dinges, P.J. Hajduk, M.K. Joseph, S. Kitada, S.J. Korsmeyer, A.R. Kunzer, A. Letai, C. Li, M.J. Mitten, D.G. Nettesheim, S. Ng, P.M. Nimmer, J.M. O'Connor, A. Oleksijew, A.M. Petros, J.C. Reed, W. Shen, S.K. Tahir, C.B. Thompson, K.J. Tomaselli, B. Wang, M.D. Wendt, H. Zhang, S.W. Fesik, S.H. Rosenberg, An inhibitor of Bcl-2 family proteins induces regression of solid tumours, Nature 435 (2005) 677−681, https://doi.org/10.1038/nature03579.

[73] S. Monaco, L.E. Tailford, N. Juge, J. Angulo, Differential epitope mapping by STD NMR spectroscopy to reveal the nature of protein-ligand contacts, Angew. Chem. Int. Ed. 56 (2017) 15289−15293, https://doi.org/10.1002/anie.201707682.

[74] R. Nepravishta, S. Walpole, L. Tailford, N. Juge, J. Angulo, Deriving ligand orientation in weak protein−ligand complexes by DEEP-STD NMR spectroscopy in the absence of protein chemical-shift assignment, ChemBioChem 20 (2019) 340−344, https://doi.org/10.1002/cbic.201800568.

[75] J.E. Watt, G.R. Hughes, S. Walpole, S. Monaco, G.R. Stephenson, P.C. Bulman Page, A.M. Hemmings, J. Angulo, A. Chantry, Discovery of small molecule WWP2 ubiquitin ligase inhibitors, Chem. Eur. J. 24 (2018) 17677−17680, https://doi.org/10.1002/chem.201804169.

[76] L.H. Lucas, C.K. Larive, Measuring ligand-protein binding using NMR diffusion experiments, Concepts Magn. Reson. 20A (2004) 24−41, https://doi.org/10.1002/cmr.a.10094.

[77] S. Vanwetswinkel, R.J. Heetebrij, J. van Duynhoven, J.G. Hollander, D.V. Filippov, P.J. Hajduk, G. Siegal, TINS, target immobilized NMR screening: an efficient and sensitive method for ligand discovery, Chem. Biol. 12 (2005) 207−216, https://doi.org/10.1016/j.chembiol.2004.12.004.

[78] D. Chen, J.C. Errey, L.H. Heitman, F.H. Marshall, A.P. IJzerman, G. Siegal, Fragment screening of GPCRs using biophysical methods: identification of ligands of the adenosine A_{2A} receptor with novel biological activity, ACS Chem. Biol. 7 (2012) 2064−2073, https://doi.org/10.1021/cb300436c.

[79] R. Rosenzweig, L.E. Kay, Bringing dynamic molecular machines into focus by methyl-TROSY NMR, Annu. Rev. Biochem. 83 (2014) 291−315, https://doi.org/10.1146/annurev-biochem-060713-035829.

[80] M.S. Chimenti, S.L. Bulfcr, R.J. Neitz, A.R. Renslo, M.P. Jacobson, T.L. James, M.R. Arkin, M.J.S. Kelly, A fragment-based ligand screen against part of a large protein machine: the ND1 domains of the AAA+ ATPase p97/VCP, J. Biomol. Screen. 20 (2015) 788−800, https://doi.org/10.1177/1087057115570550.

[81] A. Bax, Two-dimensional NMR and protein structure, Annu. Rev. Biochem. 58 (1989) 223−256, https://doi.org/10.1146/annurev.bi.58.070189.001255.

[82] D.P. Frueh, Practical aspects of NMR signal assignment in larger and challenging proteins, Prog. Nucl. Magn. Reson. Spectrosc. 78 (2014) 47−75, https://doi.org/10.1016/j.pnmrs.2013.12.001.

[83] C. Kang, S. Gayen, W. Wang, R. Severin, A.S. Chen, H.A. Lim, C.S.B. Chia, A. Schüller, D.N.P. Doan, A. Poulsen, J. Hill, S.G. Vasudevan, T.H. Keller, Exploring the binding of peptidic West Nile virus NS2B−NS3 protease inhibitors by NMR, Antivir. Res. 97 (2013) 137−144, https://doi.org/10.1016/j.antiviral.2012.11.008.

[84] M. Arai, J.C. Ferreon, P.E. Wright, Quantitative analysis of multisite protein−ligand interactions by NMR: binding of intrinsically disordered p53 transactivation subdomains with the TAZ2 domain of CBP, J. Am. Chem. Soc. 134 (2012) 3792−3803, https://doi.org/10.1021/ja209936u.

[85] B. VanSchouwen, G. Melacini, Cracking the allosteric code of NMR chemical shifts, Proc. Natl. Acad. Sci. U.S.A. 113 (2016) 9407−9409, https://doi.org/10.1073/pnas.1611068113.

[86] H. Su, Y. Xu, Application of ITC-based characterization of thermodynamic and kinetic association of ligands with proteins in drug design, Front. Pharmacol. 9 (2018) 1133, https://doi.org/10.3389/fphar.2018.01133.

[87] S. Geschwindner, J. Ulander, P. Johansson, Ligand binding thermodynamics in drug discovery: still a hot tip? J. Med. Chem. 58 (2015) 6321–6335, https://doi.org/10.1021/jm501511f.

[88] A. Schön, E. Freire, Enthalpy screen of drug candidates, Anal. Biochem. 513 (2016) 1–6, https://doi.org/10.1016/j.ab.2016.08.023.

[89] H.-Y. Yen, J.T.S. Hopper, I. Liko, T.M. Allison, Y. Zhu, D. Wang, M. Stegmann, S. Mohammed, B. Wu, C.V. Robinson, Ligand binding to a G protein–coupled receptor captured in a mass spectrometer, Sci. Adv. 3 (2017) e1701016, https://doi.org/10.1126/sciadv.1701016.

[90] D. Martinez Molina, P. Nordlund, The cellular thermal shift assay: a novel biophysical assay for in situ drug target engagement and mechanistic biomarker studies, Annu. Rev. Pharmacol. Toxicol. 56 (2016) 141–161, https://doi.org/10.1146/annurev-pharmtox-010715-103715.

[91] J.M. Dziekan, H. Yu, D. Chen, L. Dai, G. Wirjanata, A. Larsson, N. Prabhu, R.M. Sobota, Z. Bozdech, P. Nordlund, Identifying purine nucleoside phosphorylase as the target of quinine using cellular thermal shift assay, Sci. Transl. Med. 11 (2019), https://doi.org/10.1126/scitranslmed.aau3174 eaau3174.

[92] D.A. Annis, E. Nickbarg, X. Yang, M.R. Ziebell, C.E. Whitehurst, Affinity selection-mass spectrometry screening techniques for small molecule drug discovery, Curr. Opin. Chem. Biol. 11 (2007) 518–526, https://doi.org/10.1016/j.cbpa.2007.07.011.

[93] C.E. Whitehurst, Z. Yao, D. Murphy, M. Zhang, S. Taremi, L. Wojcik, J.M. Strizki, J.D. Bracken, C.C. Cheng, X. Yang, G.W. Shipps, M. Ziebell, E. Nickbarg, Application of affinity selection-mass spectrometry assays to purification and affinity-based screening of the chemokine receptor CXCR4, Comb. Chem. High Throughput Screen. 15 (2012) 473–485.

[94] T.N. O'Connell, J. Ramsay, S.F. Rieth, M.J. Shapiro, J.G. Stroh, Solution-based indirect affinity selection mass spectrometry—a general tool for high-throughput screening of pharmaceutical compound libraries, Anal. Chem. 86 (2014) 7413–7420, https://doi.org/10.1021/ac500938y.

[95] K.P. Imaduwage, E.P. Go, Z. Zhu, H. Desaire, HAMS: high-affinity mass spectrometry screening. A high-throughput screening method for identifying the tightest-binding lead compounds for target proteins with No false positive identifications, J. Am. Soc. Mass Spectrom. 27 (2016) 1870–1877, https://doi.org/10.1007/s13361-016-1472-3.

[96] M.J. Chalmers, S.A. Busby, B.D. Pascal, Y. He, C.L. Hendrickson, A.G. Marshall, P.R. Griffin, Probing protein ligand interactions by automated hydrogen/deuterium exchange mass spectrometry, Anal. Chem. 78 (2006) 1005–1014, https://doi.org/10.1021/ac051294f.

[97] T. Lavold, R. Zubarev, J. Astorga-Wells, Hydrogen-deuterium exchange mass spectrometry in drug discovery — theory, practice and future, in: D. Huddler, E.R. Zartler (Eds.), Applied Biophysics for Drug Discovery, John Wiley & Sons, Ltd, Chichester, UK, 2017, pp. 61–71, https://doi.org/10.1002/9781119099512.ch4.

[98] D.A. Erlanson, J.A. Wells, A.C. Braisted, Tethering: fragment-based drug discovery, Annu. Rev. Biophys. Biomol. Struct. 33 (2004) 199–223, https://doi.org/10.1146/annurev.biophys.33.110502.140409.

[99] D.A. Erlanson, A.C. Braisted, D.R. Raphael, M. Randal, R.M. Stroud, E.M. Gordon, J.A. Wells, Site-directed ligand discovery, Proc. Natl. Acad. Sci. U.S.A. 97 (2000) 9367–9372, https://doi.org/10.1073/pnas.97.17.9367.

[100] J.M. Ostrem, U. Peters, M.L. Sos, J.A. Wells, K.M. Shokat, K-Ras, G12C) inhibitors allosterically control GTP affinity and effector interactions, Nature 503 (2013) 548−551, https://doi.org/10.1038/nature12796.

[101] D.A. Erlanson, J.W. Arndt, M.T. Cancilla, K. Cao, R.A. Elling, N. English, J. Friedman, S.K. Hansen, C. Hession, I. Joseph, G. Kumaravel, W.-C. Lee, K.E. Lind, R.S. McDowell, K. Miatkowski, C. Nguyen, T.B. Nguyen, S. Park, N. Pathan, D.M. Penny, M.J. Romanowski, D. Scott, L. Silvian, R.L. Simmons, B.T. Tangonan, W. Yang, L. Sun, Discovery of a potent and highly selective PDK1 inhibitor via fragment-based drug discovery, Bioorg. Med. Chem. Lett. 21 (2011) 3078−3083, https://doi.org/10.1016/j.bmcl.2011.03.032.

[102] M.R. Arkin, M. Randal, W.L. DeLano, J. Hyde, T.N. Luong, J.D. Oslob, D.R. Raphael, L. Taylor, J. Wang, R.S. McDowell, J.A. Wells, A.C. Braisted, Binding of small molecules to an adaptive protein-protein interface, Proc. Natl. Acad. Sci. U.S.A. 100 (2003) 1603−1608, https://doi.org/10.1073/pnas.252756299.

[103] S.G. Kathman, Z. Xu, A.V. Statsyuk, A fragment-based method to discover irreversible covalent inhibitors of cysteine proteases, J. Med. Chem. 57 (2014) 4969−4974, https://doi.org/10.1021/jm500345q.

[104] J.M. Rainard, G.C. Pandarakalam, S.P. McElroy, Using microscale thermophoresis to characterize hits from high-throughput screening: a european lead factory perspective, SLAS Discov. 23 (2018) 225−241, https://doi.org/10.1177/2472555217744728.

[105] M. Ehlers, J.-N. Grad, S. Mittal, D. Bier, M. Mertel, L. Ohl, M. Bartel, J. Briels, M. Heimann, C. Ottmann, E. Sanchez-Garcia, D. Hoffmann, C. Schmuck, Rational design, binding studies, and crystal-structure evaluation of the first ligand targeting the dimerization interface of the 14-3-3ζ adapter protein, ChemBioChem 19 (2018) 591−595, https://doi.org/10.1002/cbic.201700588.

[106] M.H. Moon, T.A. Hilimire, A.M. Sanders, J.S. Schneekloth, Measuring RNA−ligand interactions with microscale thermophoresis, Biochemistry 57 (2018) 4638−4643, https://doi.org/10.1021/acs.biochem.7b01141.

[107] J. Schiebel, S.G. Krimmer, K. Röwer, A. Knörlein, X. Wang, A.Y. Park, M. Stieler, F.R. Ehrmann, K. Fu, N. Radeva, M. Krug, F.U. Huschmann, S. Glöckner, M.S. Weiss, U. Mueller, G. Klebe, A. Heine, High-throughput crystallography: reliable and efficient identification of fragment hits, Structure 24 (2016) 1398−1409, https://doi.org/10.1016/j.str.2016.06.010.

[108] I. Tickle, A. Sharff, M. Vinkovi?, J. Yon, H. Jhoti, High-throughput protein crystallography and drug discovery, Chem. Soc. Rev. 33 (2004) 558, https://doi.org/10.1039/b314510g.

[109] T.L. Blundell, H. Jhoti, C. Abell, High-throughput crystallography for lead discovery in drug design, Nat. Rev. Drug Discov. 1 (2002) 45−54, https://doi.org/10.1038/nrd706.

[110] V. Cherezov, M.A. Hanson, M.T. Griffith, M.C. Hilgart, R. Sanishvili, V. Nagarajan, S. Stepanov, R.F. Fischetti, P. Kuhn, R.C. Stevens, Rastering strategy for screening and centring of microcrystal samples of human membrane proteins with a sub-10 μm size X-ray synchrotron beam, J. R. Soc. Interface 6 (2009), https://doi.org/10.1098/rsif.2009.0142.focus.

[111] S. Tsujino, T. Tomizaki, Ultrasonic acoustic levitation for fast frame rate X-ray protein crystallography at room temperature, Sci. Rep. 6 (2016) 25558, https://doi.org/10.1038/srep25558.

[112] D. Lyumkis, Challenges and opportunities in cryo-EM single-particle analysis, J. Biol. Chem. 294 (2019) 5181–5197, https://doi.org/10.1074/jbc.REV118.005602.

[113] G. Scapin, C.S. Potter, B. Carragher, Cryo-EM for small molecules discovery, design, understanding, and application, Cell Chem. Biol. 25 (2018) 1318–1325, https://doi.org/10.1016/j.chembiol.2018.07.006.

[114] X. Li, P. Mooney, S. Zheng, C.R. Booth, M.B. Braunfeld, S. Gubbens, D.A. Agard, Y. Cheng, Electron counting and beam-induced motion correction enable near-atomic-resolution single-particle cryo-EM, Nat. Methods 10 (2013) 584–590, https://doi.org/10.1038/nmeth.2472.

[115] A. Boland, L. Chang, D. Barford, The potential of cryo-electron microscopy for structure-based drug design, Essays Biochem. 61 (2017) 543–560, https://doi.org/10.1042/EBC20170032.

[116] G. Fan, M.R. Baker, Z. Wang, A.B. Seryshev, S.J. Ludtke, M.L. Baker, I.I. Serysheva, Cryo-EM reveals ligand induced allostery underlying InsP3R channel gating, Cell Res. 28 (2018) 1158–1170, https://doi.org/10.1038/s41422-018-0108-5.

[117] A. Merk, A. Bartesaghi, S. Banerjee, V. Falconieri, P. Rao, M.I. Davis, R. Pragani, M.B. Boxer, L.A. Earl, J.L.S. Milne, S. Subramaniam, Breaking cryo-EM resolution barriers to facilitate drug discovery, Cell 165 (2016) 1698–1707, https://doi.org/10.1016/j.cell.2016.05.040.

[118] X. Bai, C. Yan, G. Yang, P. Lu, D. Ma, L. Sun, R. Zhou, S.H.W. Scheres, Y. Shi, An atomic structure of human γ-secretase, Nature 525 (2015) 212–217, https://doi.org/10.1038/nature14892.

[119] J.C. Tsai, L.E. Miller-Vedam, A.A. Anand, P. Jaishankar, H.C. Nguyen, A.R. Renslo, A. Frost, P. Walter, Structure of the nucleotide exchange factor eIF2B reveals mechanism of memory-enhancing molecule, Science 359 (2018), https://doi.org/10.1126/science.aaq0939 eaaq0939.

[120] S. Banerjee, A. Bartesaghi, A. Merk, P. Rao, S.L. Bulfer, Y. Yan, N. Green, B. Mroczkowski, R.J. Neitz, P. Wipf, V. Falconieri, R.J. Deshaies, J.L.S. Milne, D. Huryn, M. Arkin, S. Subramaniam, 2.3 Å resolution cryo-EM structure of human p97 and mechanism of allosteric inhibition, Science 351 (2016) 871–875, https://doi.org/10.1126/science.aad7974.

[121] E. Sijbesma, K.K. Hallenbeck, S. Leysen, P.J. de Vink, L. Skóra, W. Jahnke, L. Brunsveld, M.R. Arkin, C. Ottmann, Site-directed fragment-based screening for the discovery of protein–protein interaction stabilizers, J. Am. Chem. Soc. 141 (2019) 3524–3531, https://doi.org/10.1021/jacs.8b11658.

[122] D. Bier, P. Thiel, J. Briels, C. Ottmann, Stabilization of Protein–Protein Interactions in chemical biology and drug discovery, Prog. Biophys. Mol. Biol. 119 (2015) 10–19, https://doi.org/10.1016/j.pbiomolbio.2015.05.002.

[123] A. Ballone, F. Centorrino, C. Ottmann, 14-3-3: A case study in PPI modulation, Molecules 23 (2018) 1386, https://doi.org/10.3390/molecules23061386.

[124] J. Ohkanda, A. Kusumoto, L. Punzalan, R. Masuda, C. Wang, P. Parvatkar, D. Akase, M. Aida, M. Uesugi, Y. Higuchi, N. Kato, Structural effects of fusicoccin upon upregulation of 14-3-3-phospholigand interaction and cytotoxic activity, Chem. Eur. J. 24 (2018) 16066–16071, https://doi.org/10.1002/chem.201804428.

Principles and applications of small molecule peptidomimetics

6

Andrea Trabocchi

Department of Chemistry "Ugo Schiff", University of Florence, Sesto Fiorentino, Florence, Italy

6.1 Introduction

The term "peptidomimetics" [1,2] is now a well-established concept in drug discovery for addressing those bioactive peptides that are being taken into account for developing new drugs, also called "peptidomimetic drugs." This concept dates back some decades ago, since the biomedical research got interested to developing new therapeutics based on peptides and proteins, taking advantage of their features responsible for biological activity and introducing both structural and functional specific modifications.

The importance of taking into account bioactive peptides as a starting point for drug discovery and development is due to the fact that an important number of biologically active peptides has been discovered and characterized over the years, including hormones, vasoactive peptides, and neuropeptides. It is now well understood that a wide array of peptides influence cell-cell communication and control a series of vital functions by interacting with membrane-bound receptors and enzymes. Thus, they are of great interest in the biomedical field, and the number of native and modified peptides used as therapeutics is ever increasing, many of them having been prepared on large scale and tested both in pharmacology and clinic, already. The continuous interest of considering peptides in a drug discovery program for developing new drugs is justified by some strengths that peptides possess, including their good efficacy, safety, and tolerability, their high selectivity and potency, as well as favourable issues in the development stages, including the knowledge of a predictable metabolism, shorter time to market and lower attrition rates, and the existence of standard synthetic protocols [3].

Nevertheless, the need of modifying bioactive peptides and ultimately developing peptidomimetics as drugs arises from the limitations of peptides as potential therapeutics due to several factors, including:

1. Limited stability toward proteolysis by peptidases in the gastrointestinal tract and in serum, with $t_{1/2}$ on the order of minutes
2. Poor transport properties to the blood and across the blood-brain barrier due to high molecular weight and lack of specific transport systems
3. Rapid excretion through the liver and kidneys

4. Intrinsic flexibility enables the interaction with multiple proteins besides the target, and thus resulting in undesired side effects

The flexibility issue is the most relevant together with poor stability in serum. It is mainly due to the rotatable bonds present in each amino acids, that in medium-sized polypeptides ($<$30 amino acids) determine multiple conformations that are energetically accessible for each residue constituting the peptide. Specifically, the flexibility of each residue constituting a peptide is due to two degrees of conformational freedom addressed by N-C_α and C_α-CO rotational bonds described by ϕ and ψ dihedral angles, respectively, which result in a population of local conformations, all contributing to the overall flexibility of the peptide backbone chain in dynamically interconverting equilibria in aqueous solution.

6.2 Definition and classification

A peptidomimetic compound is defined as "a substance having a secondary structure, besides other structural features similar to native peptide, such that it binds to enzymes or receptors with higher affinity than the starting peptide. As an overall result, the native peptide effects are inhibited (antagonist or inhibitor) or increased (agonist)." Since their introduction, peptidomimetics have shown great interest among the organic and medicinal chemistry community. This is due to the significant improvement that peptidomimetics deliver to the parent starting peptide, including higher selectivity, thus resulting in less side effects, and greater oral bioavailability and biological activity, that are prolonged due to lowered enzymatic degradation [4].

Peptidomimetics are historically divided into three classes depending on their structural and functional characteristics [5]:

1. **type I mimetics**, or *structural* mimetics, show an analogy of a local topography with the native substrate, and they carry all the functionalities responsible of the interaction with an enzyme or a receptor in a well-defined spatial orientation;
2. **type II mimetics**, or *functional* mimetics, the analogy with the native compound is based on the interaction with the target receptor or enzyme, without apparent structural analogies;
3. **type III mimetics**, or *functional-structural* mimetics, are generally conceived possessing a scaffold having a structure different from the substrate, in which all the functional groups needed for biological interactions are mounted in a well-defined spatial orientation [6]. Many examples are reported in the literature in which an unnatural framework substitutes the peptide backbone and carries the required functional groups for biological activity.

A new classification has been proposed recently, in view of taking account of recent advances in the field and to allow clear assignment of all approaches. Specifically, Grossmann et al. [7] introduced a new classification of peptidomimetics based on the degree of their similarity to the natural peptide precursor, thereby resulting in four different classes A–D, where A features the most and D the least similarities (Fig. 6.1).

1. **Class A mimetics** are defined as peptides that mainly consist of the parent peptide amino acid sequence, only a limited number of modified amino acids are introduced to stabilize the bioactive conformation, and the backbone and side chains of a class A mimetic align closely with the bioactive conformation of the precursor peptide.

   This class also includes stapled peptides, and it is closely related to the classic type I peptidomimetics.

2. **Class B mimetics** consist of peptides modified with various nonnatural amino acids, isolated small molecule building blocks, and/or major backbone alterations.

Peptides

natural peptide sequences derived from proteins and (non) ribosomal peptides

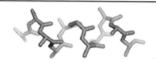

Class A — modified peptides

peptides mainly formed by α-amino acids with minor side chain or backbone alterations

Class B — modified peptides/foldamers

peptides with various backbone and side chain alterations also including foldamers

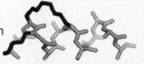

peptidic character

Class C — structural mimetics

small molecule-like scaffolds that project substituents in analogy to peptide side chains

Class D — mechanistic mimetics

molecules that mimic the mode of action of a peptide without a direct link to its side chains

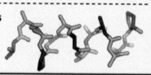
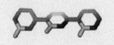
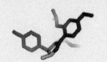

small molecules

FIGURE 6.1

Recent classification of peptidomimetics proposed by Grossmann et al.

Reproduced from M. Pelay-Gimeno, A. Glas, O. Koch, T.N. Grossmann, Structure-based design of inhibitors of protein–protein interactions: mimicking peptide binding epitopes, Angew. Chem. Int. 54 (2015) 8896–8927. Copyright (2015), with permission from John Wiley and Sons.

This class also includes foldamers, such as β- and α/β-peptides as well as peptoids, which contain a modified backbone and retain their side chains topologically similar to the parent peptide.

3. **Class C mimetics** are characterized by highly modified structures with small molecule character that replace the peptide backbone completely. The central scaffold projects substituents in analogy to the orientation of key residues (e.g., hot spots) in the bioactive conformation of the parent peptide.

    This class of molecules are related to the traditional type III peptidomimetics, also called scaffold peptidomimetics.

4. **Class D mimetics** are molecules that mimic the mode of action of a bioactive peptide without a direct link to its side chain functionalities.

    These molecules are generated through a hit-to-lead process starting from class C peptidomimetics, or they result from screenings of compound libraries.

Two classic examples of peptidomimetics are reported in Fig. 6.2. The thyrotropin-releasing hormone (TRH) mimetic is based on a cyclohexane scaffold **1** replacing the peptide backbone, and the three functional groups that constitute

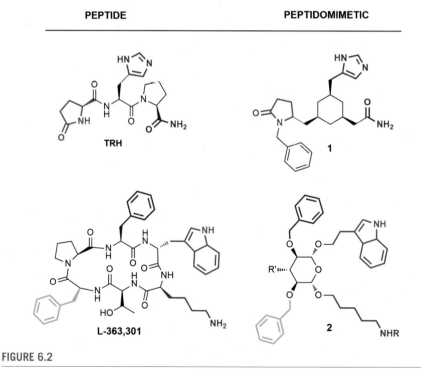

FIGURE 6.2

Peptidomimetic compounds consisting of cyclohexane (**1**) or glucose (**2**) scaffolds.

the pharmacophore are placed on the scaffold with the same spatial orientation of amino acid side chains found in TRH hormone [8]. Another classic entry includes the replacement of a peptidic fragment in the somatostatin receptor binding cyclo-peptide with D-glucose scaffold **2**, which proved to be an attractive mimic of a β-turn resulting from an inexpensive carbohydrate starting material with well-defined stereochemistry [9,10] (Fig. 6.2).

6.3 Strategic approaches to peptidomimetic design

As previously anticipated, the development of peptidomimetics in drug discovery is taken into account when the hit compounds toward an identified target are of peptide nature [11]. Specifically, the first step in a drug discovery process is hit identification, and for peptide hit compounds this is generally carried out by scanning peptide libraries for binding affinity (i.e., by phage display or combinatorial chemistry of synthetic peptide libraries). Molecular biology techniques, such as sequencing, cloning, and site-directed mutagenesis experiments, in combination to molecular modeling calculations, are necessary to achieve structural information about the protein target, and specifically to address the receptor residues responsible for peptide recognition. This early-stage research in drug discovery is instrumental in delivering the information required for selecting a bioactive peptide to be successively processed in a hierarchical way.

Since the generation of peptidomimetics is generally focused on the structural features of the native peptide and its protein target (i.e., receptor or enzyme), the development of peptidomimetics must take account of some basic principles [12], including:

1. replacement of peptide backbone with a nonpeptide framework: this is particularly important for improving bioavailability. If an amide bond substitution does not change the biological activity or amide bonds are not exposed to the active site, then the template may be designed in order to eliminate peptide bonds;
2. preservation of side chain involved in biological activity, as they constitute the structural elements of the pharmacophore. Taking advantage of the structural determinants for molecular recognition, several modifications are introduced to improve biological activity, including chain length modification, introduction of constraints and of isosteric replacements [13];
3. maintenance of flexibility in first-generation peptidomimetics: if a biological activity is observed for a flexible mimetic, then the introduction of elements of rigidity to side chains is a rational approach to improve the preliminary activity observed;
4. selection of proper targets based on a pharmacophore hypothesis. The knowledge of the structure-activity relationship or the three-dimensional structure of bioactive conformation is an efficient approach to rapidly achieve the lead

compound, thus reducing the number of compounds being generated with poor biological activity.

The initial steps in developing a lead peptidomimetic compound starting from a bioactive peptide are connected to the comprehension of the key residues of the parent peptide for bioactivity and also of potential weak sites for guiding the rational design of chemical modifications to improve enzymatic stability. This is particularly important when no structural data corresponding to the interaction with the target receptor are available, and currently, it is a relevant issue in the panorama of protein-protein interactions. The identification of potential cleavage sites has been undertaken experimentally by subjecting peptide to tissue homogenates, plasma and serum, and cocktails of purified proteases, followed by the characterization of the corresponding fragments by HPLC or modern LC-MS techniques.

The hierarchical approach in progressively transforming a bioactive peptide into a peptidomimetic compound takes advantage of several steps, which are important in giving insight to structure-activity relationships:

1. size reduction
2. alanine scanning
3. D-amino acid scanning
4. *N*-methylation
5. local and global constraints to define the bioactive conformation

The **reduction of the size** of the parent peptide is carried out with aim of identifying the key peptide residues interacting directly with the active site of the target enzyme or receptor. The result of this process is important in view of developing small molecule peptidomimetics possessing improved pharmacokinetics. This step is generally approached by assaying an array of peptides generated from the systematic removal of amino acids from either *N*- or *C*-termini, ultimately resulting in the minimal peptide sequence carrier of bioactivity, which is considered as the hit compound encompassing the pharmacophore, which must be maintained during subsequent modifications of the peptide structure.

Alanine scanning is the process consisting of the systematic synthesis and biological evaluation of an array of peptides containing an alanine residue replacing in place of any amino acid of the original bioactive peptide sequence. The peptides containing an alanine residue in place of key amino acids as in the original sequence result in the loss of bioactivity as a consequence of the lack of side chains interacting with the target receptor, thus allowing for the identification of pharmacophoric amino acids.

The introduction of **D-amino acids** systematically in the parent peptide sequence is a similar approach as of the alanine scanning in selecting the key residues carriers of bioactivity. This is also carried out in order to assess any possible modification as such to improve the metabolic stability of the peptide. Moreover, as the shift of chirality results in a different arrangement of the side chains, this approach gives insight on the structural organization of the bioactive conformation.

The *N*-methylation of amino groups and assay of the corresponding *N*-methylated peptides is another well-established approach for understanding the role of each amino acid constituting the bioactive peptide. Specifically, the alkylation of amide bonds is strategic for identifying which amino acids act as hydrogen bonding donors in the interaction with the target enzyme/receptor. Additionally, the *N*-methylation generates a tertiary amide bond, which is prone to establish *cis/trans* equilibrium at the amide bond, ultimately contributing to provide insight about the relationship between the conformational preferences of the peptide and bioactivity. Specifically, the strategy of systematically applying *N*-methylation to bioactive peptides was conceived by taking into account the role of proline in modulating the conformational profile of proteins, thus considering proline as a "mimic for *N*-methylation" in proteins [14]. In this view, *N*-methylation introduces another dimension to this "spatial screening" [15] due to the remarkable property of conformational modulation. *N*-methylation augments the amount of *cis*-peptide bonds and allows for the study of the role of amide protons in establishing potential hydrogen bonds, resulting in the optimization of the conformational and structural profile of a peptidomimetic.

Once the minimal peptide sequence carrier of bioactivity is identified together with the role of each amino acid constituting the parent peptide, structural modifications are introduced hierarchically in order to develop a compound with reduced peptide character, and possibly improving both the intrinsic binding affinity to the target receptor or enzyme and the pharmacokinetics/pharmacodynamics (PK/PD) profile from a pharmacological point of view. The achieved first-generation peptidomimetics are then subjected to further conformational studies aimed to defining the rationale for receptor-ligand or enzyme-inhibitor key interactions. The results are then applied for the optimization of hit peptidomimetics toward improved compounds possessing a nonpeptide framework.

6.3.1 Local modifications

The modification of the peptide through the introduction of local structural changes is the most conservative approach, which produce minimal modifications to the parent peptides although introducing varying degree of conformational bias. Generally, such modifications are restricted to single amino acids, and these local alterations of the peptide structure can be grouped into backbone and side-chain modifications. A wide range of variations at the moieties constituting the single amino acid have been reported [2], particularly addressing: the amino group being replaced with isosteric atoms or groups, such as oxygen, keto-methylene or *N*-hydroxyl moieties; the alpha carbon that can be replaced with nitrogen atoms, C-alkyl to achieve quaternary amino acids, or boron atoms; the carbonyl group that has been reported being replaced with thiol, methylene, phosphinic, and boronic groups. Unnatural amino acids are of great interest for this purpose, and their use as building blocks for the development of peptidomimetics has been taken into account extensively over the years. As well as homologation, the array of building blocks useful for developing peptidomimetics as locally modified peptides includes α-amino acids carrying modified side chains and rigid cyclic compounds as amino acid and dipeptide isosteres with added conformational restriction, which are important for modulating the conformational preferences of peptide leads.

6.3.1.1 Amide bond surrogates

The modification of the peptide backbone is generally addressed by introducing amide bond surrogates in view of improving the stability of the peptide in vivo, and several amide bond isosteres have been proposed since the advent of peptidomimetic concepts (Fig. 6.3), which mimic the structural features of the peptide bond and modulate the conformational profile and the hydrogen bonding capability [16]. For example, the hydrogen bonding capabilities are modulated by applying diverse amide bond isosteres, such as sulfonamide, phosphinic, or peptoids, depending on the accessibility of donors or acceptors for such interactions.

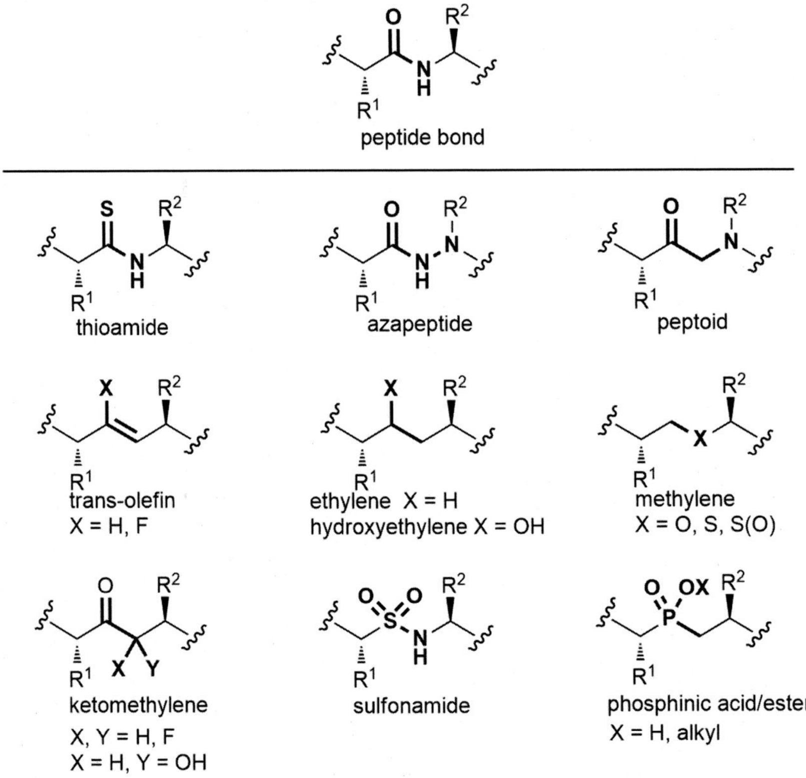

FIGURE 6.3

Array of peptide bond isosteres.

6.3.1.2 Side chain modifications

The introduction of local modifications around side chains mainly modulates the conformational profile of the peptide, specifically regulating the flexibility on all the rotatable bonds present in the amino acid unit (Fig. 6.4A). Several tethering approaches have been proposed in order to restrict the conformational freedom of selected dihedrals. Moreover, side chain modifications have been introduced for the exploration of steric and electronic interactions around pharmacophoric regions, for example, modulating the hydrophobic content by adding aromatic moieties or

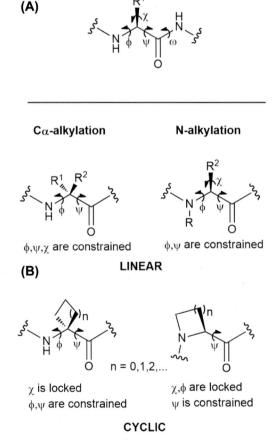

(A)

Cα-alkylation   N-alkylation

ϕ, ψ, χ are constrained ϕ, ψ are constrained

(B)   LINEAR

n = 0,1,2,...

χ is locked χ, ϕ are locked
ϕ, ψ are constrained ψ is constrained

CYCLIC

FIGURE 6.4

(A) Dihedral angle descriptors of rotatable bonds taken into account for modulating the conformational flexibility around the peptide chain; (B) C$_\alpha$- and N-alkylation to give acyclic (top) or cyclic structures (bottom) provide the introduction of chemical constraints in the amino acid moiety.

introducing polar appendages to address any polar or hydrogen bonding interactions with the target receptor.

6.3.1.3 C_α- and N-alkylation of amino acids

The approach of modifying a single amino acid unit within a peptide sequence is generally achieved by introducing constraining elements so as to reduce the conformational flexibility, and such constraints through C_α- and N-alkylation can be applied to acyclic or cyclic structures (Fig. 6.4B), resulting to varying degree of constraints to the torsion angles φ, ψ, and χ.

α-substituted amino acids are a common approach to reduce the torsional freedom around the backbone of such amino acid. α-methyl α-amino acids possess a methyl at the C_α, which greatly reduce the rotational freedom around N-C_α and C_α-CO_2H bonds. α-Me-alanine is the most widely studied α-alkylated amino acid, which is able to restrict the dihedrals φ, ψ to angles present in α or 3_{10} helices [17], as when inserted into a peptide it contributes to reducing the rotational freedom around its backbone bonds of about 90%.

6.3.1.4 N_α-C_α cyclized amino acids

The amino acids belonging to this class are generally considered as "proline mimetics," as it is unique among proteinogenic amino acids in bearing a cyclic structure and possessing the capability of giving *cis/trans* isomerization due to reduced energetic barrier of interconversion (about 2 kcal/mol). Thus, the structural properties of proline and its derivatives have been largely exploited in the development of peptidomimetics [18]. Indeed, much effort has been devoted toward the exploration of structural variants carrying higher conformational constraint and chemical diversity, and numerous mimetics and analogues of proline have been developed and applied to the synthesis of biologically active compounds, with the aim of modulating the *cis/trans* ratio of acyl-Pro bonds, constraining the conformation of the peptide bond, and producing proline-like reverse turn inducers [19–22].

Major structural features of these cyclic molecules are as follows:

1. the presence of a bond connecting N_α to C_α, that forms a cycle responsible for the local reduced conformational freedom;
2. the rotation around the C_α-C=O bond is partially impaired for the nonbonded interaction between the carbonyl group and the ring; and
3. the steric hindrance between the proline mimetic and adjacent residues, thus affecting the overall conformation locally.

Specifically, proline and all the corresponding peptidomimetics possess the capability of constraining the conformational freedom related to the torsional angle ϕ, and some approaches are generally followed:

1. modulation of the ring size, ranging from aziridines to omoproline;
2. inclusion of heteroatoms, such as azaproline or silaproline; and

3. introduction of substituents at positions 3, 4, 5 to improve the conformational restriction.

The structural diversity of proline mimetics is given by different ring size, the inclusion of heteroatoms in the ring, the presence of substituents on the ring, and the polycyclic nature as found in some scaffolds (Fig. 6.5).

Two cyclic amino acid derivatives found in nature and possessing the pyrrolidine ring as of proline are the pyroglutamic acid, derived from the lactamization of glutamic acid, and 4-hydroxy-proline, which is key in stabilizing helical structures of collagen (Fig. 6.6).

6.3.1.5 Constraining the side chain rotational freedom

The inclusion of rigidifying elements in amino acids and peptidomimetics with aim of reducing the conformational freedom of side chains has been envisaged by taking advantage of double bonds or cycles (Fig. 6.7). The application of α,β-unsaturated α-amino acids, also called dehydro-amino acids, allows for blocking the C_α-C_β rotation at the χ dihedral angle, and selecting an E or Z isomer depending on the desired position of the R side-chain. This chemical bias locks the χ angle to 0° or 180° depending on the stereochemistry of the olefin moiety.

The tethering approach has been also popular in blocking the conformational freedom of side chains by including cyclopropane rings at C_α-C_β positions, similarly to the conformational result as of dehydro amino acids, as they do not possess any conformational flexibility.

6.3.2 Global restrictions

Cyclization of a bioactive peptide is very effective in producing improved in vivo stability and higher potency as compared to their linear analogues, and specifically being more resistant to both exo- and endoproteases [23], which explains the

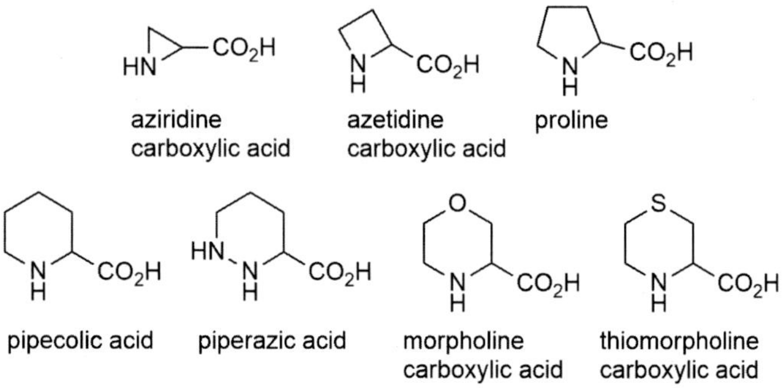

FIGURE 6.5

Cyclic α-amino acids as proline surrogates in peptidomimetic chemistry.

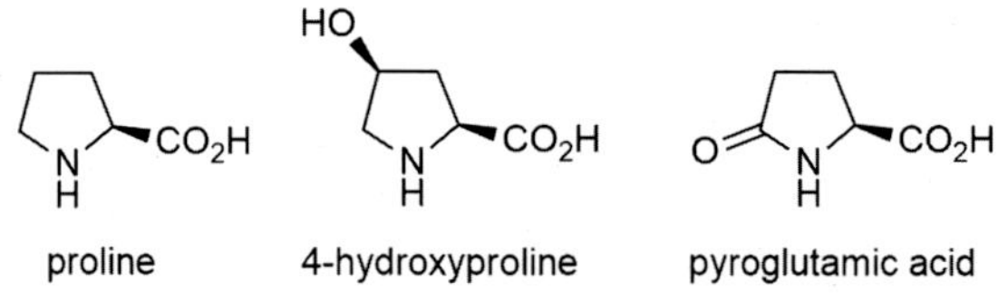

FIGURE 6.6

Cyclic pyrrolidine-based amino acids found in nature.

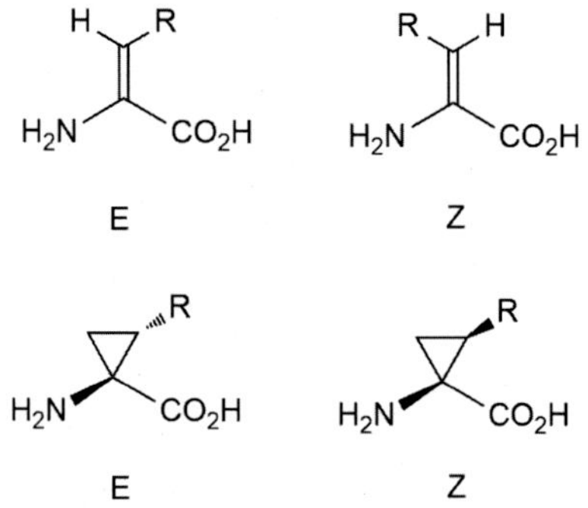

FIGURE 6.7

Dehydro- and cyclopropyl-amino acids allow for blocking the C_α–C_β rotation defined by the χ angle.

significant therapeutic potential of this class of molecules [24]. This is due to the high proportion of *cis* amide bonds and the absence of free *C*- and *N*-termini that confer higher metabolic resistance. Also, the limited conformational freedom results in higher receptor selectivity and binding affinity, by reducing unfavorable entropic effects. Indeed, many biologically active macrocyclic peptides are found in nature, like somatostatin, that is a macrocyclic peptide hormone formed in the hypothalamus and regulating the release of growth-hormone. It also acts in the pancreas, preventing the release of glucagon and insulin, leading to a lowering of blood glucose concentrations [25]. The regulatory action of somatostatin is due to interaction of its macrocyclic loop with the receptor site.

Thus, the generation of cyclic peptidomimetics is very attractive in view of constraining a native peptide structure into a conformationally reduced molecule. This approach involves the modification of the overall conformational profile of the target peptide compound; thus, it is complementary to the introduction of cyclic scaffolds for adding constraints locally.

The cyclization strategies can be classified with respect to backbone and side chains depending on the chemical moieties used for the introduction of the constraint [24,26].

Cyclization between backbone elements is approached in several ways:

1. by tethering two amide nitrogen atoms with a linker (backbone to backbone);
2. by introducing a chemical junction between a C_α and a nitrogen atom (backbone to backbone);
3. by linking an *N*-terminal amino group with an amide nitrogen atom with a spacer (head to backbone); and
4. by cyclizing the two *N*- and *C*-terminal ends of a peptidomimetic structure with an amide bond (head-to-tail).

Backbone cyclization combines cyclization with *N*-alkylation to enhance the stability of peptides [27]. These are generally carried out in solution phase, and the most popular approach is the head-to-tail generation of cyclic peptide via amide bond formation under standard peptide chemistry conditions [28]. The formation of the macrocyclic peptide is carried out under high dilution to promote the intramolecular reaction. A remarkable example of the synthesis of cyclic peptides via head-to-tail approach is given by the generation of cyclic peptidomimetics containing the Arg-Gly-Asp (RGD) sequence [29]. This approach has been extensively studied, and several reaction conditions have been studied in detail, including carboxylic activators and the role of the solvent toward the outcome of the cyclization. Solid-phase approaches have been also study with aim of linking the side chains of selected amino acids, such as Asp, Glu, Lys, Ser, and Thr, directly on resin and performing the macrolactamization before the cleavage from the solid support, taking advantage of the "pseudodilution effect" of the solid support. As a third approach, cyclative cleavage methods have been proposed by developing suitable linkers to accomplish the cyclization of the peptide molecule with concomitant release from the solid support. Some examples of suitable linkers for cyclative cleavage include the Kaiser oxime resin [30], the Kenner safety catch linker [31], and the backbone amide linker (BAL) developed by Barany and Albericio [32].

Another popular approach is the generation of cyclic peptidomimetics through a chemical bond involving two side chains. Generally, the cyclization is achieved by exploiting basic and amino acid residues for the formation of an amide bond, or taking advantage of cysteine amino acids for the development of cyclic peptidomimetics through disulfide bridges between the two side chains. Generally, lysine and ornithine are taken into account as amine counterparts if the *C*-terminal backbone is used for the cyclization, whereas glutamic and aspartic acids are the side chain functional groups used for the amide bond formation with the *N*-terminal moiety. The cyclization between cysteines is a classic approach in peptide chemistry within the side chain to side chain approach to easily generate a disulfide bond. Although the introduction of disulfide bonds to peptides confers high stability toward proteases, the sensitivity to reduction is often an issue, thus other linkages to improve the metabolic stability have been developed during the years. Modern methods for the cyclization of peptides take advantage of the well-established ring-closing metathesis, that allows for the generation of macrocyclic constructs by means of high metabolically stable olefin linkages, taking advantage of alkenyl moieties which are generally introduced in the peptide compound by means of allylglycines [33,34]. Other recently reported approaches include the azide-alkyne cycloaddition (also termed "click chemistry" or CuAAC), sulfur-mediated cyclizations, multicomponent approaches, and others [24].

6.4 Peptidomimetic molecules

6.4.1 Dipeptide isosteres

The scaffold approach for the generation of dipeptide isosteres [35] resulted particularly effective to constrain the conformational freedom of a specific region of a peptide bond by blocking ϕ, ψ, or ω rotations around backbone covalent bonds.

A milestone contribution to this field has been given by Freidinger, who first introduced the concept of using dipeptide lactams as conformational constraints to restrict the peptide bond to the *trans* conformation, and the application of ring junctions to limit the ψ backbone dihedral angle [36]. The first dipeptide lactam was achieved using protected ornithine (**4**) by means of an intramolecular acylation approach (Fig. 6.8). Similarly, the protected lysine precursor **5** was taken into account to achieve the corresponding homologue ε-lactam dipeptide isostere **6**.

Then, the successful synthesis of dipeptide lactams paved the way toward the application of dipeptide lactams as peptidomimetics useful for increasing the stability toward protease degradation, potency and receptor selectivity, as well as being instrumental as molecular tools to decipher the biologically active conformation of the parent peptide [37,38]. Indeed, a wide array of dipeptide isosteres were reported over the years capable of constraining the conformation around a dipeptide unit, including azabicycloalkanes, lactams, piperazinones and imidazolinones, and bridged [39] and spiranic molecular scaffolds (Fig. 6.9A−D). A common approach to dipeptide isosteres is the design and synthesis of conformationally biased δ-amino acids. δ-Amino acids are isosteric replacements of dipeptide units, and their application in the field of peptidomimetics has been extensively reported, particularly for developing β-turn mimetics. A popular approach focused on proline-based bicyclic δ-amino acids, that can be grouped into the class of azabicyclo[x.3.0]alkanes (Fig. 6.9A).

Particularly constrained dipeptide isosteres are those formed by spiro compounds, that are characterized by the combination of two cyclic moieties, that block simultaneously both ϕ and ψ, ω torsional angles, thus resulting in a blocked conformation. δ-Amino acids based on the spiro-γ-lactam structure have been applied for the development of neurotensin C-terminal hexapeptide peptidomimetics [40], taking advantage of the isosteric resemblance to constrained dipeptide motifs (Fig. 6.9D).

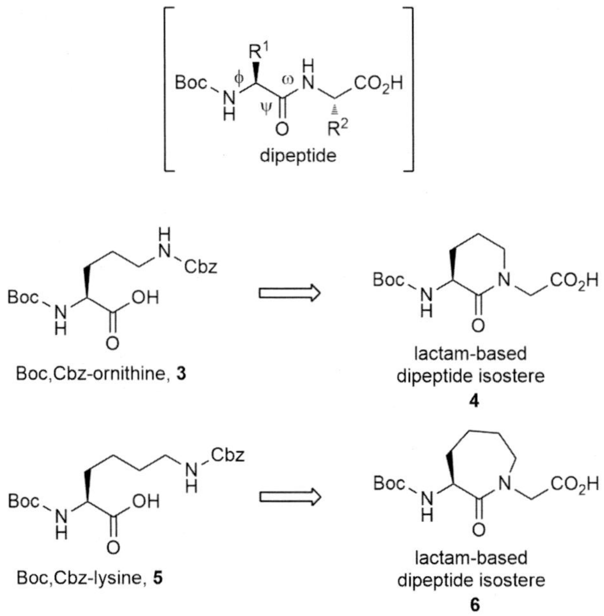

FIGURE 6.8

Six- and seven-membered ring lactam-based dipeptide isosteres synthesized from ornithine and lysine, respectively.

6.4.2 Transition-state isosteres

Transition-state isosteres can be defined as a stable and nonhydrolyzable functional group suitable for mimicking the tetrahedral transition state of peptide bond hydrolysis. Such chemical moieties are a special group of amide bond isosteres that have been widely exploited for developing peptidomimetic protease inhibitors. Specifically, the incorporation of a transition-state analogue into a peptide structure proved to be very successful for the design of efficient protease inhibitors, as these enzymes catalyze hydrolytic reactions via a tetrahedral transition state resulting from the nucleophilic attack by a water molecule on the scissile peptide bond carbonyl group [41,42]. This strategic approach consists of replacing the hydrolysable peptide linkage with nonhydrolysable transition-state isosteres [43]. Most of the known inhibitors are peptide substrate analogues characterized by the presence of a nonhydrolyzable transition-state isostere at position P_1/P_1', close to the scissile bond, in place of the normal amide bond. Statine [44,45] is a well-known peptidomimetic moiety of this class, as it represents the nonproteinogenic β-hydroxy-γ-amino acid found in the natural protease inhibitor pepstatin. Other examples of transition-state analogues being included in the structure of peptidomimetic protease inhibitors are hydroxyethylamine isosteres [46−49], phosphinates [50−52], hydroxymethylcarbonyls [53−55], and hydroxyethylenes [56−58] (Fig. 6.10).

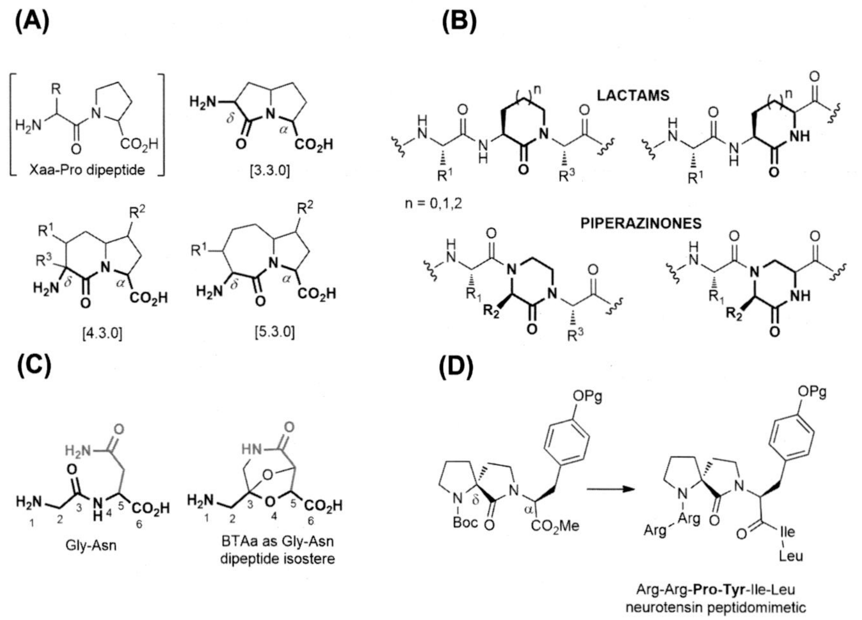

FIGURE 6.9

Molecular scaffolds (A–D) developed as constrained dipeptide isosteres.

6.4.3 Retroinverso peptides

A logic approach for improving the resistance of a peptide compound toward protease degradation is to conceive chemical modifications around amide bonds that are subjected to hydrolysis. Retroinverso isomerization is a brilliant method to modify the structure of the backbone to prevent the protease recognizing the peptide-based inhibitor as a substrate. This can be achieved by replacing one or more L-amino acids with the parent enantiomer, and by reversing the backbone direction from N→C to C→N. The retroinverso modification does not alter the conformational flexibility of the molecule, but improves highly the in vivo stability due to the modification of amide bonds, which are recognized by proteases for their hydrolysis. This approach is well-represented by the retroinverso peptidomimetic of the key tetrapeptide sequence found in gastrin (Fig. 6.11, left) [59], and by the retroinverso peptidomimetic of tuftsin (Fig. 6.11, right), which is an immune system stimulator that is completely degraded in vivo in about 8 min [60]. The retroinverso peptidomimetic of tuftsin showed less than 2% hydrolysis after 50 min and retention of bioactivity.

6.4.4 Azapeptides

Azapeptides are an interesting and synthetically easy to approach class of peptidomimetics consisting of the Cα atom of the backbone being replaced isoelectronically by a nitrogen atom (Fig. 6.12). The synthesis of azapeptides is generally approached

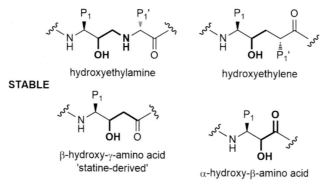

FIGURE 6.10

Common transition-state isosteres used in the design peptidomimetic protease inhibitors.

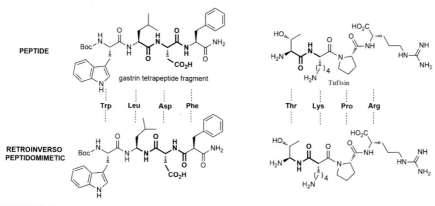

FIGURE 6.11

Retroinverso peptidomimetic of the key tetrapeptide sequence found in gastrin (left) and the immunomodulator tuftsin (right).

using substituted hydrazines or hydrazides easily [61], through the acylation of hydrazines [62], and the incorporation of aza-amino acid esters into a peptide chain. Azapeptides have been shown to be therapeutically relevant particularly in the case of serine and cysteine proteases inhibitors [63].

6.4.5 Peptoids

Oligomers of *N*-substituted glycine, or peptoids, were firstly reported by Bartlett and coworkers in 1992 [64]. They were initially proposed as an accessible class of molecules from which lead compounds could be identified for drug discovery. Peptoids can be easily defined as α-peptide mimetics in which the side chain is switched from the α-carbon to the backbone amide nitrogen (Fig. 6.13). This modification results in the formal shift of the position of the side chain with respect to the parent peptide backbone.

Sequence-specific peptoid oligomers are easily assembled from primary amines by the solid-phase submonomer approach. Relevant applications of peptoids are oriented to the exploration of peptoid secondary structures and drug design. Major advantages of peptoids as research and pharmaceutical tools include the ease and economy of synthesis, highly variable backbone and modularity of side chain chemistry.

6.4.6 The triazole ring as a peptide bond isostere

The triazole ring has become very popular in chemical biology since the introduction of the "click chemistry concept," allowing for the facile bioconjugation of biomolecules with suitable molecules through the Cu-catalyzed azide-alkyne cycloaddition (CuAAC). The reasons for the high popularity of the Huisgen 1,3-dipolar cycloaddition reaction are connected to the easy synthesis of the alkyne and azide functionalities, coupled with their kinetic stability and tolerance to a wide variety of functional groups and reaction conditions, which make these coupling partners particularly attractive. Nevertheless, the Huisgen reaction was not exploited a synthetic tool until the regioselectivity issues were solved with the introduction of copper catalysis. In 2002, Sharpless and Meldal groups reported independently the discovery of copper(I) catalysis in Huisgen cycloaddition between

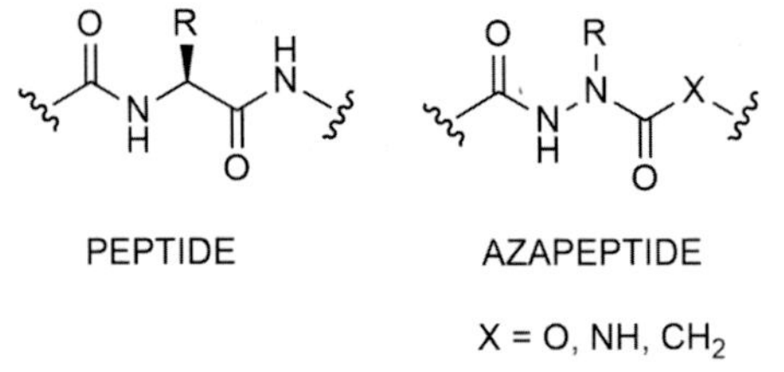

PEPTIDE   AZAPEPTIDE

X = O, NH, CH$_2$

FIGURE 6.12

Structural features of azapeptides as compared to the peptide moiety.

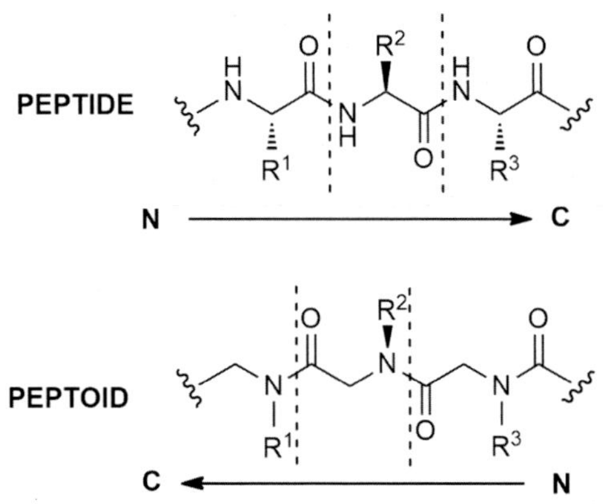

FIGURE 6.13

Atom-by-atom correlation between peptide and peptoid.

azides and terminal alkynes [65–67]. This kind of catalysis leads to high yield often in mild conditions (room temperature) and allows to achieve complete regioselectivity toward the 1,4 disubstituted adduct. Consequently, the CuAAC reaction has been applied as a conjugation strategy in the design and synthesis of complex biomimetic architectures in which the triazole linkage replaces peptide and phosphodiester bonds. Moreover, several reports demonstrated that the CuAAC reaction can be used to generate potent biologically active compounds via chemical ligation, and that 1,2,3-triazoles can be used as surrogates for biologically relevant amide bonds.

An additional important feature of the triazole ring is connected to its intrinsic capability of working as a peptidomimetic scaffold. Indeed, the triazole ring possess a peculiar property is being exploited in medicinal chemistry, as it is considered a good nonclassical bioisostere of the peptide bond with improved chemical stability in biological systems (Fig. 6.14) [68].

The triazole ring is planar and displays similar electronic content and dipole moment as of the amide bond [69,70]. It possesses analogous hydrogen bonding profile, too, with the C-2 atom playing as a hydrogen bond donor, and nitrogen atoms at 4 and 5 positions as hydrogen bonding acceptors with their lone pairs, in the same way to nitrogen and the oxygen atoms of the amide bond, respectively.

Specifically, the CuAAC has several attributes in peptidomimetic chemistry, as

1. it gives the 1,2,3-triazole nucleus that is a likely candidate for small molecule drugs;
2. it is compatible with the side chains of all the amino acids, at least in the protected form;
3. it can potentially give high yields of products with few by-products;
4. the molecular dimensions of the 1,4-disubstituted-1,2,3-triazoles are somewhat similar to amide bonds in terms of distance and planarity; and

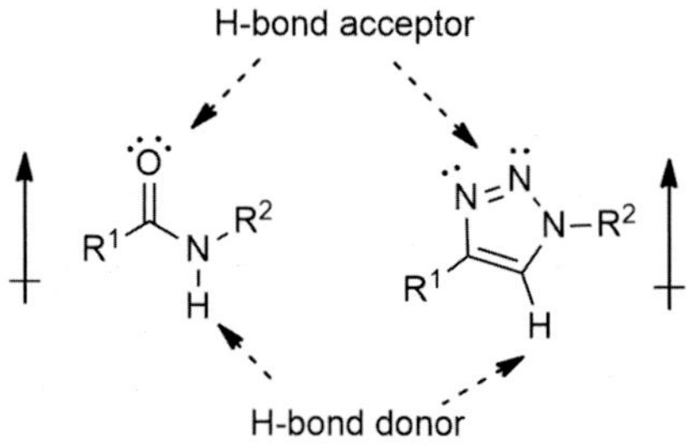

H-bond acceptor

R^1-R^2 = 3.9 ang R^1-R^2 = 5.0 ang

H-bond donor

FIGURE 6.14

Bioisosteric relationship between the 1,4-disubstituted triazole ring and the amide bond.

5. the triazole unit is resistant to enzymatic degradation, hydrolysis, and oxidation, making it an attractive moiety in biologically active compounds.

6.5 Secondary structure peptidomimetics

Protein-protein interactions (PPIs) are essential for numerous biological processes, including cell proliferation, cell division, signal transduction, transcription, translation, and programmed cell death. PPIs also play a critical role in various diseases and pathological conditions such as neurodegeneration, cardiovascular diseases, and cancer. Drugs targeting PPI possess a huge therapeutic potential, nevertheless addressing PPIs with small molecules is challenging. This is due to the fact that binding surfaces between proteins are usually large and involve many polar and hydrophobic interactions, and also they are typically flat, mostly not possessing a defined binding pocket for binding of a small molecule. Peptides can be rationally designed based on the natural sequences that mediate PPI in the proteins. As short peptides are often flexible, the potency of inhibitors can be increased by inducing the desired secondary structure. Protein-protein interfaces can involve all three main protein structural motifs: alpha-helices, beta-strands, and loop or turn regions. Accordingly, peptidomimetics mimicking such motifs aim to fix those bioactive conformations characterized to a high degree by such structural elements, which are essential conformational components for peptides and proteins (Fig. 6.15).

6.5.1 α-Helix mimetics

The α-helix is the most common peptide secondary structure, constituting almost half of the polypeptide structure in proteins. First proposed by Hamilton, a notable entry to α-helix mimetics consisted of molecular templates based on the terphenyl (**7**) [72] and terpyridyl (**8**) scaffolds [73] (Fig. 6.16). Successively, Boger described 4-aminobenzoic acid-based oligoamides **9** with the aromatic building block varying

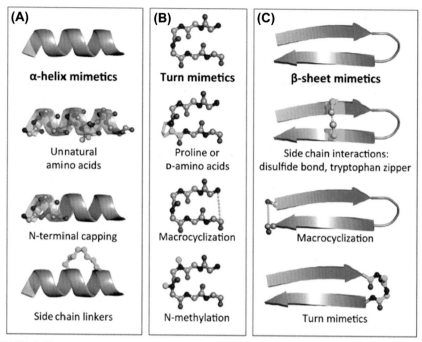

FIGURE 6.15

Chemical approaches for stabilizing peptide mimetics of alpha-helices, turns, and beta-hairpins. (A) alpha-helices can be stabilized by using unnatural amino acids, *N*-terminal capping, or sidechain linkers (stapling). (B) Turn mimetics may be stabilized by using proline or D-amino acids, macrocyclization, or *N*-methylation of the backbone amide. (C) Beta-sheet mimetics can be stabilized by side chain interactions or by using turn mimetics to induce hairpin formation.

Reprinted from A.D. Cunningham, N. Qvit, D. Mochly-Rosen, Peptides and peptidomimetics as regulators of protein-protein interactions, Curr. Opin. Struct. Biol. 44 (2017) 59–66. Copyright (2017), with permission from Elsevier.

in number from 1 to 3 units, respectively [74] (Fig. 6.16). Another entry to helix mimetics composed by aromatic rings was proposed by Koenig and collab. reporting a 1,4-dipiperazinobenzene (**10**) as a short helix mimetic containing side-chain isosteres at the two piperazines [75] (Fig. 6.16).

6.5.2 β-Sheets mimetics

β-sheets are characterized by a regular array of intramolecular hydrogen bonds connecting adjacent β-strands and represent key structural elements in the three-dimensional structure and biological activity of proteins. Their interest is growing especially in the field of CNS diseases involving aggregation of oligomeric species possessing flat structures of this type [77]. As the structure and stability of β-sheets

α-helix motif   terphenyl **7**   **8**: X = N, terpyridyl scaffold   1,4-dipiperazinobenzene **10**

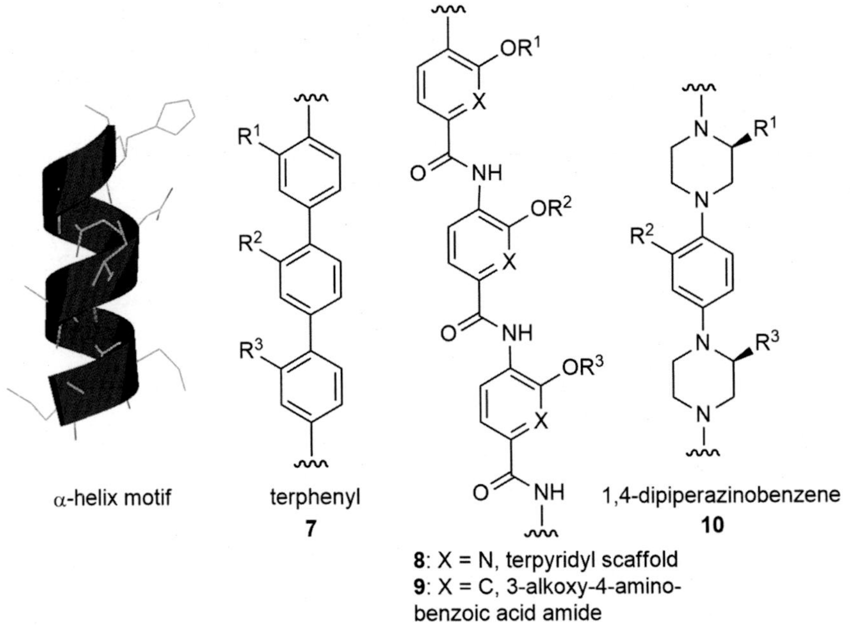

8: X = N, terpyridyl scaffold
9: X = C, 3-alkoxy-4-amino-
benzoic acid amide

FIGURE 6.16

Representative entries to α-helix mimetics.

Reproduced from A. Trabocchi, A. Guarna, Peptidomimetics in Organic and Medicinal Chemistry: The Art of Transforming Peptides in Drugs, John Wiley & Sons, UK, 2014 (Chapter 1). Copyright (2014), with permission from John Wiley and Sons.

are still not as well understood as α-helices, the interest of β-sheet-like peptidomimetics is connected to developing model structures to study such folding propensities in peptides and proteins. The design of β-sheet structures takes advantage of specific moieties, such as urea bonds, or designed molecular scaffolds, possessing a flat structure and the capability of mimicking β-strands, specifically establishing parallel hydrogen bonds both as donors and acceptors [78]. Some key examples include β-strand mimetics formed by 5-amino-2-methoxybenzamides and hydrazides (**11**), that have been proposed for stabilizing antiparallel β-sheets [79,80], and methoxypyrrole-based amino acids (**12**) [81], all serving as a central strand of the β-sheet to orient the hydrogen bonding functionality appropriately (Fig. 6.17).

6.5.3 β-Turn mimetics

β-turns are the most explored protein secondary structure mimetics using small molecules [82]. A β-turn is defined as a tetrapeptide sequence where the distance between $C_{\alpha i}$ and $C_{\alpha i+3}$ is $\leq 7 \text{Å}$. The turn can be stabilized by chelation of a cation, or intramolecular hydrogen bonds. The general requirements in designing suitable turn mimetics consist of identifying a rigid scaffold that orients the side-chain residues in the same direction as the natural peptide, while conferring better solubility and/or resistance to enzymatic degradation. Different approaches have been devised for the generation of β-turn mimetics by synthesizing either scaffolds mimicking the whole peptide motif (**13**) or developing dipeptide isosteres capable of inducing a turn in a peptide motif (**14**). Also, chemical tethers have been introduced as constraining elements to stabilize β-turn structures within a macrocyclic molecule (**15**) (Fig. 6.18).

6.6 Application of peptidomimetics as protease inhibitors

The most successful application of the concept of peptidomimetics in drug discovery is represented by the development of enzyme inhibitors. More specifically, proteases have been found as an attractive therapeutic target for a number of pathologies, as they are crucial for a number of processes, including the regulation of peptide hormones and neuromodulators through proteolytic activation of inactive precursors. In the course of designing peptidomimetic protease inhibitors, the nomenclature proposed by Schechter and Berger is often considered as a standard in describing the cleavage site of a substrate (P_1–P_1' at the cleavage site and P_n–P_n' for the flanking

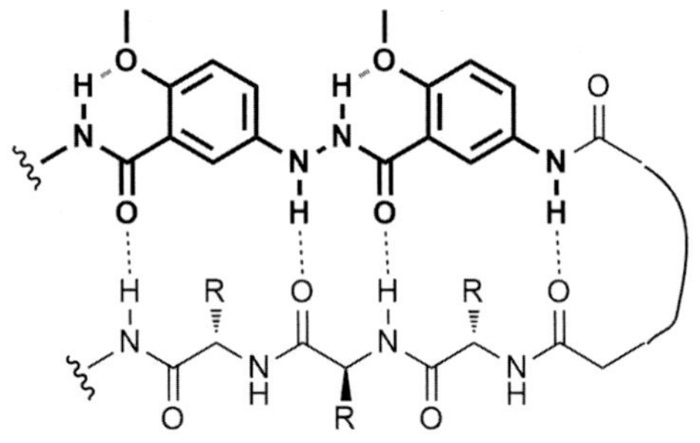

11

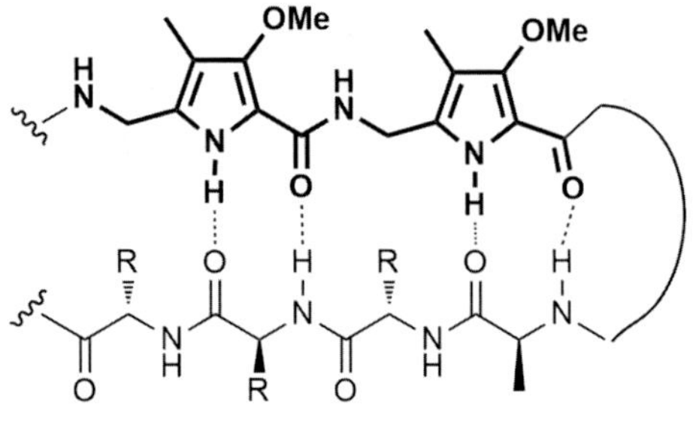

12

FIGURE 6.17

Representative entries to β-sheet mimetics.

amino acids) with respect of amino acid sequence, and the corresponding enzyme sites (S_1–S_1' at the cleavage site and S_n–S_n' for the flanking sites), as represented in Fig. 6.19 [83].

6.6.1 A case study of peptidomimetic drugs: HIV protease inhibitors

The generation of peptidomimetic drugs has been particularly effective for the treatment of viral infections. There are nine peptidomimetic drugs on the market for the treatment of AIDS, and at least four in clinical development for treatment of hepatitis C virus (HCV) infections. The array of drugs against AIDS, which is caused by HIV infection, includes peptidomimetic compounds that target the virally encoded aspartic protease enzyme. This enzyme is crucial for the maturation process during the replication of HIV and for promoting the infectivity.

The general strategy to develop transition-state analogues of aspartic peptidases was developed during 1972–1983 [84]. HIV-1 aspartic protease represents the most extensively studied therapeutic target in the history of structure-based drug design. The initial design of aspartic protease inhibitors was based resembling of the transition-state intermediate, which is formed by the enzyme during the catalytic mechanism (see Section 6.4.2) [85]. Starting from statine, which is the nonproteinogenic β-hydroxy-γ-amino acid found in the natural protease inhibitor pepstatin, a

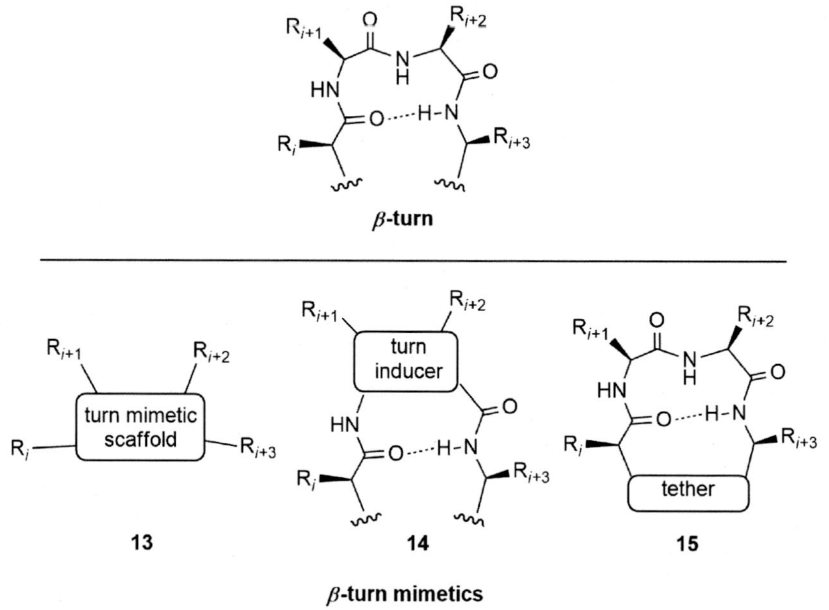

FIGURE 6.18

Chemical approaches to β-turn mimetics.

variety of different transition-state analogue moieties were developed and exploited to create inhibitors of different aspartic proteases [86–90]. The modifications of the side-chains at P_1 and P_1' positions and of the amino acids within the peptide sequence allowed for the development of selective inhibitors.

HIV-1 protease inhibitors were developed taking into account both the hydrolytic mechanism and the tridimensional structure of the target enzyme. These can be subdivided in two main groups, first- and second-generation protease inhibitors: (Fig. 6.20)

• First-generation inhibitors are peptidomimetics showing similarities with the substrates, including saquinavir (SQV), ritonavir (RTV), indinavir (IDV), and amprenavir (APV)
• Second-generation HIV PIs include fosamprenavir (FPV), darunavir (DRV) among the others

First approaches toward the design of HIV aspartic protease inhibitors conceived the presence of a hydroxyl moiety mimicking the transition state of the amide bond hydrolysis, which establishes hydrogen bonding interactions with the two catalytic aspartic acids in the protease active site. Such inhibitors were characterized by hydroxyethylamine or hydroxypropylamine transition-state isosteres in the backbone and by the presence of a phenylalanine isostere as the P_1 group close to the scissile bond [91].

Specifically, the hydroxyethylamine isostere was taken into account for the development of saquinavir as the first approved HIV protease inhibitor in 1995. This inhibitor, possessing a K_i value of 0.12 nM, was designed taking into account the specificity of the protease in recognizing a Pro residue at P_1', and the Phe-Asn dipeptide on the other side of the molecule.

Although showing high potency and significant therapeutic profile, first-generation drugs showed some limitations due to resistance phenomena of viral strains, toxicity, and low pharmacokinetics, thus requiring too many doses to be assumed daily. Many efforts in solving these problems resulted in second-generation inhibitors, which showed improved features with respect to potency,

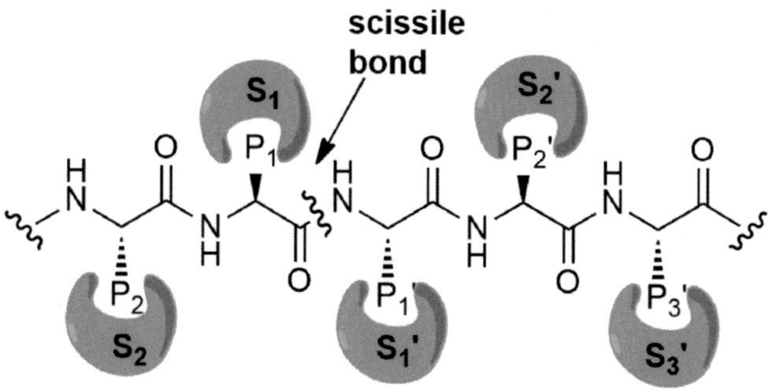

FIGURE 6.19

Peptide substrate and enzyme subsites according to Schechter and Berger notation.

activity toward mutant strains, and pharmacokinetic profiles. In particular, darunavir was designed starting from the structure of amprenavir by replacing the furan ring at P_2 with a bicyclic *bis*-tetrahydrofuran system. This group demonstrated a better ligand capacity in the S_2 subsite due to extra hydrogen bonding interactions of the two oxygen atoms of the bicyclic system [92]. The improved interactions of the bicyclic system resulted in darunavir possessing an inhibitory potency toward wild-type HIV protease of about 1000 times that of first-generation drugs.

Because of the rapid genomic evolution of viruses, an inevitable consequence in the treatment of viral infections is the advent of drug resistance. Therefore, the incomplete suppression of HIV in AIDS patients will continue to drive the search for more effective therapeutic agents that exhibit efficacy against the mutants raised by the earlier generation of protease inhibitors.

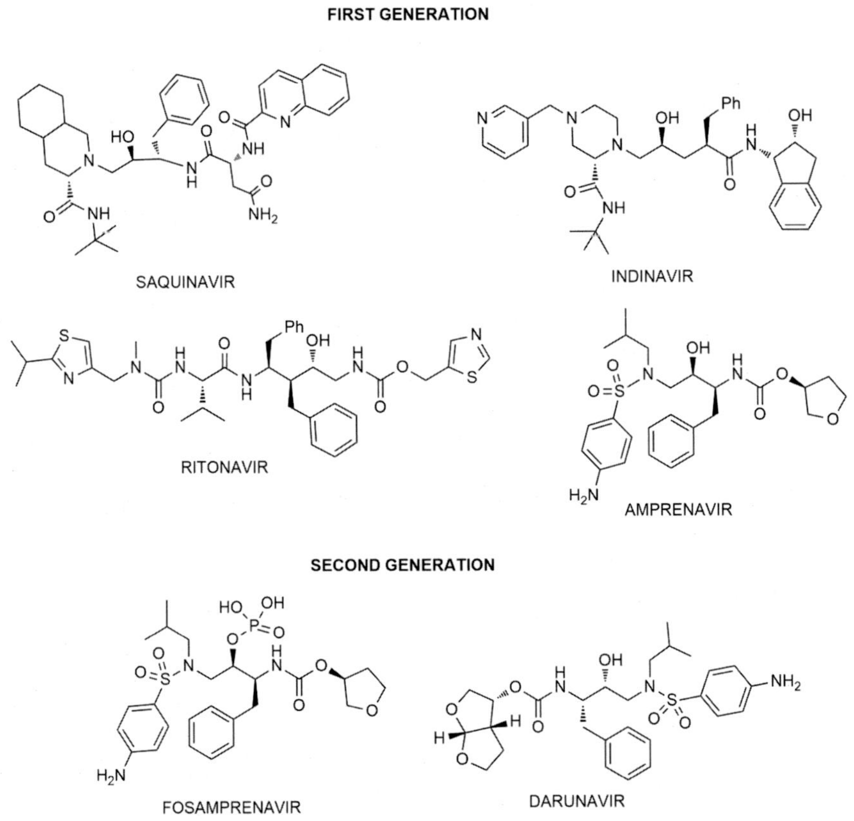

FIGURE 6.20

Representative first- and second-generation HIV protease inhibitors.

6.7 Conclusion

After several decades since the introduction of the concept of peptidomimetics, this approach in drug discovery is still timely, due to the never ending interest in new compounds based on peptides and proteins. Besides the development of biotechnological therapeutics based on antibody-derived compounds, the field of small molecules encompassing the panorama of peptide drugs still comprises by peptidomimetic approaches with aim to achieve hit compounds possessing improved bioactivity and pharmacokinetics. During last decades the basic concepts and approaches to peptidomimetic compounds have evolved to diverse compounds and synthetic strategies spanning from combinatorial chemistry to solid-phase synthesis and heterocyclic chemistry. It is expected that in the future peptidomimetics will continue leading the drug discovery workflow characterized by presence of hit compounds of peptide nature. The approach of considering peptides as drugs is experiencing a renaissance, taking advantage of modern omics sciences, thus peptidomimetic chemistry will parallel with such renewed interest to improve the druggability of those bioactive molecules.

References

[1] A. Giannis, T. Kolter, Peptidomimetics for receptor ligands-discovery, development, and medical perspectives, Angew. Chem. Int. Ed. Engl. 32 (1993) 1244–1267.

[2] J. Gante, Peptidomimetics-tailored enzyme inhibitors, Angew. Chem. Int. Ed. Engl. 33 (1994) 1699–1720.

[3] K. Fosgerau, T. Hoffmann, Peptide therapeutics: current status and future directions, Drug Discov. Today 20 (1) (2015) 122–128.

[4] R.M.J. Liskamp, Conformationally restricted amino acids and dipeptides, (non)peptidomimetics and secondary structure mimetics, Recl. Trav. Chim. Pays-Bas 113 (1994) 1–19.

[5] A.S. Ripka, D.H. Rich, Peptidomimetic design, Curr. Opin. Chem. Biol. 2 (1998) 441–452.

[6] R. Marshall, F. Ballante, Limiting assumptions in the design of peptidomimetics, Drug Dev. Res. 78 (2017) 245–267.

[7] M. Pelay-Gimeno, A. Glas, O. Koch, T.N. Grossmann, Structure-based design of inhibitors of protein–protein interactions: mimicking peptide binding epitopes, Angew. Chem. Int. 54 (2015) 8896–8927.

[8] G.L. Olson, D.R. Bolin, M.P. Bonner, M. Bös, C.M. Cook, D.C. Fry, B.J. Graves, M. Hatada, D.E. Hill, M. Kahn, V.S. Madison, V.K. Rusiecki, R. Sarabu, J. Sepinwall, G.P. Vincent, M.E. Voss, Concepts and progress in the development of peptide mimetics, J. Med. Chem. 36 (1993) 3039–3049.

[9] R. Hirschmann, K.C. Nicolaou, S. Pietranico, E.M. Leahy, J. Salvino, B. Arison, M.A. Cichy, P.G. Spoors, W.C. Shakespeare, P.A. Sprengeler, P. Hamley, A.B. Smith III, T. Reisine, K. Raynor, L. Maechler, C. Donaldson, W. Vale, R.M. Freidinger, M.R. Cascieri, C.D. Strader, De novo design and synthesis of

somatostatin non-peptide peptidomimetics utilizing .beta.-D-glucose as a novel scaffolding, J. Am. Chem. Soc. 115 (1993) 12550−12568.

[10] R.F. Hirschmann, K.C. Nicolaou, A.R. Angeles, J.S. Chen, A.B. Smith III, The β-D-glucose scaffold as a β-turn mimetic, Acc. Chem. Res. 42 (10) (2009) 1511−1520.

[11] A. Giannis, F. Rübsam, Peptidomimetics in drug design, Adv. Drug Res. 29 (1997) 1−78.

[12] P.S. Farmer, in: E.J. Ariëns (Ed.), Drug Design, vol. X, Academic Press, New York, 1980, p. 119.

[13] G.R. Marshall, A hierarchical approach to peptidomimetic design, Tetrahedron 49 (1993) 3547−3558.

[14] J. Chatterjee, F. Rechenmacher, H. Kessler, N-methylation of peptides and proteins: an important element for modulating biological functions, Angew. Chem. Int. Ed. 52 (2013) 254−269.

[15] H. Kessler, R. Gratias, G. Hessler, M. Gurrath, G. Müller, Conformation of cyclic peptides. Principle concepts and the design of selectivity and superactivity in bioactive sequences by 'spatial screening', Pure Appl. Chem. 68 (1996) 1201−1205.

[16] A. Choudhary, R.T. Raines, An evaluation of peptide-bond isosteres, ChemBioChem 12 (2011) 1801−1807.

[17] I.L. Karle, P. Balaram, Structural characteristics of .alpha.-helical peptide molecules containing Aib residues, Biochemistry 29 (1990) 6747−6756.

[18] M. Breznik, S. Golič Grdadolnik, G. Giester, I. Leban, D. Kikelj, Influence of chirality of the preceding acyl moiety on the cis/trans ratio of the proline peptide bond, J. Org. Chem. 66 (2001) 7044−7050.

[19] M.J. Genin, R.L. Johnson, Design, synthesis, and conformational analysis of a novel spiro-bicyclic system as a type II .beta.-turn peptidomimetic, J. Am. Chem. Soc. 114 (1992) 8778−8783.

[20] S. Hanessian, G. McNaughton-Smith, H.-G. Lombart, W.D. Lubell, Design and synthesis of conformationally constrained amino acids as versatile scaffolds and peptide mimetics, Tetrahedron 53 (1997) 12789−12854.

[21] B.E. Fink, P.R. Kym, J.A. Katzenellenbogen, Design, synthesis, and conformational analysis of a proposed type I β-turn mimic, J. Am. Chem. Soc. 120 (1998) 4334−4344.

[22] L. Halab, W.D. Lubell, Use of steric interactions to control peptide turn geometry. Synthesis of type VI β-turn mimics with 5-tert-butylproline, J. Org. Chem. 64 (1999) 3312−3321.

[23] J.D.A. Tyndall, T. Nall, D.P. Fairlie, Proteases universally recognize beta strands in their active sites, Chem. Rev. 105 (2005) 973−999.

[24] C.J. White, A.K. Yudin, Contemporary strategies for peptide macrocyclization, Nat. Chem. 3 (2011) 509−524.

[25] R. Burgus, N. Ling, M. Butcher, R. Guillemin, Primary structure of somatostatin, a hypothalamic peptide that inhibits the secretion of pituitary growth hormone, Proc. Natl. Acad. Sci. U.S.A. 70 (1973) 684−688.

[26] A. Grauer, B. König, Peptidomimetics − a versatile route to biologically active compounds, Eur. J. Org. Chem. (2009) 5099−5111.

[27] C. Gilon, D. Halle, M. Chorev, Z. Selinger, G. Byk, Backbone cyclization: a new method for conferring conformational constraint on peptides, Biopolymers 31 (1991) 745−750.

[28] P. Li, P.P. Roller, J. Xu, Current synthetic approaches to peptide and peptidomimetic cyclization, Curr. Org. Chem. 6 (2002) 411−440.

[29] R. Haubner, W. Schmitt, G. Hoelzemann, S.L. Goodman, A. Jonczyk, H. Kessler, Cyclic RGD peptides containing β-turn mimetics, J. Am. Chem. Soc. 118 (34) (1996) 7881–7891.

[30] W.F. DeGrado, E.T. Kaiser, Polymer-bound oxime esters as supports for solid-phase peptide synthesis. The preparation of protected peptide fragments, J. Org. Chem. 45 (1980) 1295–1300.

[31] G.W. Kenner, J.R. McDermott, R.C. Sheppard, The safety catch principle in solid phase peptide synthesis, J. Chem. Soc. Chem. Commun. (1971) 636–637.

[32] K.J. Jensen, J. Alsina, M.F. Songster, J. Vagner, F. Albericio, G. Barany, Backbone amide linker (BAL) strategy for solid-phase synthesis of C-terminal-modified and cyclic peptides, J. Am. Chem. Soc. 120 (1998) 5441–5452.

[33] S.J. Miller, H.E. Blackwell, R.H. Grubbs, Application of ring-closing metathesis to the synthesis of rigidified amino acids and peptides, J. Am. Chem. Soc. 118 (1996) 9606–9614.

[34] C.J. Creighton, A.B. Reitz, Synthesis of an eight-membered cyclic pseudo-dipeptide using ring closing metathesis, Org. Lett. 3 (2001) 893–895.

[35] R.M. Freidinger, Design and synthesis of novel bioactive peptides and peptidomimetics, J. Med. Chem. 46 (2003) 5553–5566.

[36] R.M. Freidinger, D.F. Veber, R. Hirschmann, L.M. Paege, Lactam restriction of peptide conformation in cyclic hexapeptides which alter rumen fermentation, Int. J. Pept. Protein Res. 16 (1980) 464–470.

[37] R.M. Freidinger, D.S. Perlow, W.C. Randall, R. Saperstein, B.H. Arison, D.F. Veber, Conformational modifications of cyclic hexapeptide somatostatin analogs, Int. J. Pept. Protein Res. 23 (1984) 142–150.

[38] M.A. Cascieri, G.G. Chicchi, R.M. Freidinger, C.D. Colton, D.S. Perlow, B. Williams, N.R. Curtis, A.T. McKnight, J.J. Maguire, D.F. Veber, T. Liang, Conformationally constrained tachykinin analogs which are selective ligands for the eledoisin binding site, Mol. Pharmacol. 29 (1986) 34–38.

[39] A. Trabocchi, G. Menchi, E. Danieli, A. Guarna, Synthesis of a bicyclic delta-amino acid as a constrained Gly-Asn dipeptide isostere, Amino Acids 35 (2008) 37–44.

[40] H. Bittermann, J. Einsiedel, H. Hübner, P. Gmeiner, Evaluation of lactam-bridged neurotensin analogues adjusting psi(Pro10) close to the experimentally derived bioactive conformation of NT(8-13), J. Med. Chem. 47 (2004) 5587–5590.

[41] A. Brik, C. Wong, HIV-1 protease: mechanism and drug discovery, Org. Biomol. Chem. 1 (2003) 5–14.

[42] F. Wångsell, P. Nordeman, J. Sävmarker, R. Emanuelsson, K. Jansson, J. Lindberg, A. Rosenquist, B. Samuelsson, M. Larhed, Investigation of α-phenylnorstatine and α-benzylnorstatine as transition state isostere motifs in the search for new BACE-1 inhibitors, Bioorg. Med. Chem. 19 (2011) 145–155.

[43] E.D. Clercq, Toward improved anti-HIV chemotherapy: therapeutic strategies for intervention with HIV infections, J. Med. Chem. 38 (1995) 2491–2517.

[44] M. Müller, Chemoenzymatic synthesis of building blocks for statin side chains, Angew. Chem. Int. Ed. 44 (2005) 362–365.

[45] D. Gupta, R.S. Yedidi, S. Varghese, L.C. Kovari, P.M. Woster, Mechanism-based inhibitors of the aspartyl protease plasmepsin II as potential antimalarial agents, J. Med. Chem. 53 (2010) 4234–4247.

[46] K. Akaji, K. Teruya, S. Aimoto, Solid-phase synthesis of HTLV-1 protease inhibitors containing hydroxyethylamine dipeptide isostere, J. Org. Chem. 68 (2003) 4755–4763.

[47] A. Gautier, D. Pitrat, J. Hasserodt, An unusual functional group interaction and its potential to reproduce steric and electrostatic features of the transition states of peptidolysis, Bioorg. Med. Chem. 14 (2006) 3835−3847.

[48] L.R. Marcin, M.A. Higgins, F.C. Zusi, Y. Zhang, M.F. Dee, M.F. Parker, J.K. Muckelbauer, D.M. Camac, P.E. Morin, V. Ramamurthy, A.J. Tebben, K.A. Lentz, J.E. Grace, J.A. Marcinkeviciene, L.M. Kopcho, C.R. Burton, D.M. Barten, J.H. Toyn, J.E. Meredith, C.F. Albright, J.J. Bronson, J.E. Macor, L.A. Thompson, Synthesis and SAR of indole-and 7-azaindole-1,3-dicarboxamide hydroxyethylamine inhibitors of BACE-1, Bioorg, Med. Chem. Lett. 21 (2011) 537−541.

[49] S.S. Kale, S.T. Chavan, S.G. Sabharwal, V.G. Puranik, G.J. Sanjayan, Bicyclic amino acid-carbohydrate-conjugates as conformationally restricted hydroxyethylamine (HEA) transition-state isosteres, Org. Biomol. Chem. 9 (2011) 7300−7302.

[50] M. Drag, R. Grzywa, J. Oleksyszyn, Novel hydroxamic acid-related phosphinates: inhibition of neutral aminopeptidase N (APN), Bioorg. Med. Chem. Lett 17 (2007) 1516−1519.

[51] T. Yamagishi, H. Ichikawa, T. Haruki, T. Yokomatsu, Diastereoselective synthesis of alpha,beta'-disubstituted aminomethyl(2-carboxyethyl)phosphinates as phosphinyl dipeptide isosteres, Org. Lett. 10 (2008) 4347−4350.

[52] R. Grzywa, J. Oleksyszyn, First synthesis of alpha-aminoalkyl-(N-substituted) thiocarbamoyl-phosphinates: inhibitors of aminopeptidase N (APN/CD13) with the new zinc-binding group, Bioorg. Med. Chem. Lett 18 (2008) 3734−3736.

[53] A. Stoeckel-Maschek, B. Stiebitz, R. Koelsch, K. Neubert, Novel 3-amino-2-hydroxy acids containing protease inhibitors. Part 1: synthesis and kinetic characterization as aminopeptidase P inhibitors, Bioorg. Med. Chem. 13 (2005) 4806−4820.

[54] S. Weik, T. Luksch, A. Evers, J. Bçttcher, C.A. Sotriffer, A. Hasilik, H. Lçffler, G. Klebe, J. Rademann, The potential of P1 site alterations in peptidomimetic protease inhibitors as suggested by virtual screening and explored by the use of C-C-coupling reagents, ChemMedChem 1 (2006) 445−457.

[55] S. Nakatani, K. Hidaka, E. Ami, K. Nakahara, A. Sato, J. Nguyen, Y. Hamada, Y. Hori, N. Ohnishi, A. Nagai, T. Kimura, Y. Hayashi, Y. Kiso, Combination of non-natural D-amino acid derivatives and allophenylnorstatine-dimethylthioproline scaffold in HIV protease inhibitors have high efficacy in mutant HIV, J. Med. Chem. 51 (2008) 2992−3004.

[56] S. Hanessian, G. Yang, J. Rondeau, U. Neumann, C. Betschart, M. Tintelnot-Blomley, Structure-based design and synthesis of macroheterocyclic peptidomimetic inhibitors of the aspartic protease beta-site amyloid precursor protein cleaving enzyme (BACE), J. Med. Chem. 49 (2006) 4544−4567.

[57] Y. Yamaguchi, K. Menear, N. Cohen, R. Maha, F. Cumin, C. Schnell, J.M. Wood, J. Maibaum, The P1N-isopropyl motif bearing hydroxyethylene dipeptide isostere analogues of aliskiren are in vitro potent inhibitors of the human aspartyl protease renin, Bioorg. Med. Chem. Lett 19 (2009) 4863−4867.

[58] C. Björklund, H. Adolfsson, K. Jansson, J. Lindberg, L. Vrang, A. Hallberg, A. Rosenquist, B. Samuelsson, Discovery of potent BACE-1 inhibitors containing a new hydroxyethylene (HE) scaffold: exploration of P1' alkoxy residues and an aminoethylene (AE) central core, Bioorg. Med. Chem. 18 (2010) 1711−1723.

[59] M. Rodriguez, P. Dubreuil, J.P. Bali, J. Martinez, Synthesis and biological activity of partially modified retro-inverso pseudopeptide derivatives of the C-terminal tetrapeptide of gastrin, J. Med. Chem. 30 (1987) 758−763.

[60] A.S. Verdini, S. Silvestri, C. Becherucci, M.G. Longobardi, L. Parente, S. Peppoloni, M. Perretti, P. Pileri, M. Pinori, G.C. Viscomi, L. Nencioni, Immunostimulation by a partially modified retro-inverso-tuftsin analogue containing Thr1 psi[NHCO](R,S) Lys2 modification, J. Med. Chem. 34 (1991) 3372−3379.

[61] J. Gante, Azapeptides, Synthesis 6 (1989) 405−413.

[62] T.L. Graybill, M.J. Ross, B.R. Gauvin, J.S. Gregory, A.L. Harris, M.A. Ator, J.M. Rinker, R.E. Dolle, Synthesis and evaluation of azapeptide-derived inhibitors of serine and cysteine proteases, Bioorg. Med. Chem. Lett. 2 (1992) 1375−1380.

[63] J. Magrath, R.H. Abeles, Cysteine protease inhibition by azapeptide esters, J. Med. Chem. 35 (1992) 4279−4283.

[64] R.J. Simon, R.S. Kania, R.N. Zuckermann, V.D. Huebner, D.A. Jewell, S. Banville, S. Ng, L. Wang, S. Rosenberg, C.K. Marlowe, D.C. Spellmeyer, R. Tan, A.D. Frankel, D.V. Santi, F.E. Cohen, P.A. Bartlett, Peptoids: a modular approach to drug discovery, Proc. Natl. Acad. Sci. U.S.A. 89 (1992) 9367−9371.

[65] V.V. Rostovtsev, L.G. Green, V.V. Fokin, K.B. Sharpless, A stepwise huisgen cycloaddition process: copper(I)-catalyzed regioselective "ligation" of azides and terminal alkynes, Angew. Chem. Int. Ed. 41 (2002) 2596−2599.

[66] C.W. Tornøe, C. Christensen, M. Meldal, Peptidotriazoles on solid phase: [1,2,3]-triazoles by regiospecific copper(I)-catalyzed 1,3-dipolar cycloadditions of terminal alkynes to azides, J. Org. Chem. 67 (2002) 3057−3064.

[67] H.C. Kolb, K.B. Sharpless, The growing impact of click chemistry on drug discovery, Drug Discov. Today 8 (2003) 1128−1137.

[68] G.C. Tron, T. Pirali, L.A. Billington, P.L. Canonico, G. Sorba, A.A. Genazzani, Click chemistry reactions in medicinal chemistry: applications of the 1,3-dipolar cycloaddition between azides and alkynes, Med. Res. Rev. 28 (2) (2008) 278−308.

[69] F. Himo, T. Lovell, R. Hilgraf, V.V. Rostovtsev, L. Noodleman, K.B. Sharpless, V.V. Fokin, Copper(I)-catalyzed synthesis of azoles. DFT study predicts unprecedented reactivity and intermediates, J. Am. Chem. Soc. 127 (2005) 210−216.

[70] V.D. Bock, H. Hiemstra, J.H. van Maarseveen, CuI-Catalyzed alkyne−azide "click" cycloadditions from a mechanistic and synthetic perspective, Eur. J. Org. Chem. (2006) 51−68.

[71] A.D. Cunningham, N. Qvit, D. Mochly-Rosen, Peptides and peptidomimetics as regulators of protein-protein interactions, Curr. Opin. Struct. Biol. 44 (2017) 59−66.

[72] B.P. Orner, J.T. Ernst, A.D. Hamilton, Toward proteomimetics: terphenyl derivatives as structural and functional mimics of extended regions of an alpha-helix, J. Am. Chem. Soc. 123 (2001) 5382−5383.

[73] J.T. Ernst, J. Becerril, H.S. Park, H. Yin, A.D. Hamilton, Design and application of an alpha-helix-mimetic scaffold based on an oligoamide-foldamer strategy: antagonism of the Bak BH3/Bcl-xL complex, Angew. Chem. Int. Ed. 42 (2003) 535−539.

[74] A. Shaginian, L.R. Whitby, S. Hong, I. Hwang, B. Farooqi, M. Searcey, J. Chen, P.K. Vogt, D.L. Boger, Design, synthesis, and evaluation of an alpha-helix mimetic library targeting protein-protein interactions, J. Am. Chem. Soc. 131 (2009) 5564−5572.

[75] P. Maity, B. Koenig, Synthesis and structure of 1,4-dipiperazino benzenes: chiral terphenyl-type peptide helix mimetics, Org. Lett. 10 (2008) 1473−1476.

[76] A. Trabocchi, A. Guarna, Peptidomimetics in Organic and Medicinal Chemistry: The Art of Transforming Peptides in Drugs, John Wiley & Sons, UK, 2014 (Chapter 1).

[77] J.D. Harper, P.T. Lansbury, Models of amyloid seeding in Alzheimer's disease and scrapie: mechanistic truths and physiological consequences of the time-dependent solubility of amyloid proteins, Annu. Rev. Biochem. 66 (1997) 385–407.

[78] T. Moriuchi, T. Hirao, Highly ordered structures of peptides by using molecular scaffolds, Chem. Soc. Rev. 33 (2004) 294–301.

[79] J.S. Nowick, D.L. Holmes, G. Mackin, G. Noronha, A.J. Shaka, E.M. Smith, An artificial β-sheet comprising a molecular scaffold, a β-strand mimic, and a peptide strand, J. Am. Chem. Soc. 118 (1996) 2764–2765.

[80] J.S. Nowick, J.H. Tsai, Q.-C.D. Bui, S. Maitra, A chemical model of a protein β-sheet dimer, J. Am. Chem. Soc. 121 (1999) 8409–8410.

[81] C. Bonauer, M. Zabel, B. König, Synthesis and peptide binding properties of methoxypyrrole amino acids (MOPAS), Org. Lett. 6 (2004) 1349–1352.

[82] R.V. Nair, S.B. Baravkar, T.S. Ingole, G.J. Sanjayan, Synthetic turn mimetics and hairpin nucleators: Quo Vadimus? Chem. Commun. 50 (90) (2014) 13874–13884.

[83] I. Schechter, A. Berger, On the size of the active site in proteases. I. Papain, Biochem. Biophys. Res. Commun. 27 (1967) 157–162.

[84] D.H. Rich, M.G. Bursavich, M.A. Estiarte, Discovery of nonpeptide, peptidomimetic peptidase inhibitors that target alternate enzyme active site conformations, Biopolymers 66 (2002) 115–125.

[85] D. Leung, G. Abbenante, D.P. Fairlie, Protease inhibitors: current status and future prospects, J. Med. Chem. 43 (2000) 305–341.

[86] J.P. Vacca, B.D. Dorsey, W.A. Schleif, R.B. Levin, S.L. McDaniel, P.L. Darke, J. Zugay, J.C. Quintero, O.M. Blahy, E. Roth, V.V. Sardana, A.J. Schlabach, P.I. Graham, J.H. Condra, L. Gotlib, M.K. Holloway, J. Lin, I.-W. Chen, K. Vastag, D. Ostovic, P.S. Anderson, E.A. Emini, J.R. Huff, L.-735,524: an orally bioavailable human immunodeficiency virus type 1 protease inhibitor, Proc. Natl. Acad. Sci. U.S.A. 91 (1994) 4096–4100.

[87] D.J. Kempf, K.C. Marsh, J.F. Denissen, E. McDonald, S. Vasavanonda, C.A. Flentge, B.E. Green, L. Fino, C.H. Park, X.-P. Kong, N.E. Wideburg, A. Saldivar, L. Ruiz, W.M. Kati, H.L. Sham, T. Robins, K.D. Stewart, A. Hsu, J.J. Plattner, J.M. Leonard, D.W. Norbeck, ABT-538 is a potent inhibitor of human immunodeficiency virus protease and has high oral bioavailability in humans, Proc. Natl. Acad. U.S.A. 92 (1995) 2484–2488.

[88] A.K. Patick, H. Mo, M. Markowitz, K. Appelt, B. Wu, L. Musick, V. Kalish, S. Kaldor, S. Reich, D. Ho, S. Webber, Antiviral and resistance studies of AG1343, an orally bioavailable inhibitor of human immunodeficiency virus protease, Antimicrob. Agents Chemother. 40 (1996) 292–297.

[89] B.S. Robinson, K.A. Riccardi, Y.F. Gong, Q. Guo, D.A. Stock, W.S. Blair, B.J. Terry, C.A. Deminie, F. Djang, R.J. Colonno, P.F. Lin, BMS-232632, a highly potent human immunodeficiency virus protease inhibitor that can be used in combination with other available antiretroviral agents, Antimicrob. Agents Chemoter. 44 (2000) 2093–2099.

[90] Y. Koh, H. Nakata, K. Maeda, H. Ogata, G. Bilcer, T. Devasamudram, J.F. Kincaid, P. Boross, Y.F. Wang, Y. Tie, P. Volarath, L. Gaddis, R.W. Harrison, I.T. Weber, A.K. Ghosh, H. Mitsuya, Novel bis-tetrahydrofuranylurethane-containing nonpeptidic protease inhibitor (PI) UIC-94017 (TMC114) with potent activity against multi-PI-

resistant human immunodeficiency virus in vitro, Antimicrob. Agents Chemother. 47 (2003) 3123—3129.

[91] A. Mastrolorenzo, S. Rusconi, A. Scozzafava, G. Barbaro, C.T. Supuran, Inhibitors of HIV-1 protease: current state of the art 10 years after their introduction. From antiretro-viral drugs to antifungal, antibacterial and antitumor agents based on aspartic protease inhibitors, Curr. Med. Chem. 14 (2007) 2734—2748.

[92] A.K. Ghosh, Z.L. Dawson, H. Mitsuya, Darunavir, a conceptually new HIV-1 protease inhibitor for the treatment of drug-resistant HIV, Bioorg. Med. Chem. 15 (2007) 7576—7580.

sp²-Iminosugars as chemical mimics for glycodrug design

Elena M. Sánchez-Fernández[1], M. Isabel García-Moreno[1], José M. García Fernández[2], Carmen Ortiz Mellet[1]

University of Seville, Seville, Spain[1]; Institute for Chemical Research (IIQ), CSIC — University of Seville, Seville, Spain[2]

7.1 Introduction

Glycoproteins and glycolipids, collectively known as glycoconjugates [1], and oligosaccharides are vital biomolecules that control essential biological and pathological events, including cellular recognition and adhesion, cell growth and cell differentiation [2], fertilization, inflammation, the immune response, cancer metastasis, or viral and bacterial infection [3]. Their biosynthesis and degradation is catalyzed by glycosyl transferases [4] and glycosyl hydrolases (glycosidases) [5], respectively. Not surprisingly, regulators of these enzymes, either inhibitors or activators, are highly sought to advance our knowledge on the biological routes where glycosidases and glycosyltransferases are involved in and as drug candidates for conditions resulting from their dysfunction. In the last decades, iminosugars, sugar mimics where the endocyclic oxygen typical of monosaccharides has been replaced by a nitrogen atom, have flourished as the most popular family of modulators of carbohydrate processing enzymes [6], enabling, for instance, altering the glycosylation profile of eukaryotic cells, interfering in the metabolism of glycoconjugates and carbohydrates, modifying the carbohydrate-dependent properties of glycoproteins or blocking the carbohydrate-mediated interaction of host cells with infective agents [7].

Owing to their interesting biological properties, iminosugars have been explored as a new and innovative source of therapeutic drugs in the treatment of a wide range of diseases [8]. Thus, miglitol (Glyset) (Fig. 7.1), a potent inhibitor of the α-glucosidases in the digestive tract, was licensed in 1996 as the first iminosugar-based drug for diabetic patients [9]. Outstanding results have also been obtained with various glycomimetics based on iminosugars in the field of lysosomal storage disorders (LSDs), a heterogeneous group of rare diseases characterized by a genetic defect in the genes that encode specific lysosomal enzymes, leading to the accumulation of complex nonmetabolized molecules in the lysosome. The therapeutic potential of small organic molecules against such genetic diseases is extremely attractive, especially when a neurological deterioration is revealed. In this sense, miglustat,

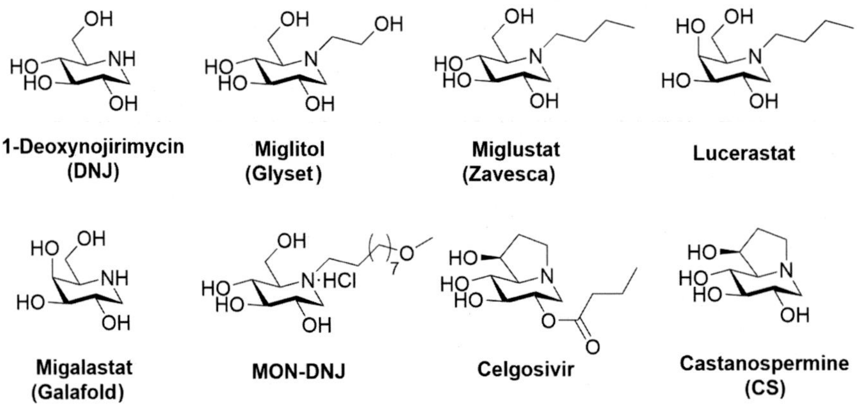

FIGURE 7.1

Chemical structures of some classical iminosugars.

N-butyl-deoxynojirimycin (OGT918, Zavesca) (Fig. 7.1), an inhibitor of the first step in the biosynthesis of many glycosphingolipids, was licensed in Europe (2002) and United States (2003) for substrate reduction therapy against nonneuronopathic Gaucher and Niemann-Pick C diseases [10]. Lucerastat, its analogue with D-*galacto* configuration, is also being tested in a Phase I trial for its potential clinical usc in paticnts suffering from Fabry disease [11]. In 2016, migalastat (Galafold, Amicus Therapeutics) obtained marketing approval by the European Medicine Agency as the first pharmacological chaperone-based therapy against this disease [12]. Currently, MON-DNJ and celgosivir, modified compounds from the naturally occurring iminosugars 1-deoxynojirimycin (DNJ) and castanospermine (CS) (Fig. 7.1), respectively, are being evaluated in clinical trials against dengue, hepatitis C, and the human immunodeficiency virus [13−15].

The therapeutic potential exhibited by the iminosugars has fueled huge synthetic efforts aimed at exploring the impact of chemical transformations in the inhibition potency and selectivity against different enzymes. Structural modifications that propitiate a substantial change in the hybridization of the endocyclic nitrogen atom, from sp³ (pyramidal) to sp² (planar), have led to different derivatives with interesting biological activity, such as glycolactams and glycoamidines [16,17] (Fig. 7.2). Among the lactams, ND2001, a potent and competitive β-D-glucuronidase inhibitor, behaves as an antimetastatic agent in a pulmonary metastatic model of mouse B16 melanoma [18]. More recently, *N*-arylated-lactam-type iminosugars have been described as potent immunosuppressive agents, highlighting the *N*-biphenyl-lactam **1** (Fig. 7.2). This D-*galacto*-type iminosugar exhibited a very promising effect on the inhibition of the proliferation of mouse splenocytes, displaying a half maximal inhibitory concentration (IC$_{50}$) value of 6.94 μM [19]. In turn, isofagomidine (Fig. 7.2), an isofagomine analogue bearing an amidine group at the

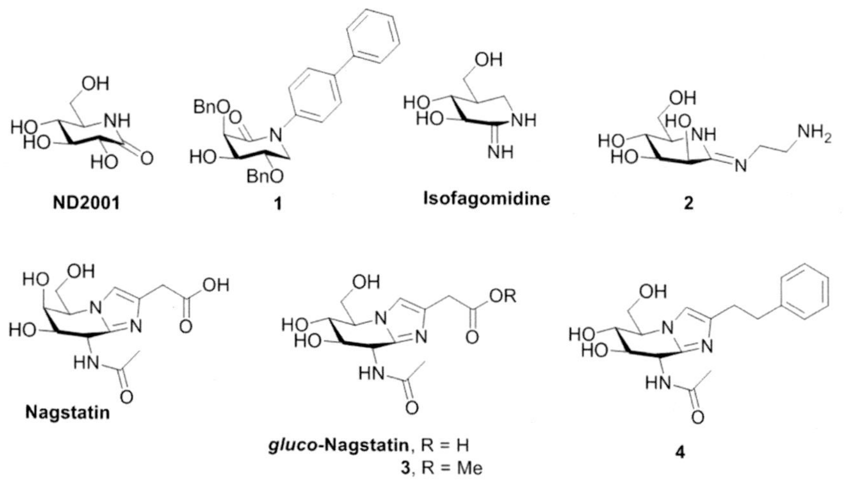

FIGURE 7.2

Chemical structures of glycolactams and glycoamidines bearing a sp^2-hybridized carbon in the piperidine ring.

pseudoanomeric position [20], behaves as a potent and selective inhibitor of Jack bean α-mannosidase, with an inhibition constant value (K_i) in the submicromolar range ($K_i = 0.75$ μM). Its structurally related analogue mannoamidine **2** (Fig. 7.2), bearing an ethylamine exocyclic substituent, significantly decreased the K_i value up to two orders of magnitude against either α- or β-mannosidase, reaching the low nanomolar range ($K_i = 6-9$ nM) [21]. The different charge distribution of these glycoamidines with regard to classical iminosugars has provided insights into the mechanism of enzymatic glycoside hydrolysis.

Within the glycoamidine-type structure, derivatives that incorporate a fused heterocyclic ring have proven very efficient at inhibiting β-hexosaminidases, a topic of great interest since dysregulation of N-acetylglucosamine levels is associated with a range of human diseases, such as cancer, diabetes, and neurodegeneration [22]. The naturally occurring glycoimidazole nagstatin (Fig. 7.2) has been known for a long time as a potent competitive and specific inhibitor of β-N-acetylglucosaminidase (IC$_{50}$ = 13.4 nM, bovine kidney) [23]. As expected, the related *gluco*-configured analogues *gluco*-nagstatin and its methyl ester **3** (Fig. 7.2) have been reported as more potent inhibitors than nagstatin, up to one order of magnitude [24]. Likewise, replacement of the carboxylic acid by a phenylmethyl moiety in the bicyclic glucoamidine scaffold (**4**, Fig. 7.2) resulted in a potent inhibitor of the human O-GlcNAcase enzyme, with K_i in the subnanomolar range (0.42 nM) [25].

Kifunensine, a natural product incorporating a cyclic oxamide (Fig. 7.3), was identified more than 3 decades ago as inhibitor of Jack bean α-mannosidase (IC$_{50}$ = 0.12 μM) [26]. Subsequent studies have allowed to confirm its inhibitory activity against the human endoplasmic reticulum (ER) α-1,2-mannosidase and mouse

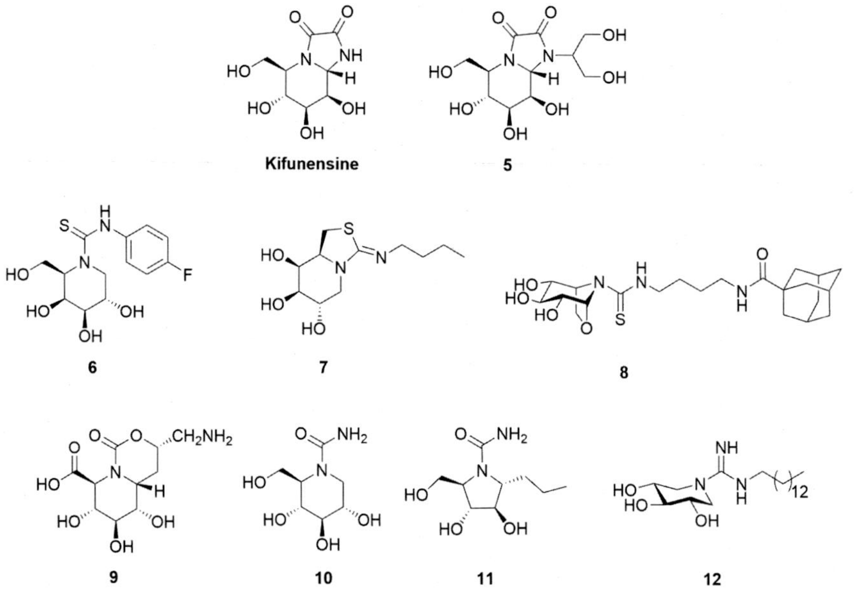

FIGURE 7.3

Chemical structures of iminosugar-type derivatives incorporating an exocyclic pseudoamide functionality.

Golgi α-mannosidase IA (IC$_{50}$ < 30 μM, in both cases) [27]. Although none of its synthetic analogues featuring different substituents at the N-1 position led to improved derivatives in terms of potency, it is worth pointing out that the bicyclic oxamide **5** (Fig. 7.3), bearing an N-bis(hydroxymethyl)methyl fragment, displayed remarkable selectivity against the human ER mannosidase [27].

In the late 1990s, Ortiz Mellet and García Fernández conceived a novel approach for the synthesis of glycomimetics sharing a pseudoamide-type functionality at the analogous position to the endocyclic oxygen of glycopyranosides. Briefly, they exploited the susceptibility of the masked aldehyde group of monosaccharides to undergo intramolecular nucleophilic addition by a nitrogen center in a carbamate, thiocarbamate, urea, thiourea, isourea, isothiourea, or guanidine functionality [28,29]. Differently from the classical iminosugars (sp³-hybridized) or from the previous examples of sp²-hybridized analogues, which cannot reproduce the chemical features and reactivity of the anomeric center in carbohydrates and their conjugates, the new compounds, for which the authors coined the term sp²-iminosugars, bear a hemiaminal function that emulates the hemiacetal group of reducing sugars [30]. The anomeric hydroxyl is stabilized by a very efficient hyperconjugation between the p orbital housing the electron pair of the endocyclic N-atom and the antibonding orbital of the contiguous pseudoanomeric bond, which fixes the axial orientation, in agreement with the anomeric effect (Fig. 7.4). This unique characteristic empowers sp²-iminosugars with the capability to participate in glycosylation reactions, the iconic

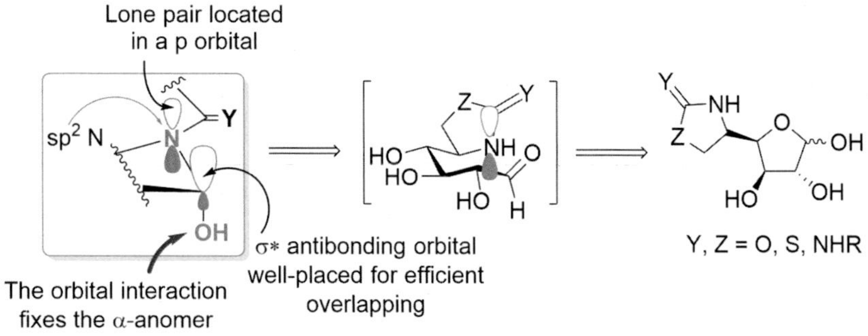

FIGURE 7.4

Retrosynthetic scheme for reducing bicyclic sp^2-iminosugars.

transformation of sugars in biology. In other words, sp^2-iminosugars are structural mimics of monosaccharides that replicate not only the glycosylation profile of their natural monosaccharides, but also their reactivity: they can be considered as "carbo-hydrate chemical mimics."

1-Deoxy (nonreducing) iminosugars have also been transformed into 1-deoxy-sp^2-iminosugars by standard chemical transformations, e.g., reaction with isocyanates or isothiocyanates. Although the possibility of further elaboration at the anomeric position is lost, the transformation of the amino group of iminosugars into pseudoamide functionalities permits the incorporation of an extremely broad diversity of substituents, which has been exploited to optimize the affinity toward complementary glycosidases. For instance, several 1-deoxy-sp^2-iminosugars incorporating thioureas and isothioureas reported by Higaki, Ortiz Mellet, García Fernández, and coworkers are under investigational or preclinical development as pharmacological chaperones for LSDs such as Fabry (**6**) [31], G_{M1}-gangliosidosis (**7**) [32] and Gaucher diseases (**8**) [33] (Fig. 7.3) [34]. Additional examples came from the laboratories of Nicotra and coworkers, who synthesized a variety of nonnatural nojirimycin-derived iminosugars with carbamate (e.g., **9**, Fig. 7.3) [35] or urea functionalities (e.g., **10** and **11**, Fig. 7.3) [36] as selective inhibitors of yeast α-glucosidase or insect trehalase, and Pieters and coworkers where the incorporation of an exocyclic N-alkylated guanidinium moiety provided a set of stable iminosugars as selective inhibitors of the human β-glucocerebrosidase (e.g., **12**, Fig. 7.3) [37,38]. Nevertheless, it is the capability of the sp^2-iminosugars to engage in conjugation strategies, thereby affording glycoconjugate mimetics, that makes their idiosyncrasy. In the last years, several approaches to access sp^2-iminosugar conjugates have been explored and remarkable biological properties have been reported. This chapter focuses specifically on the synthesis and biological activities displayed by sp^2-iminosugar-type glycoconjugate mimetics bearing O-, S-, N-, or C-anomeric aglycone groups, with an emphasis on the relationships between their chemical structure and their potential as drug leads targeting cancer, parasitic diseases, and inflammatory disorders.

7.2 sp^2-Iminosugars as antiproliferative and antimetastatic agents

In 2010, Ortiz Mellet, Ouadid-Ahidouch, García Fernández, and coworkers first reported the synthesis of a new family of nojirimycin-related bicyclic sp^2-iminosugar prototypes with *N*-, *S*-, and *C*-pseudoglycoside structure (**17**, **19** and **21**, Scheme 7.1). The pseudoanomeric substituent consisted in all cases of an *n*-octyl chain in axial orientation, analogous to that found in the aglycones in the natural α-D-gluco-pyranosides. As a result, **17**, **19** and **21** possess amphiphilic character and can be considered as glycolipid mimics (sp^2-glycolipids, sp^2-GLs). They were initially conceived as potential inhibitors of the ER α-glucosidases I and II involved in glyco-protein biosynthesis, a process that is deregulated in cancer cells. Interestingly, they exhibited potent and specific in vitro activity against breast cancer cell lines [39].

The authors developed efficient methodologies for the conjugation of the pivotal precursor (1*R*)-5*N*,6*O*-oxomethylidenenojirimycin (**14**), obtained by a sequence of nine steps from commercially available D-glucofuranurone-6,3-lactone (**13**) [40], with the appropriate glycosyl acceptor (Scheme 7.1). In the case of the *N*-pseudogly-coside derivative **17**, they adapted a procedure previously used to generate the *gem*-diamine anomeric functional group in the preparation of *N*-linked sp^2-pseudo-disaccharides [41]. It consisted in the straightforward coupling of the reducing carbamate **14** with *n*-octylamine in methanol at 65°C. For the *S*-glycosylation reaction, peracetate **15** was used as the glycosyl donor. The reaction with octanethiol in dichloromethane (DCM), using boron trifluoride-diethyl ether (BF$_3$OEt$_2$) as promo-tor, afforded the corresponding sulfide **18** that was deacetylated using sodium methoxide (NaOMe) to give the target compound **19**. *C*-Glycosylation requested the preparation of the α-pseudoglycosyl fluoride intermediate **16**, which was next coupled with the organometallic reagent tri-*n*-octylaluminum to achieve **20** and

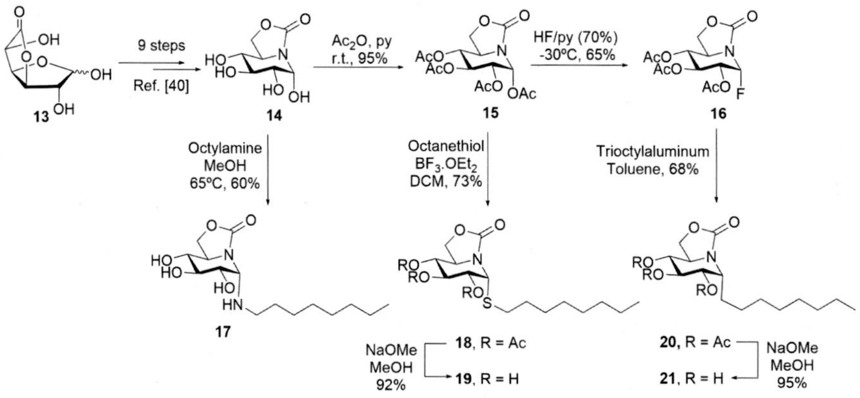

SCHEME 7.1

Stereoselective synthesis of the *N*-, *S*- and *C*-octyl pseudo-α-glycosides 17, 19, and 21, with hydroxylation profiles analogous to D-glucose.

deprotected to afford the *C*-octylglycoside **21** (Scheme 7.1). In all cases, the coupling constants supported the formation of the α-anomer in a $^4C_1$ chair conformation as the single or the main diastereomer, underlining the major influence of the orbital contribution to the anomeric effect in these sp^2-glycomimetics.

Structure-activity relationship (SAR) studies revealed that the glycosidic linkage plays a relevant role in the antiproliferative potential of these amphiphilic sp^2-iminosugars when tested in the noninvasive (MCF-7) and invasive (MDA-MB-231) human model breast cancer cell lines. Both the *S*- and *C*-glycosides (**19** and **21**) achieved remarkable IC$_{50}$ values, in the micromolar range (Table 7.1). It is worth pointing out that none of the octyl sp^2-α-pseudoglycosides affected the viability of normal breast cells (MCF-10A).

The antiproliferative activity of (1*R*)-1-octylthio-5*N*,6*O*-oxomethylidenenojirimycin (**19**) and (1*R*)-1-octyl-5*N*,6*O*-oxomethylidene-1-deoxynojirimycin (**21**) against cells with uncontrolled growth rates results from apoptosis induction, as indicated by the increase in the ratio between the proapoptotic protein Bax versus the antiapoptotic protein Bcl-2 (Bax/Bcl-2 ratio > 1) [42]. Flow cytometric assays indicated the accumulation of breast tumor cells at the G0/G1 and G2/M phases of the cell cycle. In full agreement, Western blotting analysis showed alterations in the expression of some regulatory proteins closely related to the mentioned checkpoints, such as cyclins (D1 and E), and cyclin-dependent kinases (CDK4 and CDK1). An in-depth investigation of the effect of the sp^2-α-*C*-pseudo-octyl glycoside **21** on metastatic processes also evidenced a significant decrease in the migrating malignant cells that was much more accentuated in the invasive MDA-MB-231 tumor line (60% vs. 38% for MCF-7, Fig. 7.5A–F) [43].

The anti-migratory effect exerted by **21** on malignant cells seems to be orchestrated by inhibition of the expression of β1-integrin, a protein overexpressed in breast cancer. This is accompanied by a decrease of the intracellular calcium concentration as a result of the alteration of Stim 1 expression, a glycoprotein implicated in store-operated calcium entry (SOCE) [43]. All these findings make the nojirimycin-derived bicyclic *S*- and *C*-octyl sp^2-GLs **19** and **21** promising chemotherapeutic candidates against breast carcinoma.

Table 7.1 Antiproliferative activity (IC$_{50}$ values, μM) of *N*-, *S*- and *C*-octyl nojirimycin-related derivatives (**17**, **19**, and **21**) against different human breast cell lines.

sp^2-GL	Breast cell lines		
	MCF-10A (normal)	**MCF-7 (non-invasive)**	**MDA-MB-231 (invasive)**
17 (*N*)	>100	120	—
19 (*S*)	>100	37	35
21 (*C*)	>100	26	44

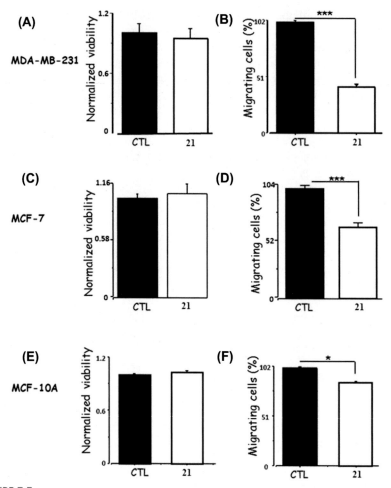

FIGURE 7.5

Effect of (1*R*)-1-octyl-5*N*,6*O*-oxomethylidene-1-deoxynojirimycin (21) on the reduction of migrating cells: (A), (C) and (E): Effect of 21 on cells viability MDA-MB-231, MCF-7 and MCF-10A, respectively, (B) invasive breast tumor line (MDA-MB-231), (D) non-invasive breast tumor line (MCF-7), (F) normal breast cell line (MCF-10A).

Reprinted (adapted) with permission from J. Cell Physiol. 232 (12) (2017) 3631–3640. Copyright (2017) John Wiley and Sons.

The three sp²-glycomimetics **17**, **19** and **21**, with α-D-*gluco* configurational profile behaved as potent and specific competitive inhibitors of the commercial enzymes neutral yeast α-glucosidase and yeast isomaltase, with K_i values in the low micromolar and submicromolar range (0.54—7.2 μM). The inhibition potency against different β-glucosidases (almonds, bovine liver) was up to two or three orders of magnitude lower ($K_i = 21-682$ μM) [39]. These data were consistent with the initial hypothesis of a mechanism of action implying inhibition of ER α-glucosidases. However, further experimental assays based on the detection of ER-free oligosaccharides [44] were not consistent with this assumption. To deepen in this question, the synthesis and biological evaluation of a wider series of N-, S-, C- and O-glycoside sp^2-GLs that incorporated compounds with D-*manno* (**22–25**) and D-*galacto* (**28–31**) hydroxylation patterns was undertaken [45] (Scheme 7.2). The methodologies previously optimized for the D-*gluco* epimers (**17, 19, 21**) were applied for the preparation of mannojirimycin (**22–24**) and galactonojirimycin N-, S-, C-octyl pseudoglycosides (**28–30**). The requested reducing carbamate precursors **26** and **32** were obtained from commercially available L-gulonic acid γ-lactone (**27**) and D-galactose (**33**), respectively [30] (Scheme 7.2).

The synthesis of O-octyl pseudoglycosides (**36, 25** and **31**) was accomplished by glycosylation of *n*-octanol, using the corresponding peracetylated α-sp^2-iminosugar glycosyl fluoride (**16**, *gluco*; **34**, *manno*; **35**, *galacto*) as the glycosyl donor and BF$_3$OEt$_2$ as promotor [45] (Scheme 7.3), following the approach reported for the synthesis of methyl isomaltoside and maltoside sp^2-iminosugar mimics [46].

The crystal structure of the only product isolated from the reaction crude obtained by heating the reducing mannojirimycin derivative **26** with *n*-octylamine in methanol (Fig. 7.6), allows to confirm its *gem*-diamine structure **22** [47] and the

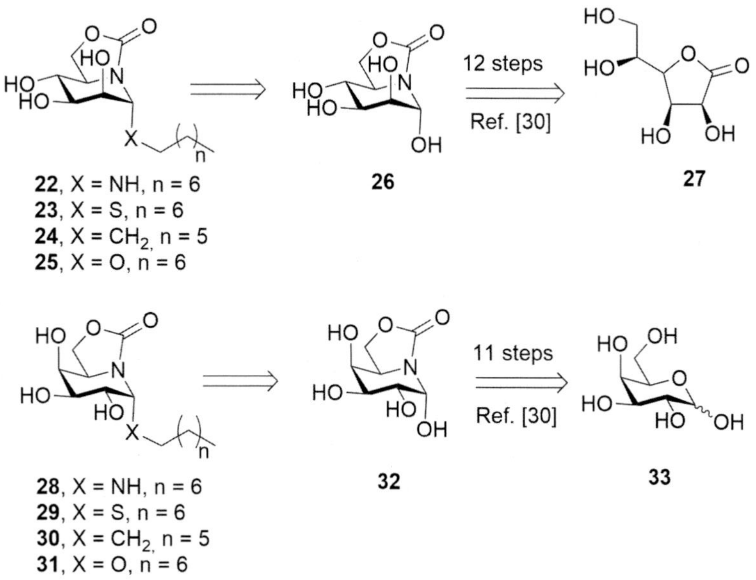

22, X = NH, n = 6
23, X = S, n = 6
24, X = CH$_2$, n = 5
25, X = O, n = 6

26

27

28, X = NH, n = 6
29, X = S, n = 6
30, X = CH$_2$, n = 5
31, X = O, n = 6

32

33

SCHEME 7.2

sp^2-Iminosugar N-, S-, C-, and O-octyl pseudo-α-glycosides with analogous hydroxylation profile to D-mannose (**22–25**) and D-galactose (**28–31**).

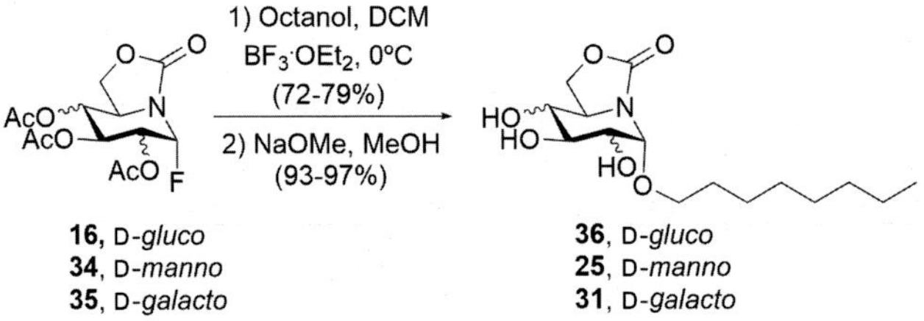

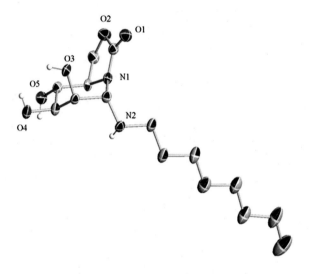

SCHEME 7.3

Representative synthetic procedure for the stereoselective synthesis of sp²-iminosugar *O*-octyl pseudo-α-glycosides (36, 25, 31) with hydroxylation profiles analogous to D-glucose, D-mannose, and D-galactose.

FIGURE 7.6

Molecular structure of (1*S*)-1-octylamino-5*N*,6*O*-oxomethylidene-1-deoxymannojirimycin (22), ORTEP (Oak Ridge Thermal Ellipsoid Plot) representation.

Reprinted with permission from J. Org. Chem. 79 (23) (2014) 11722–11728. Copyright (2014) American Chemical Society.

$^4C_1$ chair conformation for the six-membered ring, as well as the (1*S*)-stereochemistry, i.e., the α-configuration, at the pseudoanomeric center, with the octylamino substituent axially arranged.

None of the sp²-glycomimetics with D-mannose and D-galactose hydroxylation profile behaved as inhibitor of neutral α-glucosidase or isomaltase. On the other hand, high-throughput screening (HTS) against a panel of cancer cells let identify derivatives with antiproliferative activities against six different human solid tumor cell lines, namely breast (HBL-100, T-47D), lung (A549, SW1573), cervix

(HeLa), and colon (WiDr), in both configurational series [45]. The *N*- and *S*-octyl-pseudo-α-glycosides related to mannojirimycin (**22** and **23**) and the *N*-octyl-galactonojirimycin analogue (**28**) were spotted as the most promising candidates, exhibiting growth inhibition values (GI$_{50}$) in the 30—90 μM range (Table 7.2).

The acetal function in the *O*-glycoside sp^2-GLs abolishes the tumor growth inhibitory activity in any of the three configurationally different series (**25, 31, 36**; GI$_{50}$ > 100 μM). In contrast, the *gem*-diamines derivatives with either *manno* or *gal-acto* configurations (**22** and **28**) turned to be rather efficient. The *gem*-diamine segment is also present in the natural iminosugar siastatin B, which exhibits potent antimetastatic activity as reported in 1994 by Nishimura and coworkers [48,49]. The therapeutic potential of siastatin B synthetic analogues has been expanded since then, being remarkable, for instance, the inhibitory activity against influenza virus infection [50].

The body of data above discussed evidences that the influence of the hydrophobic aglycone fragment and the nature of the glycosidic linkage in amphiphilic sp^2-glycomimetics are critical for the antitumor activity, since neither of the corresponding reducing precursors (D-*gluco* **14**, D-*manno* **26,** and D-*galacto* **32**) showed an effect on any of the human tumor cell lines tested in vitro (GI$_{50}$ > 100 μM). However, the results incite to discard a direct relationship with the inhibition of ER α-glucosidases. Further studies on the biomolecular targets and the molecular mechanisms at play are still necessary. All in all, extensive research during the last 10 years has revealed the promising ability of sp^2-iminosugars that incorporate a lipidic aglycone (sp^2-GLs) to reduce the proliferation of tumor cells by arresting the cell cycle, behaving as potent antimitotic, proapoptotic, and antimetastatic agents.

7.3 sp^2-Iminosugars in cancer immunotherapy

Immunotherapy has consolidated in recent years as a revolutionary strategy for cancer treatment, complementing the conventional surgery, radiotherapy and chemotherapy options. It is characterized by a high selectivity and with minimal toxicity for healthy cells of the host, thereby less side effects [51]. Vaccination with tumor antigens can, in principle, trigger the production of antibodies that specifically bind to tumor cells and elicit their annihilation by the immune system. One of the most studied proteins associated to the design of anticancer vaccines is the mucin MUC1, an *O*-glycoprotein overexpressed in different types of cancers, including breast, ovary, lung, colon, and pancreas [52]. The sequence formed by the amino acids proline-aspartate-threonine or serine-arginine (PDT[S]R) in MUC1 is recognized by a variety of monoclonal antibodies that are specific for tumor cells. The immunogenicity of this fragment is enhanced when the amino acid threonine or serine is glycosylated with the carbohydrate *N*-acetyl-D-galactosamine (GalNAc) [53]. This glycosylamino acid, known as Tn antigen (GalNAc-α-*O*-Ser/Thr) (Fig. 7.7), is indeed one of the most specific tumor-associated structures involved

Table 7.2 Antiproliferative activity (GI$_{50}$, µM) of N-, S- and C-octyl pseudo-α-glycosides related to nojirimycin (**19**, **21**), mannojirimycin (**22**, **23**), and galactonojirimycin (**28**) against several tumor cell lines.

| sp^2-GL | | Tumor cell lines | | | | | |
| | | Breast | | Lung | | Cervix | Colon |
		HBL-100	T-47D	A549	SW1573	HeLa	WiDr
Gluco	**19** (S)	73 ± 38	95 ± 7.2	58 ± 7.1	>100	92 ± 12	89 ± 15
	21 (C)	>100	91 ± 12	79 ± 30	>100	96 ± 5.9	>100
Manno	**22** (N)	43 ± 14	88 ± 21	91 ± 16	62 ± 23	68 ± 24	>100
	23 (S)	87 ± 18	84 ± 7.7	72 ± 16	>100	80 ± 6.6	>100
Galacto	**28** (N)	42 ± 12	47 ± 18	34 ± 1.6	50 ± 9.7	30 ± 2.1	44 ± 14

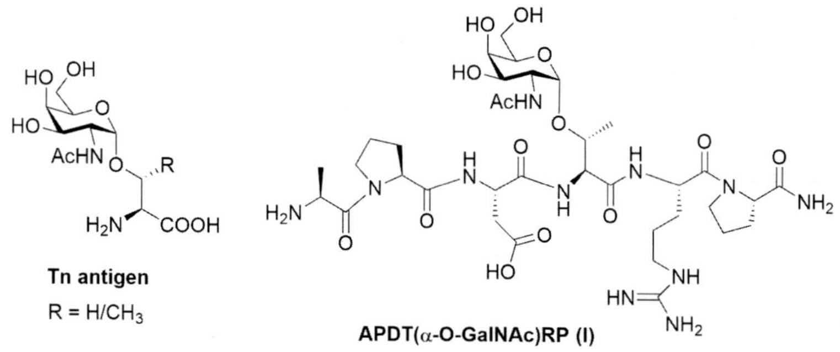

FIGURE 7.7

Chemical structures of the tumor associated carbohydrate antigen Tn and of the natural glycohexapeptide antigen APDT(α-O-GalNAc)RP (I).

in recognition processes of malignant cells. It is a cancer biomarker and an attractive target for the development of cancer vaccines [54].

In spite of the intense research conducted in the field, generating a powerful and selective immune response against tumor cells still remains an unmet goal. The main limitation is the insufficient immune response promoted by the natural glycopeptide antigenic motifs [55]. To address this issue, Ortiz Mellet, García Fernández, and Peregrina proposed the replacement of the GalNAc moiety in the Tn antigen by a bicyclic sp²-iminosugar that reproduces the substitution and stereochemical pattern of the parent monosaccharide [56]. The synthetic versatility of sp²-iminosugars, along with the high α-selectivity achieved in different glycosylation reactions, makes the approach feasible and very appealing for SAR studies. Thus, the authors prepared eight α-O-glycosylamino acid mimetics differing in the nature of sp²-iminosugar (Glc, Gal, GlcNAc, or GalNAc analogues) and the amino acid partner (Ser or Thr), namely sp²-Glc-α-O-Ser (**37**), sp²-Glc-α-O-Thr (**38**), sp²-Gal-α-O-Ser (**39**),

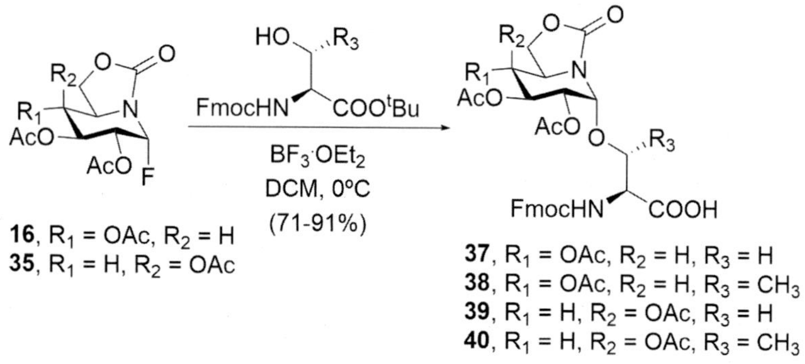

SCHEME 7.4

Preparation of sp²-iminosugar building blocks with D-*gluco* (**37, 38**) and D-*galacto* (**39, 40**) configurational profiles as Tn antigen mimics.

sp^2-Gal-α-O-Thr (**40**) (Scheme 7.4), and the corresponding 2-acetamido-2-deoxy-derivatives sp^2-GlcNAc-α-O-Ser (**52**), sp^2-GlcNAc-α-O-Thr (**53**), sp^2-GalNAc-α-O-Ser (**54**), and sp^2-GalNAc-α-O-Thr (**55**) (Scheme 7.5). The use of the 1-fluoro derivatives **16** or **35** as glycosyl donors in the O-glycosidation reactions of N-Fmoc-serine/threonine *tert*-butyl esters, promoted by BF$_3$·OEt$_2$, led exclusively to the required pseudo-α-O-glycosylamino acids with D-*gluco* (**37**, **38**) and D-*gal-acto* (**39**, **40**) configuration, respectively, avoiding tricky separations of mixtures of α- and β-anomers. Note that the acid medium elicits concomitant hydrolysis of the *tert*-butyl ester present in the Tn antigen mimic structures, affording the free carboxylic acids in one pot.

The incorporation of the acetamido group (NHAc) at C-2 in the sp^2-iminosugars was unprecedented and demanded the implementation of an appropriate synthetic strategy that employed the sp^2-D-glucal (**42**) and sp^2-D-galactal (**43**) as key intermediates (Scheme 7.5). Their preparation was achieved by adapting an efficient methodology reported by Schwartz and coworkers [57], based on the use of titanocene monochloride (Cp$_2$Ti^{III}Cl, obtained in turn, by reduction of titanocene dichloride Cp$_2$Ti^{IV}Cl$_2$ with an excess of manganese) to render protected glycals by reduction of glycosyl halides. Azidonitration of **42** and **43** with ceric ammonium nitrate (CAN) and sodium azide (NaN$_3$), following the method of Lemieux and Ratcliffe [58], was next attempted. Although the expected 2-azido-2-deoxyglycosyl nitrates could not be isolated, aqueous workup and subsequent acetylation delivered the corresponding O-acetylated 2-azido-2-deoxy derivatives **44** and **45**, which were activated by fluorination at the pseudoanomeric position (→**46** and **47**) and engaged

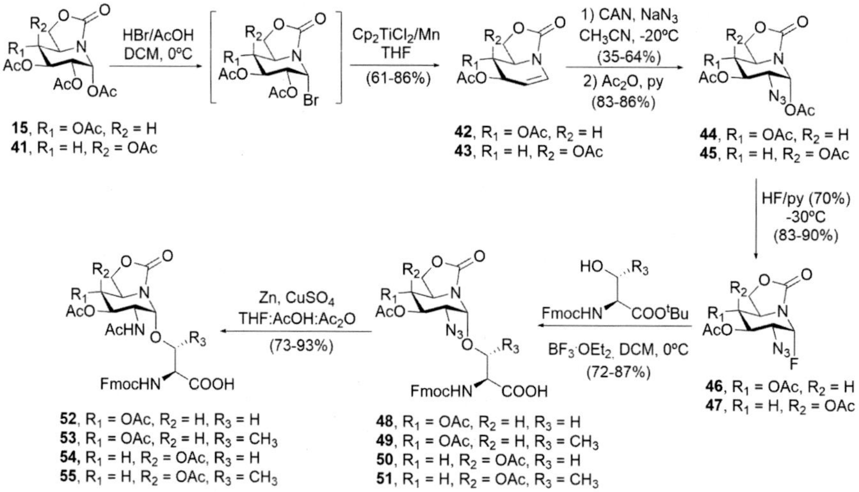

SCHEME 7.5

Preparation of pseudoglycopeptide building blocks bearing sp^2-iminosugar motifs with D-gluco (52, 53; GlcNAc) and D-galacto (54, 55; GalNAc) configurational profiles as Tn antigen mimics.

in glycosylation reactions (→**48**−**51**). Finally, treatment with Zn and CuSO$_4$ in a mixture of THF:AcOH:Ac$_2$O enabled the transformation of N$_3$ to NHAc in one pot procedure to afford the target pseudo-α-*O*-glycoamino acid Tn antigen mimics **52**−**55**.

Compounds **37**−**40** and **52**−**55** were synthesized at a relatively large scale and used as building blocks in the solid phase peptide synthesis (SPPS) of the glycosylated alanine-proline-aspartate-serine/threonine-arginine-proline (APDS/TRP) sequence recognized by anti-MUC1 antibodies (Fig. 7.8).

The dissociation constant (K_D) values of the novel nojirimycin- and galactonojirimycin-related sp²-α-*O*-glycopeptidomimetics **56**−**63** against the monoclonal antibody scFv-SM3 mAb, for which the potential in the diagnosis and treatment of breast cancer has been shown [59], were determined by bio-layer interferometry (BLI). The sp²-glycopeptidomimetics bearing the sp²-iminosugar

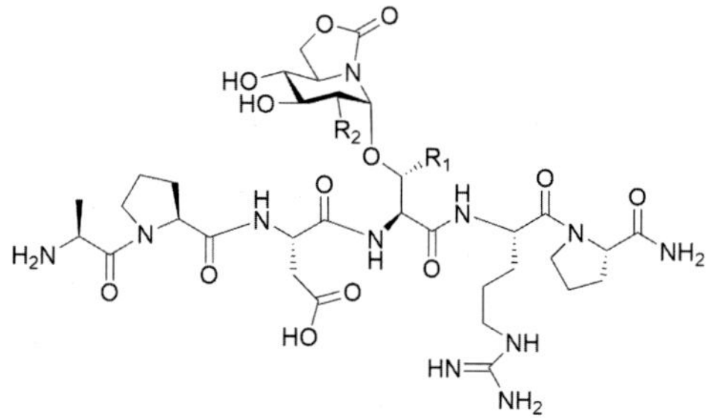

56, R$_1$ = H, R$_2$ = OH
57, R$_1$ = Me, R$_2$ = OH
58, R$_1$ = H, R$_2$ = NHAc
59, R$_1$ = Me, R$_2$ = NHAc

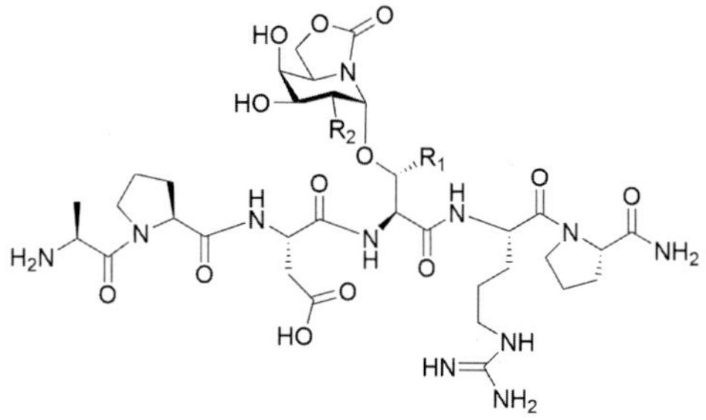

60, R$_1$ = H, R$_2$ = OH
61, R$_1$ = Me, R$_2$ = OH
62, R$_1$ = H, R$_2$ = NHAc
63, R$_1$ = Me, R$_2$ = NHAc

FIGURE 7.8

Chemical structure of sp²-α-*O*-glycohexapeptidomimetics containing a bicyclic sp²-iminosugar glycone motif with D-*gluco* (56−59) and D-*galacto* (60−63) stereochemistry and incorporating APDTRP and APDSRP sequences in the aglycone fragment.

moiety bound to the amino acid serine exhibited much lower affinity for the SM3 antigen than those incorporating threonine, irrespective of the sp^2-iminosugar stereochemistry. Gratifyingly, the best binding affinities have been achieved with APDT(α-sp²-GalNAc)RP (**63**, $K_D = 1.6 \times 10^{-6} \pm 1.6 \times 10^{-7}$ M) and its analogue with D-glucose profile APDT(α-sp²-GlcNAc)RP (**59**, $K_D = 2.0 \times 10^{-6} \pm 3.0 \times 10^{-7}$ M), which exhibited lower K_D values than the natural glycohexapeptide APDT(α-O-GalNAc)RP (**I**; Fig. 7.7), used as a positive control ($K_D = 3.3 \times 10^{-6} \pm 8.4 \times 10^{-7}$ M). The sp^2-glycohexapeptidomimetics that lacked the NHAc group showed much pronounced differences in binding affinity, the *galacto* configuration being strictly essential in this case for the efficient molecular recognition of the SM3 mAb (**61**, $K_D = 4.1 \times 10^{-6} \pm 1.1 \times 10^{-6}$ M).

In summary, the studies conducted on tumor-associated carbohydrate antigen sp^2-glycopeptidomimetics allowed the selection of APDT(α-sp²-GalNAc)RP (**63**) as a suitable candidate for its incorporation into cancer vaccines. Preclinical assays using rodent models are currently being conducted to evaluate the efficiency of anti-cancer vaccines obtained using **63** as the antigenic epitope.

7.4 sp²-Iminosugars as antileishmanial candidates

Cancer and parasitic diseases exhibit well-documented analogies that make, for instance, that several drugs show therapeutic benefit in both contexts [60–62]. Parasites and tumor cells share similar hallmarks related to the development of defense strategies, evasion of the immune surveillance as well as their capacity for fast cellular division, among others [63]. This is the reason why compounds that activate apoptotic pathways in malignant disorders frequently exhibit antiparasitic activity [64]. Hence, it is not striking that compounds approved and used as antitumor drugs, such as the alkylphospholipid analogue miltefosine, the antimetabolite 5-fluorouracil or the antimitotic diterpene taxol (Fig. 7.9), also display antileishmanial properties [65–67]. Indeed, different studies suggest an association between these two pathologies [68].

Leishmaniasis, a chronic infection caused by protozoan parasites of the genus *Leishmania*, is one of the most neglected diseases in spite of the fact that 12 million

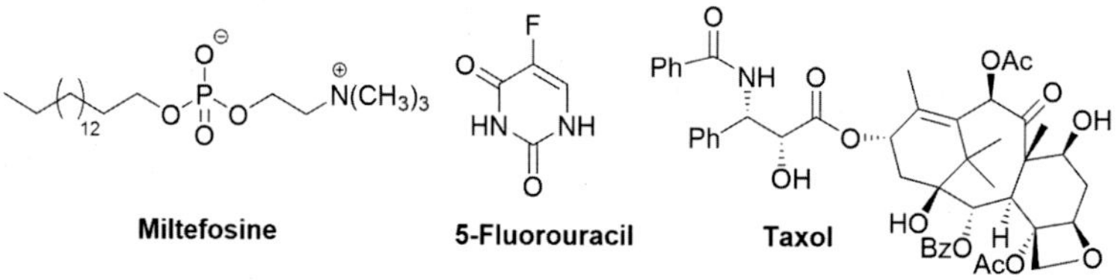

FIGURE 7.9

Chemical structures of drugs having antitumor and antileishmanial activities: miltefosine, 5-fluorouracil and taxol.

people are infected worldwide or at risk of infection, causing around 50,000 deaths per year. Global warming, environmental changes, socioeconomic factors, and population movements, among other factors, have favored an increase of the incidence with the spread of the disease at an alarming rate of 1.5-2 million new cases occurring annually, placing itself at the head of emerging diseases. Pharmacotherapy is the only fighting tool currently available against this disease, given that results about the use of prophylactic/therapeutic vaccines against any form of leishmaniasis are inconclusive [69]. However, the chemoresistance of the parasite has provoked a decrease of the effectiveness of conventional medication. The development of new and potent leishmanicidal drugs that complement the current available therapeutic strategies is, therefore, a priority.

The remarkable antitumor activity of amphiphilic sp²-iminosugars let envisage the possibility that these compounds could also bear potential as antiparasitic drugs. This hypothesis was substantiated in 2015, when the first sp²-glycomimetic derivatives with antileishmanial activity, namely the nojirimycin-related α-S-dodecyl pseudoglycoside **65**, the corresponding α-glycosylsulfoxides **72** (S_S), **73** (S_R), and the α-glycosylsulfone **77**, were reported by Ortiz Mellet, García Fernández, and Gamarro [70] (Scheme 7.6). These compounds were evaluated as inhibitors of the growth of intracellular amastigotes of *Leishmania donovani* Dd8, responsible for the visceral form of the disease, together with the homologous octyl pseudoglycosides (**19**, **70**, **71**, **76**), in the context of an SAR study (Scheme 7.6). The pseudoanomeric sulfinyl and sulfonyl groups were readily generated by oxidation of the sulfide group in the corresponding α-acetylated octyl(dodecyl)sulfides (**18**, **64**) by treatment with one or two equivalents of the oxidizing agent *m*-chloroperoxybenzoic acid (*m*CPBA), respectively. Final deacetylation reaction with NaOMe in methanol

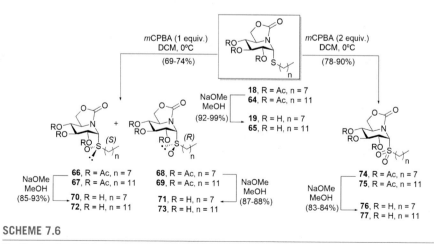

SCHEME 7.6

Synthesis of octyl(dodecyl) pseudothioglycosides (**19**, **65**) and the related sulfinyl (**70**—**73**) and sulfonyl (**76**, **77**) nojirimycin derivatives.

afforded the fully unprotected oxidized compounds **70, 71, 72, 73, 76,** and **77** (Scheme 7.6).

The length of the lipophilic chain in this series of sp²-GLs was found critical for the antiparasitic activity. Thus, the representatives bearing the dodecyl aglycone attached to the sp²-iminosugar moiety either by sulfide (**65**), sulfinyl (**72, 73**), or sulfonyl linkers (**77**), behaved as potent *Leishmania* growth inhibitors, displaying half maximal effective concentration values (EC_{50}) ranging from 10 to 33 μM (Fig. 7.10). By contrast, the partners equipped with an *n*-octyl chain were much less efficient: the sulfoxide and sulfone derivatives (**70, 71,** and **76**) did not show any significant effect ($EC_{50} > 100$ μM) and the pseudo *N*- and *S*-octyl glycosides (**17** and **19**; Scheme 7.1) were only moderately active (EC_{50} values 47 and 73 μM, respectively). This differential behavior may be ascribed to the superior cellular permeability conferred by the longer aliphatic chain. Determination of the EC_{50} values against human leukemia cell line THP-1 confirmed the relatively low cytotoxicity of these antiparasitic agents (Fig. 7.10).

All the assayed sp²-GLs in this study failed to inhibit the growth of flagellated extracellular forms of the parasite, the infective promastigote forms Dd8 ($EC_{50} > 100$ μM), regardless of the length of the aliphatic chain (octyl/dodecyl) or the nature of the glycosidic linkage (N, C, S, SO, SO₂). Altogether, the results highlight the favorable selectivity of action of the sp²-iminosugar glycolipids incorporating the longer dodecyl aliphatic chain toward the intracellular forms of the parasite. It is worth noting that the antileishmanial activity was not correlated to the neutral α-glucosidase inhibitory potency, which was higher for the octyl as compared to the dodecyl conjugates, similarly to that observed for the anticancer

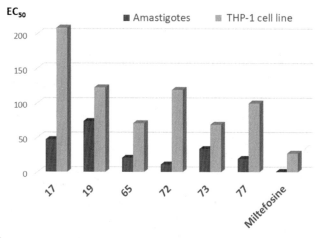

FIGURE 7.10

EC_{50} values (μM) against amastigotes of *Leishmania donovani* Dd8 and EC_{50} values (μM) for cytotoxicity (MTT colorimetric assay) against human leukemia cell line THP-1 for the sp²-GLs 17, 19, 65, 72, 73, and 77, compared to miltefosine.

activity. Alternative mechanisms pertaining the immune response probably operate in both cases.

The World Health Organization (WHO) recommends combined therapy approaches to deal with the drug resistance developed by the parasites over time and the side effects of conventional medication. In this line, the use of the S_S sulfinyl diastereomer **72**, the most potent antileishmanial candidate identified among the sp^2-GLs in terms of efficacy (EC$_{50}$ = 10.80 ± 0.27 μM) and safety (EC$_{50}$ in THP-1 leukemia cells = 118.83 ± 3.37 μM), in amalgamation with the antileishmanial reference drug miltefosine was investigated. As reflected in Fig. 7.11, the EC$_{50}$ value of **72** for amastigote growth inhibition decreased to about one order of magnitude, from 10.80 to 1.31 μM, in the presence of 0.3 μM of miltefosine. Significant synergistic interactions were determined by isobolographic analysis, supporting the effectiveness of this combination therapy in this family of sp^2-glycomimetics.

The findings achieved with this family of amphiphilic sp^2-iminosugars as antiparasitic agents are highly relevant since they not only illustrate the potential of the sp^2-GLs as prototypes for the development of new therapeutic strategies against leishmaniasis, but also justify the search for new pseudoglycolipid prototypes useful against other emerging/re-emerging diseases caused by parasites, bacteria, or viruses.

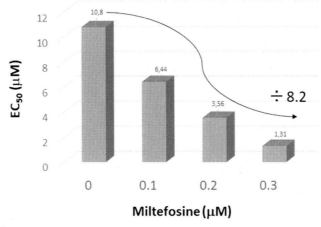

FIGURE 7.11

EC$_{50}$ values (μM) of (1R)-1-dodecylsulfinyl-5N,6O-oxomethylidenenojirimycin (72) against amastigotes in infected THP-1 cells as result of its combination treatment with miltefosine.

7.5 sp^2-Iminosugars and Inflammation

Since R. Virchow speculated more than 150 years ago on the relationship between cancer and inflammation [71], a strong body of evidences supports the use of anti-inflammatory therapies for the prevention and treatment of cancer. It is now well-established that a large number of tumors are triggered by pathological processes leading to chronic inflammation [72], which provides the appropriate environment for the transformation of normal cells into malignant cells.

The inflammatory pathway is characterized by two types of cellular responses, proinflammatory (M1) and anti-inflammatory (M2), activated by entirely different mediators. An imbalance of both responses may cause the dysregulation of the immune system [73]. Diabetic retinopathy (DR) [74], one of the main complications of diabetes, displays the perfect scenario of imbalance between pro- and anti-inflammatory mediators where the loss of vision coexists with a marked polarization toward the M1 proinflammatory state [75]. Motivated by the indications pointing to the role of the sp^2-GLs as modulators of the immune response, Ortiz Mellet, Arroba, and Valverde investigated their potential as anti-inflammatory agents in the context of DR. Very encouraging results were initially obtained with the pseudo-α-glycosides **73** [76] and **77** [77] in vitro assays using the microglia Bv.2 mouse cell line (immune cells located in the central nervous system) stimulated with bacterial lipopolysaccharide (LPS) as a model. Both sp^2-iminosugar glycolipids significantly decreased a number of M1 response markers, including nitrite (NO$_2^-$) accumulation, expression of inducible nitric oxide synthase (iNOS) and proinflammatory cytokines such as tumor necrosis factor-α (TNF-α) and interleukins IL1β and IL6. In parallel, anti-inflammatory M2 response mediator levels, such as the expression of arginase-1, the antioxidant enzyme heme oxygenase-1 (HO-1) or the powerful anti-inflammatory interleukin IL10, were greatly enhanced. Treatment of Bv.2 cells with the sp^2-GLs **73** or **77** further prevented the translocation of the proapototic and proinflammatory p65 subunit of the transcription factor nuclear factor-kappa B (NF-κB) to the nucleus, as determined by confocal immunofluorescence, following LPS stimulation. Most interestingly, additional experiments with the most efficient sp^2-GL candidate, (1*R*)-1-dodecylsulfonyl-5*N*,6*O*-oxomethylidene-nojirimycin (**77**), elicited a notable increase of the phosphorylation levels of the serine-threonine p38α mitogen activated protein kinase (MAPK) (Fig. 7.12), a master regulator of inflammatory signaling also investigated as a therapeutic target against cancer [78], parasitic diseases [79], and neurodegenerative disorders [80]. This effect was evidenced even in the absence of an external proinflammatory stimulus.

This result demonstrates the ability of the sp^2-α-glycosylsulfone **77** to intervene in key signaling pathways of the innate immune system and points to p38α MAPK as a plausible target. Computational docking supported that **77** can indeed bind p38α MAPK at the lipid binding allosteric site (Fig. 7.13B and C), analogously to the potent anti-inflammatory phosphatidylinositol ether lipid analogues (PIAs) or the alkyl phospholipid perifosine [81] (Fig. 7.13A). Compounds binding at this site

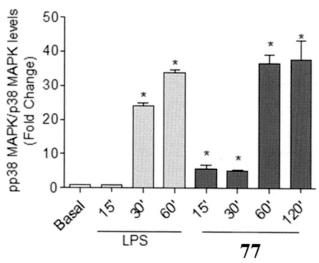

FIGURE 7.12

Direct activation of p38α MAPK in the presence of the α-glycosylsulfone 77 (10 μM).

Reprinted (adapted) with permission from Food Chem. Toxicol. 111 (2018) 456–466. Copyright (2018) Elsevier.

are able to induce self-phosphorylation of p38α MAPK by an alternative activation mechanism to the canonical mode (classical activation of the p38α cascade mediated by upstream kinases), probably after promoting local conformational changes in the protein.

The ability of the sp^2-iminosugar glycolipid **77** to induce an anti-inflammatory response by activation of p38α was also validated by further ex vivo assays in retinal explants from diabetic mice at 8 weeks of age that suffered from DR (C57BL/KsJ-*db/db* mouse model). A significant increase of HO-1 levels along with reduction of both iNOS expression and reactive gliosis (proliferation/hypertrophy of glial cells) confirmed a neuroprotective effect over the mouse retinas in the presence of **77** (20 μM for 24 h). The ensemble of data collected identifies sp^2-iminosugar glyco-lipids as a new family of activators of the autophosphorylation of p38α protein ki-nase able to provoke an anti-inflammatory response and revert the progression of DR. The anti-inflammatory activity of the sp^2-GL **77** has very recently been confirmed in human dendritic cells and in a mouse model (C57BL/6J) of LPS-induced acute inflammation [82].

7.6 Concluding remarks

Originally developed as a subfamily of iminosugar glycomimetics for fundamental studies in glycosidase mechanisms and biomedical applications in the context of

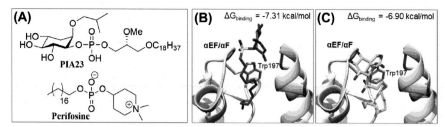

FIGURE 7.13

(A) Chemical structure of PIA23 and perifosine. (B and C) Predicted binding modes from docking experiments of 77 into the p38α structure.

Reprinted (adapted) with permission from Food Chem. Toxicol. 111 (2018) 456–466. Copyright (2018)
Elsevier.

glycosidase dysfunction-associated diseases [34], the sp²-iminosugars have nowadays consolidated as an original group of carbohydrate look-alikes with the unique property of engaging in glycosylation reactions [39,41,45]. A variety of conjugates have been prepared in the last decade taking advantage of this "chemical mimic" character that emulate the structure and function of the natural glycoconjugates, including oligosaccharides, glycolipids, and glycopeptides. The new prototypes have considerably enlarged the potential interactions with biomolecular partners beyond carbohydrate processing enzymes to carbohydrate-binding receptors (lectins) [83,84], tumor-associated carbohydrate antigens [56], and lipid-binding kinases

Table 7.3 Selected sp²-iminosugars as leading candidates against different pathologies.

Pathology	sp²-Iminosugar	Biological activity
Cancer	*21*, *63* structures	**21**; IC$_{50}$ (MCF-7) = 26 μM IC$_{50}$ (MDA-MB-231) = 44 μM [42] **63**; K_D (antibody SM3) = 1.6 ± 0.16 μM [56]
Parasitic infection	*72* structure	Amastigotes Dd8 of *Leishmania donovani* EC$_{50}$ = 10.80 ± 0.27 μM [70]
Inflammation	*77* structure	- Diabetic retinopathy [77] Bv.2 microglial cells Retinal explants *db/db* mouse model - Acute inflammation in mice: significant reduction of IL6 levels [82]

[77]. Specifically, the experimental evidence accumulated over the last decade regarding the antitumor, antileishmanial, and anti-inflammatory of sp^2-iminosugar glycolipids discard the inhibition of ER α-glucosidases and place p38α MAPK in the spotlight as a common molecular target against these pathologies.

As a summary, the following table shows at a glance the chemical structures and the potency of the leading compounds for each of the biological activities discussed in this chapter. Further developments are expected in the medium term in fields where carbohydrates and their conjugates are key players, which expands from neurological and aging diseases to bacterial or viral infection (Table 7.3).

Acknowledgments

The authors thank the funding from the Spanish Ministry of Economy and Competitiveness: SAF2016-76083-R (MINECO-FEDER).

References

[1] T.M. Gloster, D.J. Vocadlo, Developing inhibitors of glycan processing enzymes as tools for enabling glycobiology, Nat. Chem. Biol. 8 (2012) 683−694.

[2] P.O. Seitz, Glycopeptide synthesis and the effects of glycosylation on protein structure and activity, ChemBioChem 1 (4) (2000) 214−246.

[3] R.G. Spiro, Protein glycosylation: nature, distribution, enzymatic formation, and disease implications of glycopeptide bonds, Glycobiology 12 (4) (2002) 43R−56R.

[4] L.L. Lairson, B. Henrissat, G.J. Davies, S.G. Withers, Glycosyltransferases: structures, functions, and mechanisms, Annu. Rev. Biochem. 77 (2008) 521−555.

[5] T.M. Gloster, G.J. Davies, Glycosidase inhibition: assessing mimicry of the transition state, Org. Biomol. Chem. 8 (2) (2010) 305−320.

[6] G. Horne, F.X. Wilson, J. Tinsley, D.H. Williams, R. Storer, Iminosugars past, present and future: medicines for tomorrow, Drug Discov. Today 16 (2011) 107−118.

[7] P. Compain, O.R. Martin, Iminosugars: From Synthesis to Therapeutic Applications, Wiley-VCH, Weinheim, 2007.

[8] B.G. Winchester, Iminosugars: from botanical curiosities to licensed drugs, Tetrahedron Asymmetry 20 (2009) 645−651.

[9] M.L. Drent, A.T. Tollefsen, F.H. van Heusden, E.B. Hoenderdos, J.J. Jonker, E.A. van der Veen, Dose-dependent efficacy of miglitol, an alpha-glucosidase inhibitor, in type 2 diabetic patients on diet alone: results of a 24-week double-blind placebo-controlled study, Diabetes Nutr. Metab. 15 (3) (2002) 152−159.

[10] L. L Bennett, C. Fellner, Pharmacotherapy of Gaucher disease: current and future options, Pharm. Therapeut. 43 (5) (2018) 274−280.

[11] N. Guérard, O. Morand, J. Dingemanse, Lucerastat, an iminosugar with potential as substrate reduction therapy for glycolipid storage disorders: safety, tolerability, and pharmacokinetics in healthy subjects, Orphanet J. Rare Dis. 12 (9) (2017) 1−10.

[12] A. Markham, Migalastat: first global approval, Drugs 76 (11) (2016) 1147−1152.

[13] S. Watanabe, K.W.-K. Chan, G. Dow, E.E. Ooi, J.G. Low, S.G. Vasudevan, Optimizing Celgosivir therapy in mouse models of dengue virus infection of serotypes 1 and 2: the search for a window for potential therapeutic efficacy, Antivir. Res. 127 (2016) 10—19.

[14] D.S. Alonzi, K.A. Scott, R.A. Dwek, N. Zitzmann, Iminosugar antivirals: the therapeutic sweet spot, Biochem. Soc. Trans. 45 (2) (2017) 571—582.

[15] K.L. Warfield, D.L. Barnard, S.G. Enterlein, D.F. Smee, M. Khaliq, A. Sampath, M.V. Callahan, U. Ramstedt, C.W. Day, The iminosugar UV-4 is a broad inhibitor of influenza A and B viruses ex vivo and in mice, Viruses 8 (3) (2016) 1—9.

[16] I. Arora, A.K. Shaw, Ketenimine mediated synthesis of lactam iminosugars: development of one-pot process via tandem hydrative amidation of amino-alkynes and intermolecular transamidation, Tetrahedron 72 (35) (2016) 5479—5487.

[17] G. Papandreou, M.K. Tong, B. Ganem, Amidine, amidrazone, and amidoxime derivatives of monosaccharide aldonolactams: synthesis and evaluation as glycosidase inhibitors, J. Am. Chem. Soc. 115 (25) (1993) 11682—11690.

[18] Y. Kuramitsu, J. Hamada, T. Tsuruoka, K. Morikawa, H. Kobayashi, M. Hosokawa, A new anti-metastatic drug, ND-2001, inhibits lung metastases in rat hepatoma cells by suppressing haptotaxis of tumor cells toward laminin, Anti Canccer Drugs 9 (1) (1998) 88—92.

[19] H.-Q. Liu, C.-C. Song, Y.-H. Niu, T. Li, Q. Li, X.-S. Ye, Synthesis and biological evaluation of N-arylated-lactam-type iminosugars as potential immunosuppressive agents, Org. Biomol. Chem. 15 (28) (2017) 5912—5919.

[20] E. Lindbäck, O. López, J.G. Fernández-Bolaños, S.P.A. Sauer, M. Bols, An isofagomine analogue with an amidine at the pseudoanomeric position, Org. Lett. 13 (11) (2011) 2908—2911.

[21] M.-P. Heck, S.P. Vincent, B.W. Murray, F. Bellamy, C.-H. Wong, C. Mioskowski, Cyclic amidine sugars as transition-state analogue inhibitors of glycosidases: potent competitive inhibitors of mannosidases, J. Am. Chem. Soc. 126 (7) (2004) 1971—1979.

[22] M.S. Mcauley, D.J. Vocadlo, Increasing O-ClcNAc levels: an overview of small-molecule inhibitors of O-GlcNAc, Biochim. Biophys. Acta 1800 (2) (2010) 107—121.

[23] T. Aoyagi, H. Suda, K. Uotani, F. Kojima, T. Aoyama, K. Horiguchi, M. Hamada, T. Takeuchi, Nagstatin, a new inhibitor of N-acetyl-β-d-glucosaminidase produced by Streptomyces amakusaensis MG846-fF3. Taxonomy, production, isolation, physicochemical properties and biological activities, J. Antibiot. 45 (9) (1992) 1404—1408.

[24] M. Terinek, A. Vasella, Synthesis of N-Acetylglucosamine-derived Nagstatin analogues and their evaluation as glycosidase inhibitors, Helv. Chim. Acta 88 (1) (2005) 10—22.

[25] H.C. Dorfmueller, V.S. Borodkin, M. Schimpl, D.M.F. van Aalten, GlcNAc statins are nanomolar inhibitors of human O-GlcNAcase inducing cellular hyper-O-GlcNAcylation, Biochem. J. 420 (2) (2009) 221—227.

[26] H. Kayakiri, S. Takase, T. Shibata, M. Okamoto, H. Terano, M. Hashimoto, Structure of Kifunensine, a new immunomodulator isolated from an Actinomycete, J. Org. Chem. 54 (17) (1989) 4015—4016.

[27] K.W. Hering, K. Karaveg, K.W. Moremen, W.H. Pearson, A practical synthesis of kifunensine analogues as inhibitors of endoplasmic reticulum α-mannosidase I, J. Org. Chem. 70 (24) (2005) 9892—9904.

[28] J.L. Jiménez Blanco, V.M. Díaz Pérez, C. Ortiz Mellet, J. Fuentes, J.M. García Fernández, J.C. Díaz Arribas, F.J. Cañada, N-Thiocarbonyl azasugars: a new family of carbohydrate mimics with controlled anomeric configuration, Chem. Commun. 20 (1997) 1969—1970.

[29] M.I. García-Moreno, C. Ortiz Mellet, J.M. García Fernández, Polyhydroxylated *N*-(thio)carbamoyl piperidines: nojirimycin-type glycomimetics with controlled anomeric configuration, Tetrahedron: Asymmetry 10 (22) (1999) 4271−4275.

[30] P. Díaz Pérez, M.I. García-Moreno, C. Ortiz Mellet, J.M. García Fernández, Synthesis and comparative glycosidase inhibitory properties of reducing castanospermine analogues, Eur. J. Org. Chem. 14 (2005) 2903−2913.

[31] Y. Yu, T. Mena-Barragan, K. Higaki, J.L. Johnson, J.E. Drury, R.L. Lieberman, N. Nakasone, H. Ninomiya, T. Tsukimura, H. Sakuraba, Y. Suzuki, E. Nanba, C. Ortiz Mellet, J.M. García Fernandez, K. Ohno, Molecular basis of 1-deoxygalactonojirimycin arylthiourea binding to human α-Galactosidase A: pharmacological chaperoning eficacy on Fabry disease mutants, ACS Chem. Biol. 9 (7) (2014) 1460−1469.

[32] T. Takai, K. Higaki, M. Aguilar-Moncayo, T. Mena-Barragán, Y. Hirano, K. Yura, L. Yu, H. Ninomiya, M.I. García-Moreno, Y. Sakakibara, K. Ohno, E. Nanba, C. Ortiz Mellet, J.M. García Fernández, Y. Suzuki, A bicyclic 1-deoxygalactonojirimycin derivative as a novel pharmacological chaperone for GM1 gangliosidosis, Mol. Ther. 21 (3) (2013) 526−532.

[33] T. Mena-Barragán, M.I. García-Moreno, A. Sevšek, T. Okazaki, E. Nanba, K. Higaki, N.I. Martin, R.J. Pieters, J.M. García Fernández, C. Ortiz Mellet, Probing the inhibitor versus chaperone properties of sp²-iminosugars towards human β-glucocerebrosidase: a picomolar chaperone for Gaucher disease, Molecules 23 (4) (2018) 1−18.

[34] E.M. Sánchez-Fernández, J.M. García Fernández, C. Ortiz Mellet, Glycomimetic-based pharmacological chaperones for lysosomal storage disorders: lessons from Gaucher, GM1-gangliosidosis and Fabry diseases, Chem. Commun. 52 (32) (2016) 5497−5515.

[35] L. Cipolla, M.R. Fernandes, M. Gregori, C. Airoldi, F. Nicotra, Synthesis and biological evaluation of a small library of nojirimycin-derived bicyclic iminosugars, Carbohydr. Res. 342 (2007) 1813−1830.

[36] D. Bini, F. Cardona, M. Forcella, C. Parmeggiani, P. Parenti, F. Nicotra, L. Cipolla, Synthesis and biological evaluation of nojirimycin- and pyrrolidine-based trehalase inhibitors, Beilstein J. Org. Chem. 8 (2012) 514−521.

[37] A. Sevšek, L. Šrot, J. Rihter, M. Čelan, L. Quarles van Ufford, E. Moret, N.I. Martin, R.J. Pieters, *N*-Guanidino derivatives of 1,5-dideoxy-1,5-imino-D-xylitol are potent, selective, and stable inhibitors of β-glucocerebrosidase, ChemMedChem 12 (7) (2017) 483−486.

[38] R. Kooij, H.M. Branderhorst, S. Bonte, S. Wieclawska, N.I. Martin, R.J. Pieters, Glycosidase inhibition by novel guanidinium and urea iminosugar derivatives, Med. Chem. Commun. 4 (2) (2013) 387−393.

[39] E.M. Sánchez-Fernández, R. Rísquez-Cuadro, M. Chasseraud, A. Ahidouch, C. Ortiz Mellet, H. Ouadid-Ahidouch, J.M. García Fernández, Synthesis of *N*-, *S*-, and *C*-glycoside castanospermine analogues with selective neutral α-glucosidase inhibitory activity as antitumour agents, Chem. Commun. 46 (2010) 5328−5330.

[40] V.M. Díaz Pérez, M.I. García-Moreno, C. Ortiz Mellet, J. Fuentes, J.C. Díaz Arribas, F.J. Cañada, J.M. García Fernández, Generalized anomeric effect in action: synthesis and evaluation of stable reducing indolizidine glycomimetics as glycosidase inhibitors, J. Org. Chem. 65 (1) (2000) 136−143.

[41] E.M. Sánchez-Fernández, R. Rísquez-Cuadro, M. Aguilar-Moncayo, M.I. García-Moreno, C. Ortiz Mellet, J.M. García Fernández, Generalized anomeric effect in

gem-diamines: stereoselective synthesis of α-*N*-linked disaccharide mimics, Org. Lett. 11 (15) (2009) 3306–3309.

[42] G. Allan, H. Ouadid-Ahidouch, E.M. Sánchez-Fernández, R. Rísquez-Cuadro, J.M. García Fernández, C. Ortiz Mellet, A. Ahidouch, New castanospermine glycoside analogues inhibit breast cancer cell proliferation and induce apoptosis without affecting normal cells, PLoS One 8 (10) (2013) e76411.

[43] N. Gueder, G. Allan, M.-S. Telliez, F. Hague, J.M. García Fernández, E.M. Sánchez-Fernández, C. Ortiz Mellet, A. Ahidouch, H. Ouadid-Ahidouch, sp^2-Iminosugar α-glucosidase inhibitor 1-*C*-octyl-2-oxa-3-oxocastanospermine specifically affected breast cancer cell migration through Stim1, β1-integrin, and FAK signaling pathways, J. Cell. Physiol. 232 (12) (2017) 3631–3640.

[44] T.S. Rasmussen, S. Allman, G. Twigg, T.D. Butters, H.H. Jensen, Synthesis of *N*-alkylated noeurostegines and evaluation of their potential as treatment for Gaucher's disease, Bioorg. Med. Chem. Lett. 21 (5) (2011) 1519–1522.

[45] E.M. Sánchez-Fernández, R. Gonçalves-Pereira, R. Rísquez-Cuadro, G.B. Plata, J.M. Padrón, J.M. García Fernández, C. Ortiz Mellet, Influence of the configurational pattern of sp^2-iminosugar pseudo *N*-, *S*-, *O*- and *C*-glycosides on their glycoside inhibitory and antitumor properties, Carbohydr. Res. 429 (2016) 113–122.

[46] E.M. Sánchez-Fernández, R. Rísquez-Cuadro, C. Ortiz Mellet, J.M. García Fernández, P.M. Nieto, J. Angulo, sp^2-Iminosugar *O*-, *S*-, and *N*-glycosides as conformational mimics of α-linked disaccharides; implications for glycosidase inhibition, Chem. Eur. J. 18 (27) (2012) 8527–8539.

[47] E.M. Sanchez-Fernandez, E. Alvarez, C. Ortiz Mellet, J.M. García Fernandez, Synthesis of multibranched australine derivatives from reducing castanospermine analogues through the Amadori rearrangement of *gem*-diamine intermediates: selective inhibitors of β-glucosidase, J. Org. Chem. 79 (23) (2014) 11722–11728.

[48] Y. Nishimura, T. Kudo, S. Kondo, T. Takeuchi, T. Tsuruoka, H. Fukuyasu, S. Shibahara, Totally synthetic analogues of siastatin B. III. Trifluoroacetamide analogues having inhibitory activity for tumor metastasis, J. Antibiot. 47 (1) (1994) 101–107.

[49] Y. Nishimura, T. Satoh, S. Kondo, T. Takeuchi, M. Azetaka, H. Fukuyasu, Y. Iizuka, S. Shibahara, Effect on spontaneous metastasis of mouse Lewis lung carcinoma by a trifluoroacetamide analogue of siastatin B, J. Antibiot. 47 (7) (1994) 840–842.

[50] Y. Nishimura, *Gem*-diamine 1-*N*-iminosugars as versatile glycomimetics: synthesis, biological activity and therapeutic potential, J. Antibiot. 62 (8) (2009) 407–423.

[51] S. Farkona, E.P. Diamandis, I.M. Blasutig, Cancer immunotherapy: the beginning of the end of cancer? BMC Med. 14 (73) (2016) 1–18.

[52] R.E. Beatson, J. Taylor-Papadimitriou, J.M Burchell, MUC1 immunotherapy, Immunotherapy 2 (3) (2010) 305–327.

[53] N. Martínez-Sáez, J. Castro-López, J. Valero-González, D. Madariaga, I. Compañón, V.J. Somovilla, M. Salvadó, J.L. Asensio, J. Jiménez-Barbero, A. Avenoza, J.H. Busto, G.J.L. Bernardes, J.M. Peregrina, R. Hurtado-Guerrero, F. Corzana, Deciphering the non-equivalence of serine and threonine *O*-glycosylation points: implications for molecular recognition of the Tn antigen by an anti-MUC1 antibody, Angew. Chem. Int. Ed. 54 (34) (2015) 9830–9834.

[54] L.R. Loureiro, M.A. Carrascal, A. Barbas, J.S. Ramalho, C. Novo, P. Delannoy, P.A. Videira, Challenges in antibody development against Tn and sialyl-Tn antigens, Biomolecules 5 (3) (2015) 1783–1809.

[55] K. Yaddanapudi, R.A. Mitchell, J.W. Eaton, Cancer vaccines. Looking to the future, OncoImmunology 2 (3) (2013) e23403.

[56] E.M. Sanchez-Fernandez, C.D. Navo, N. Martínez-Saez, R. Goncalves-Pereira, V.J. Somovilla, A. Avenoza, J.H. Busto, G.J.L. Bernardes, G. Jimenez-Oses, F. Corzana, J.M. García Fernandez, C. Ortiz Mellet, J.M. Peregrina, Tn antigen mimics based on sp^2-iminosugars with affinity for an anti-MUC1 antibody, Org. Lett. 18 (15) (2016) 3890—3893.

[57] R.P. Spencer, C.L. Cavallaro, J. Schwartz, Rapid preparation of variously protected glycals using titanium(III), J. Org. Chem. 64 (11) (1999) 3987—3995.

[58] R.U. Lemieux, R.M. Ratcliffe, The azidonitration of tri-O-acetyl-D-galactal, Can. J. Chem. 57 (10) (1979) 1244—1251.

[59] P. Dokurno, P.A. Bates, H.A. Band, L.M.D. Stewart, J.M. Lally, J.M. Burchell, J. Taylor-Papadimitriou, D. Snary, M.J.E. Sternberg, P.S. Freemont, Crystal structure at 1.95 Å resolution of the breast tumour-specific antibody SM3 complexed with its peptide epitope reveals novel hypervariable loop recognition, J. Mol. Biol. 284 (3) (1998) 713—728.

[60] S. Benamrouz, V. Conseil, C. Creusy, E. Calderon, E. Dei-Cas, G. Certad, Parasites and malignancies, a review, with emphasis on digestive cancer induced by *Cryptosporidium parvum* (Alveolata: apicomplexa), Parasite 19 (2) (2012) 101—115.

[61] M.A.N. Al-Kamel, Leishmaniasis and malignancy: a review and perspective, Clin. Skin Cancer 2 (1—2) (2017) 54—58.

[62] A. Schwing, C. Pomares, A. Majoor, L. Boyer, P. Marty, G. Michel, *Leishmania* infection: misdiagnosis as cancer and tumor-promoting potential, Acta Trop. (2018), https://doi.org/10.1016/j.actatropica.2018.12.010.

[63] M.A. Fuertes, P.A. Nguewa, J. Castilla, C. Alonso, J.M. Pérez, Anticancer compounds as leishmanicidal drugs: challenges in chemotherapy and future perspectives, Curr. Med. Chem. 15 (5) (2008) 433—439.

[64] W. Moreira, P. Leprohon, M. Ouellette, Tolerance to drug-induced cell death favours the acquisition of multidrug resistance in *Leishmania*, Cell Death Dis. 2 (9) (2011) e201.

[65] R. Alves Moreira, S.A. Mendanha, D. Hansen, A. Alonso, Interaction of miltefosine with the lipid and protein components of the erythrocyte membrane, J. Pharm. Sci. 102 (5) (2013) 1661—1669.

[66] J.-F. Ritt, F. Raymond, P. Leprohon, D. Légaré, J. Corbeil, M. Ouellette, Gene amplification and point mutations in pyrimidine metabolic genes in 5-fluorouracil resistant *Leishmania infantum*, PLoS Neglected Trop. Dis. 7 (11) (2013), e2564.

[67] K. Georgopoulou, D. Smirlis, S. Bisti, E. Xingi, L. Skaltsounis, K. Soteriadou, In vitro activity of 10-deacetylbaccatin III against *Leishmania donovani* promastigotes and intracellular amastigotes, Planta Med. 73 (10) (2007) 1081—1088.

[68] P. Kopterides, E.G. Mourtzoukou, E. Skopelitis, N. Tsavaris, M.E. Falagas, Aspects of the association between leishmaniasis and malignant disorders, Trans. R. Soc. Trop. Med. Hyg. 101 (12) (2007) 1181—1189.

[69] H. Rezvan, M. Moafi, An overview on *Leishmania* vaccines: a narrative review article, Vet. Res. Forum 6 (1) (2015) 1—7.

[70] E.M. Sánchez-Fernández, V. Gómez-Pérez, R. García-Hernández, J.M. García Fernández, G.B. Plata, J.M. Padrón, C. Ortiz Mellet, S. Castanys, F. Gamarro, Antileishmanial activity of sp^2-iminosugar derivatives, RSC Adv. 5 (28) (2015) 21812—21822.

[71] F. Balkwill, A. Mantovani, Inflammation and cancer: back to Virchow? The Lancet 357 (9255) (2001) 539—545.

[72] G. Landskron, M. De la Fuente, P. Thuwajit, C. Thuwajit, M.A. Hermoso, Chronic inflammation and cytokines in the tumor microenvironment, J. Immunol. Res. 2014 (2014) 1—19.

[73] U. Saqib, S. Sarkar, K. Suk, O. Mohammad, M.S. Baig, R. Savai, Phytochemicals as modulators of M1-M2 macrophages in inflammation, Oncotarget 9 (25) (2018) 17937—17950.

[74] J. Tang, T.S. Kern, Inflammation in diabetic retinopathy, Prog. Retin. Eye Res. 30 (5) (2011) 343—358.

[75] A.I. Arroba, A.M. Valverde, Modulation of microglia in the retina: new insights into diabetic retinopathy, Acta Diabetol. 54 (6) (2017) 527—533.

[76] A.I. Arroba, E. Alcalde-Estévez, M. García-Ramírez, D. Cazzoni, P. de la Villa, E.M. Sánchez-Fernández, C. Ortiz Mellet, J.M. García Fernández, C. Hernández, R. Simó, A.M. Valverde, Modulation of microglia polarization dynamics during diabetic retinopathy in *db/db* mice, BBA-Mol. Basis Dis. 1862 (9) (2016) 1663—1674.

[77] E. Alcalde-Estévez, A.I. Arroba, E.M. Sánchez-Fernández, C. Ortiz Mellet, J.M. García Fernández, L. Masgrau, A.M. Valverde, The sp^2-iminosugar glycolipid 1-dodecylsulfonyl-5*N*,6*O*-oxomethylidenenojirimycin (DSO2-ONJ) as selective anti-inflammatory agent by modulation of hemeoxygenase-1 in Bv.2 microglial cells and retinal explants, Food Chem. Toxicol. 111 (2018) 456—466.

[78] A. Igea, A.R. Nebreda, The stress kinase p38α as a target for cancer therapy, Cancer Res. 75 (19) (2015) 3997—4002.

[79] M.J. Brumlik, S. Nkhoma, M.J. Kious, G.R. Thompson III, T.F. Patterson, J.J. Siekierka, T.J.C. Anderson, T.J. Curiel, Human p38 mitogen-activated protein kinase inhibitor drugs inhibit *Plasmodium falciparum* replication, Exp. Parasitol. 128 (2) (2011) 170—175.

[80] E. Kyung Kim, E.-J. Choi, Compromised MAPK signaling in human diseases: an update, Arch. Toxicol. 89 (6) (2015) 867—882.

[81] N. Tzarum, Y. Eisenberg-Domovich, J.J. Gills, P.A. Dennis, O. Livnah, Lipid molecules induce p38α activation via a novel molecular switch, J. Mol. Biol. 424 (5) (2012) 339—353.

[82] E. Schaeffer, E.M. Sánchez-Fernández, R. Gonçalves-Pereira, V. Flacher, D. Lamon, M. Duval, J.-D. Fauny, J.M. García Fernández, C.G. Mueller, C. Ortiz Mellet, sp^2-Iminosugar glycolipids as inhibitors of lipopolysaccharide-mediated human dendritic cell activation in vitro and of acute inflammation in mice in vivo, Eur. J. Med. Chem. 169 (2019) 111—120.

[83] M. Abellán Flos, M.I. García-Moreno, C. Ortiz Mellet, J.M. García Fernández, J.-F. Nierengarten, S.P. Vincent, Potent glycosidase inhibition with heterovalent fullerenes: unveiling the binding modes triggering multivalent inhibition, Chem. Eur. J. 22 (32) (2016) 11450—11460.

[84] M.I. García-Moreno, F. Ortega-Caballero, R. Rísquez-Cuadro, C. Ortiz Mellet, J.M. García Fernández, The impact of heteromultivalency in lectin recognition and glycosidase inhibition: an integrated mechanistic study, Chem. Eur. J. 23 (26) (2017) 6295—6304.

Synthesis and biological properties of spiroacetal-containing small molecules

Elena Lenci

Department of Chemistry "Ugo Schiff", University of Florence, Sesto Fiorentino, Florence, Italy

8.1 Introduction

Spiroacetals, also called spiroketals, are present in a great variety of natural products, drugs, and bioactive compounds [1—8]. Structurally, they are substituted spiranes in which the oxygen atoms belonging to different rings are linked through a common carbon atom. The three most common ring systems, 1,6-dioxaspiro[4.4] nonane (also [5,5]-spiroacetal), 1,6 dioxaspiro[4.5]decane (also [5,6]-spiroacetal), and 1,7-dioxaspiro[5.5]undecane (also [6,6]-spiroacetal), are present in more than 40% of all the acetal-containing natural products, as identified by a recent chemoinformatic study [9].

Spiroacetals can be found in nature as simple volatile molecules, such as (*R*)-olean (**1**) and (*S*)-olean (**2**) (Fig. 8.1) that act as pheromones for the olive fruit fly (*Bactrocera oleae*) males and females, respectively [10], or embedded in complex macrocycles, as in the case of the antitumoral spongistatin 2 (**3**), [11-12]. Spiroacetal are also found fused to steroidal systems, as in cephalostatins (see cephalostatin 1 (**4**)) [13—15], or to aryl rings, as in β-rubromycin (**5**), a compound characterized by a naphthoquinone and an isocoumarin moiety linked through a *bis*-benzannulated [5,6]-spiroacetal [16]. Finally, some natural products, although relatively rare, such as the marine toxin spirolide A (**6**) [17,18], have been found to possess a tricyclic *bis*-spiroacetal moiety [19].

The systematic classification of natural products containing spiroacetals, as well as their total synthesis, have been extensively discussed in previously published reviews and book chapters [1,2,20—23]. Herein, an overview of the biological profile of spiroacetal-containing small molecules is provided, showing the relevance of this moiety as a privileged structure able to bind different biological targets. Also, a panel of versatile synthetic methodologies to access spiroacetals is reported, with a focus on the feasibility and versatility of the chemistry, as required for the production of large chemical collections. Finally, few selected case studies on the synthesis of spiroacetal-containing small molecules collections are presented, together with their application in biological studies.

Small Molecule Drug Discovery. https://doi.org/10.1016/B978-0-12-818349-6.00008-X

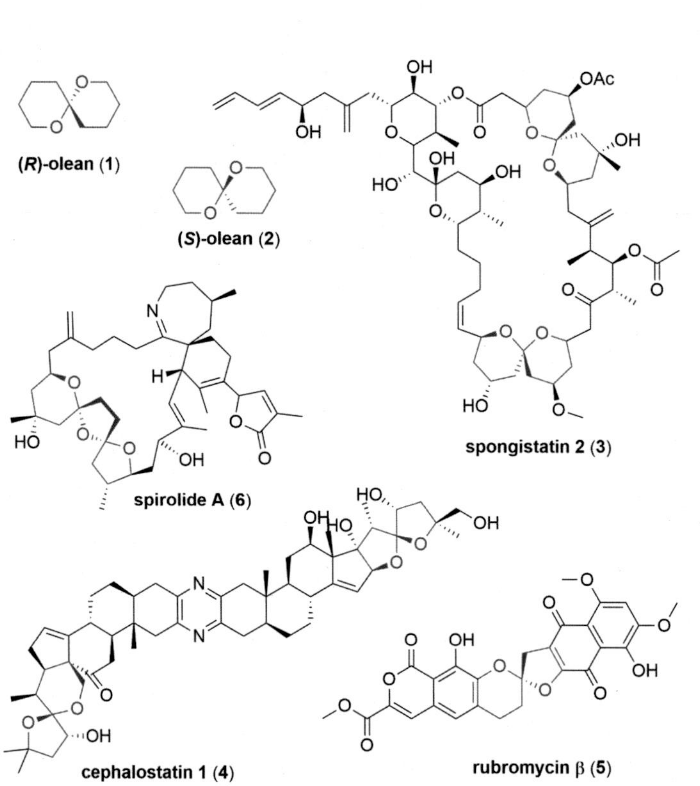

(R)-olean (1)

(S)-olean (2)

spongistatin 2 (3)

spirolide A (6)

cephalostatin 1 (4)

rubromycin β (5)

FIGURE 8.1

Representative bioactive natural products containing the [5,5]-, [6,5]-, and [6,6]-spiroacetal moiety.

8.2 Biological relevance of the spiroacetal moiety

Spiroacetals, both when they are a substructure in a highly functionalized system and when they are relatively unsubstituted, play a key role in the biological outcome of small molecules (see Table 8.1 for a list of the bioactivities of spiroacetal-containing natural products and drugs discussed along this chapter). This is mainly due to the intrinsic conformational equilibrium of these moieties that can assist the molecule in assuming the three-dimensional profile necessary for the binding, while retaining the backbone flexibility beneficial for transport and solubility. Many factors influence this equilibrium: steric strain, electrostatic interactions (such as the anomeric stabilizing effects) [24−26], and hydrogen bonds between ring-acetal oxygen and hydroxyl groups [27]. This varying conformational profile can explain why natural products coming from the same biogenetic pathway can have very different

Table 8.1 Spiroacetal-containing natural products and drugs discussed along this chapter.

Compound name	Spiroacetal moiety	Bioactivity	Ref.
R-Olean (**1**) S-Olean (**2**)	[6,6]	Pheromone for the olive fruit fly (Bactrocera oleae)	[10]
Spongistatin 2 (**3**)	[6,6]	Antitumoral, modulates mitosis	[11,12]
Cephalostatin 1 (**4**)	[5,5], [6,5]	Antitumoral, induces apoptosis	[13–15]
β-Rubromycin (**5**)	[6,5]	Antitumoral, inhibit the telomerase activity	[16]
Spirolide A (**6**)	[6,5,5]	Toxin	[17,18]
Ivermectin B$_{1a}$ (**7**)	[6,6]	Insecticidal, block GluCl channels	[29]
Bistramide A (**8**)	[6,6]	Antitumoral, activates protein kinase C	[30–32]
γ-Rubromycin (**10**)	[6,5]	Antitumoral, inhibit the telomerase activity, inhibit HIV reverse transcriptase	[33–36]
Compound **17**	[6,5]	NK1 receptor antagonist	[38]
Tofoglizon (**18**)	[6,5]	SGLT2 inhibitor	[39]
Compound **19**	[6,6]	RXR receptor antagonist	[40]
Salinomycin (**20**)	[6,6,5]	Antibacterial, antitumoral, activates ferroptosis	[41–43]
Ironomycin (**21**)	[6,6,5]	Antitumoral, activates ferroptosis	[41]
Okadaic acid (**30**)	[6,6]	Cytotoxic, inhibit phosphatases	[70,71]
Tautomycin (**31**)	[6,6]	Antibiotic, inhibit phosphatases	[72,73]
Acortartarin B (**45**)	[6,5]	Antioxidant, reduces hyperglycemia	[82,83]
Acortartarin A (**46**)	[6,5]	Antioxidant, reduces hyperglycemia	[82,83]
Pollenopyrroside A (**48**)	[6,6]	Antioxidant, reduces hyperglycemia	[84]
Shensongine A (**49**)	[6,6]	Antioxidant, reduces hyperglycemia	[85]

biological outcomes, while being structurally similar, as for the case of the nine different avermectins [28].

In many cases the specific interaction of the spiroacetal moiety of the small molecule within the binding site of the target has been extensively studied. On one hand, the spiroacetal itself can be the element directly interacting with key amino acids in the active site. This is the case of the [6,6]-spiroacetal moiety in the insecticidal ivermectin B$_{1a}$ (**7**), which is shown to bind the glutamate-gated chloride channels (GluCl), thanks to a hydrogen bond with Thr285 (Fig. 8.2A) [29]. On the other hand, the spiroacetal moiety can act as a rigid scaffold to direct the interacting appendages of the molecule in the appropriate position. As an example, in Fig. 8.2B, the antitumoral bistramide A (**8**) is reported, which possesses a [6,6]-spiroacetal moiety, behaving as a turn element able to direct the long hydrophobic chains into a deep cleft between subdomains 11 and 33 of G-actin [30−32].

Similarly, the importance of the spiroacetal moiety has been demonstrated by comparing the ability of a class of rubromycin compounds in inhibiting the telomerase activity. This enzyme is a promising target for cancer therapy, as it is responsible for compensating the telomeres shortening [33−36]. β-Rubromycin (**5**) and γ-rubromycin (**10**) are able to inhibit the telomerase activities with an IC_{50} of 3 μM (Fig. 8.3). In contrast, α-rubromycin, that does not contain the spiroacetal moiety, shows a substantially decreased potency ($IC_{50} > 200$ μM) [33]. The mode of interaction between the catalytic subunit of telomerase and the spiroacetal system still needs to be defined, and other molecular partners, such as the human telomerase RNA (hTR), might be involved.

However, the essential role of the spiroacetal moiety in inducing the antitelomerase activity has been evinced by the work of Brimble and coworkers [37]. They synthesized 29 novel rubromycin analogues with modified spiroacetal motives by introducing heteroatoms and substituents at different positions (see compounds

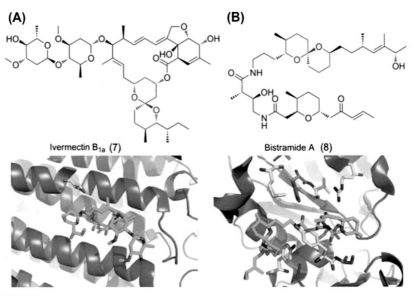

(A)

Ivermectin B$_{1a}$ (**7**)

(B)

Bistramide A (**8**)

FIGURE 8.2

(A) View from the extracellular site of ivermectin B1a into the binding site of GluCl subunit (PDB: 3RHW); (B) selected amino acid side chains of actin subdomains 1 and 3 (PDB: 2FXU) interacting with bistramide A.

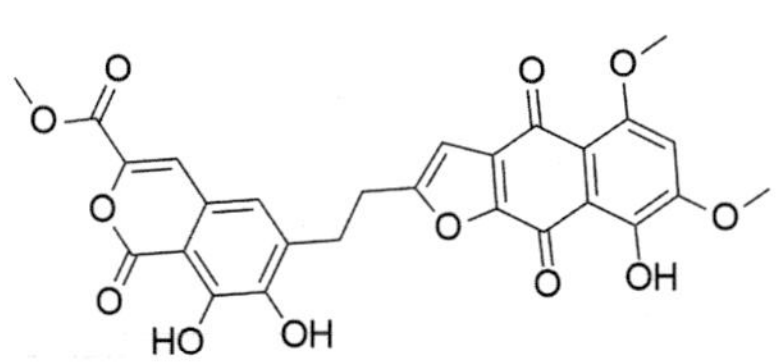

rubromycin α (**9**), IC$_{50}$ > 200 uM

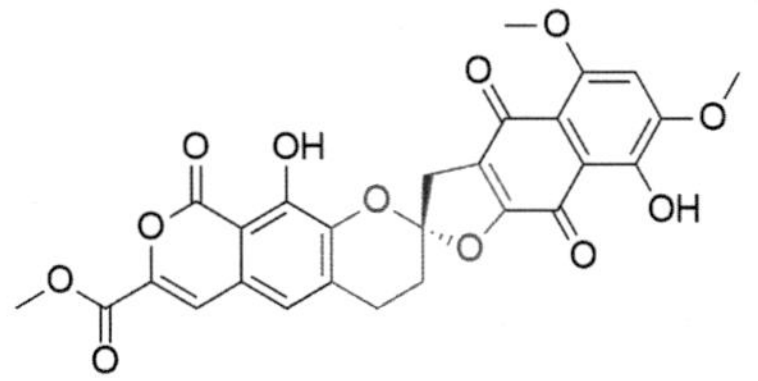

rubromycin β (**5**), IC$_{50}$ = 3 uM

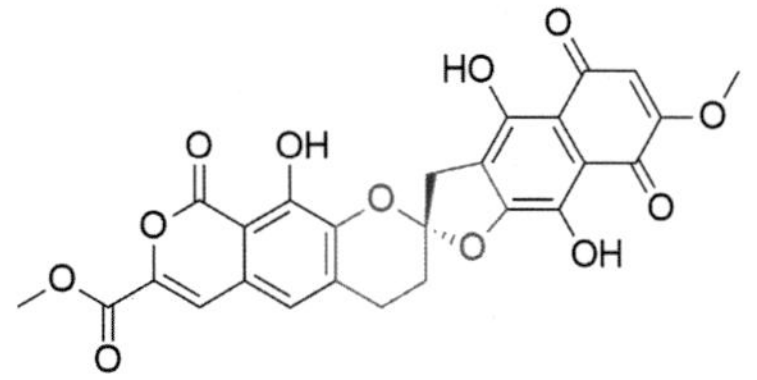

rubromycin γ (**10**), IC$_{50}$ = 3 uM

FIGURE 8.3

Structure of α-, β-, γ-rubromycin natural products and evaluation of their antitelomerase activity.

11−16 in Fig. 8.4 as representative examples). The evaluation of their antitelomerase activity showed that γ-rubromycin still possesses the highest potency, demonstrating that both the spiroacetal and the isocoumarin subunits are relevant for inducing the inhibition of this enzyme.

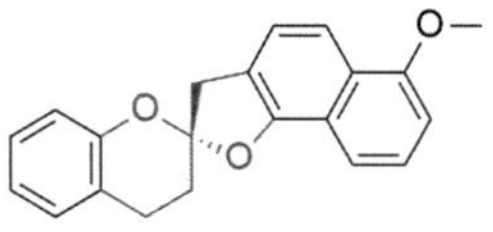

11, 31%$_{In}$ at 30 uM

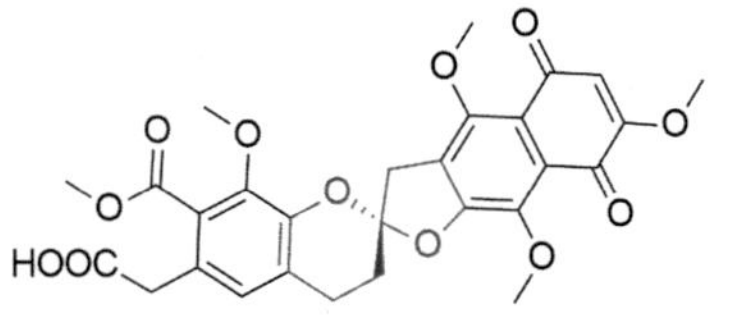

12, 40%$_{In}$ at 30 uM

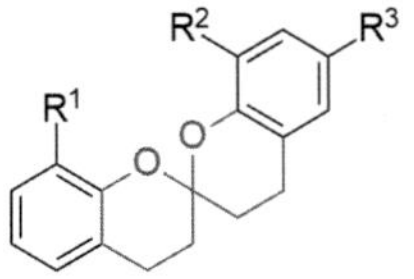

13, R^1 = R^2 = H, R^3 = OMe,
41%$_{In}$ at 30 uM
14, R^1 = R^2 = OMe, R^3 = H,
45%$_{In}$ at 30 uM

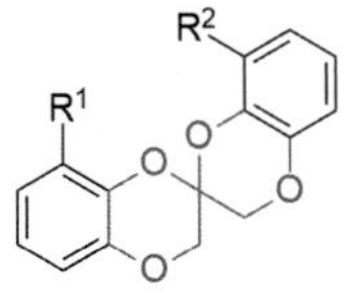

15, R^1 = R^2 = H,
41%$_{In}$ at 30 uM
16, R^1 = OMe, R^2 = H,
40%$_{In}$ at 30 uM

FIGURE 8.4

Structure of rubromycin synthetic analogues and evaluation of their antitelomerase activity.

Even though the use of spiroacetals in medicinal chemistry and drug discovery is still underexplored, acetal-containing drugs and synthetic inhibitors have appeared in the literature (Fig. 8.5). For example, a novel neurokinin 11 receptor (NK1) antagonist (compound **17**) containing a rigid 1,7-dioxa-4-azaspiro[5.5]undecane framework was developed by Williams and coworkers [38]. Tofogliflozin (**18**), designed using a 3D-pharmacophore modeling and based on the *O*-spiroacetal *C*-arylglucoside scaffold, is an active sodium glucose cotransporter 2 (SGLT2) inhibitor and a potential therapeutic agent for the treatment of type 2 diabetes [39]. Finally, Milroy and coworkers have reported the first example of a complex formed by the novel bis-benzannulated spiroacetal **19** and the retinoid X receptor (RXR), a potential therapeutic target for the Alzheimer disease [40].

Recently, the work of Rodriguez and Mehrpour gave further insight into the biological relevance of bicyclic acetals, by deciphering the mechanism by which salinomycin (**20**, Fig. 8.6) selectively attacks cancer stem cells [41–43]. Accordingly, they firstly improved the potency of salinomycin by synthesizing different analogues. From this library, they selected ironomycin (**21**), possessing a propargyl amino group as substituent of the [6,6,5]-spiroacetal moiety, that is able to enhance the activity against breast cancer stem cells, both in culture (HMLER CD24^{low}) and in mice [41]. The propargyl amino group was also used to conjugate this molecule to a fluorescent dye, thus allowing to see that ironomycin is able to induce the activation of the cell death by ferroptosis through the accumulation of cellular iron atoms into lysosomes, causing the degradation of ferritin and of the lysosomal membrane [44].

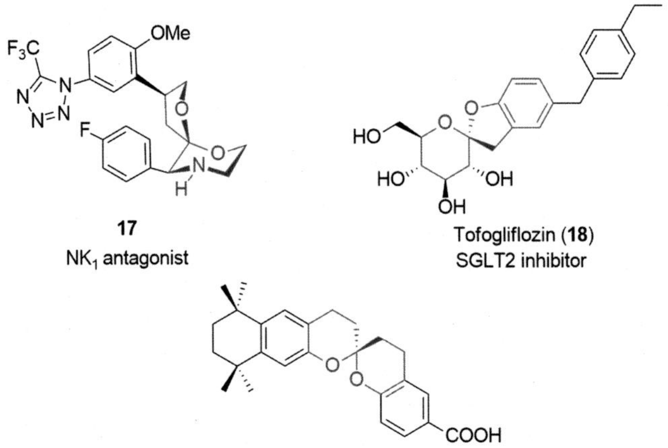

17
NK$_1$ antagonist

Tofogliflozin (**18**)
SGLT2 inhibitor

19
RXR ligand

FIGURE 8.5

Representative examples of designed acetal-containing inhibitors and drugs.

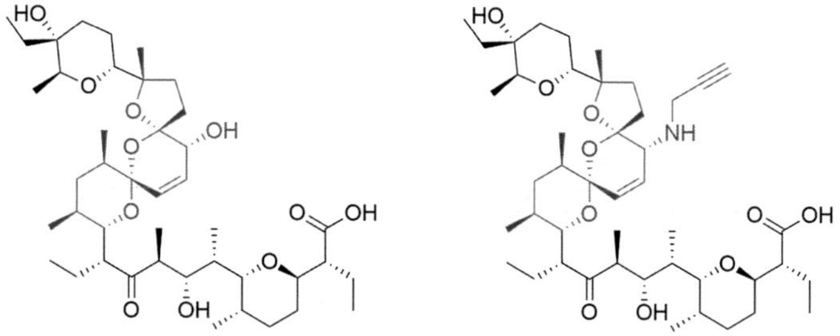

salinomycin (20)

IC_{50} = 1.0 μM
(HMLER CD24^{low} cells)
IC_{50} = 8.1 μM (ctrl)

ironomycin (21)

IC_{50} = 0.1 μM
(HMLER CD24^{low} cells)
IC_{50} = 2.1 μM (ctrl)

FIGURE 8.6

Structure of salinomycin (**20**) and ironomycin (**21**) and their activity against breast cancer stem cells.

8.3 Versatile synthetic methods for accessing spiroacetals

Given their biological relevance and their peculiar structural features, spiroacetals are useful molecular frameworks for the synthesis of chemical libraries. The variety of different configuration that can arise from spiroacetalization reactions can increase the skeletal diversity within small molecule libraries, in agreement with the principles of combinatorial chemistry [45–47] and diversity-oriented synthesis [48,49].

In addition, bicyclic acetal skeletons are attractive for the high density of polar functional groups, which offer many possibilities in the scaffold decoration and further chemical manipulations [50]. Modern advances in the efficient and selective synthesis of these skeletons are giving a great impulse to the use of acetal chemistry for the production of high-quality compound collections. Several review papers are present in the literature, describing strategies to build spiroacetals, especially in the context of natural products total synthesis [1,14,51]. Herein, we highlight the most versatile strategies to access spiroacetal molecular frameworks, keeping in mind the efficiency criteria required for the synthesis of small molecule collections, with a preference for cascade, one-pot, tandem, and/or multicomponent processes [52–54].

Many distinct approaches have been developed toward these targets during the years. However, the majority of them still involve the acid-catalyzed condensation of dihydroxyketones or their synthetic analogues (Scheme 8.1A) [19,55], as this approach leads directly to the most thermodynamically stable configuration, with the largest substituents being in an equatorial orientation, which is usually the prevalent configuration present in nature [56]. However, the stereocontrolled synthesis of complimentary configuration of spiroacetals, independently from the intrinsic thermodynamic preferences, has long attracted the interest of synthetic chemists to give an insight into the stereochemistry-activity relationship of bicyclic acetals [57]. In this field, Tan and coworkers reported several advances for the kinetically stereocontrolled spirocyclization reactions starting from glycal epoxides (Scheme 8.1B), providing systematic diversity-oriented libraries around this core.

Also, to overcome the issue of using harsh acidic conditions and to favor the stereocontrolled synthesis of spiroacetals, transition-metal-catalyzed methodologies have been increasingly applied in this area over the last 3 decades [21]. The use of transition metal catalysts has enabled the access of a variety of different modes of reactivity to provide spiroacetals. Noteworthy examples are the dihydroalkoxylation of alkynediols, catalyzed by gold, palladium, mercury, iridium, and/or rhodium complexes (Scheme 8.1C) [58,59]. In similar ways, spiroacetalization can be performed starting from monopropargylic triols [60] and propargylic vinyl ethers [61].

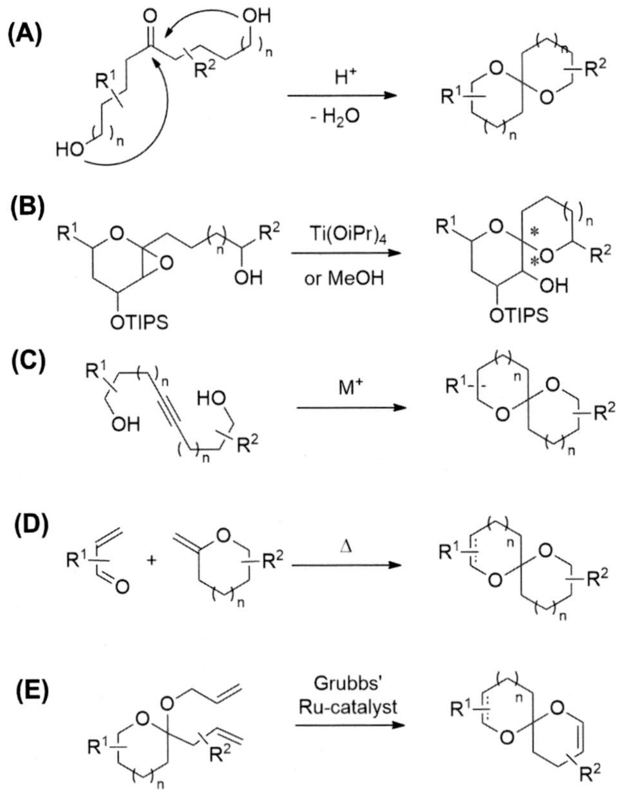

SCHEME 8.1

General synthetic strategies toward spiroacetals (n = 0,1,2).

Finally, spiroacetals can be achieved by using hetero-Diels—Alder reactions [62], as in the case of the [4 + 2]-cycloaddition between an enone and an α-methylenepyran (or furan) (Scheme 8.1D), or by using ring-closing metathesis, as in the case of the cyclization of C2-functionalized cyclic acetal diene substrates (Scheme 8.1E) [63,64].

8.4 Synthesis of libraries of spiroacetal-containing small molecules

Considering the biological relevance of the spiroacetal motifs and the easy access to their synthesis, these structures represent valuable molecular scaffolds for the generation of chemical libraries of spiroacetal-containing small molecules. In fact, some work reporting the synthesis of structurally simplified spiroacetal-containing small molecule collections for drug discovery programs and phenotypic screening have appeared in the literature [65]. The three selected examples herein reported illustrate how the application of the acetal chemistry in diversity-oriented synthesis and in combinatorial chemistry have led to the discovery of novel bioactive compounds against multiple targets, involved in a variety of diseases.

8.4.1 Spiroacetal derivatives as potential chemotherapeutic agents for the treatment of CLL leukemia

B-cell chronic lymphocytic leukemia (CLL) is the most common blood cancer in the Western world. Although some small molecules have emerged as potential therapeutic agents, this type of leukemia remains incurable [66]. Today, only the hematopoietic cell transplantation is a curative therapy for CLL, but the procedure is not generally applicable, due to high costs and the advanced age of many patients. For this reason, there is a critical need of finding new molecules that can act as alternative chemotherapeutic approaches to treat this disease. In this context, Milroy and Ley reported the biological assessment of a small molecule collection based on [6,6]- and [6,5]-spiroacetals against primary CLL cell lines [67]. The synthesis of these molecules was achieved starting from aldehyde **22** (Scheme 8.2) [68]. Briefly, the addition of the Grignard reagent, followed by oxidation of **23** with OsO_4, and treatment with 1,2-ethanedithiol gave the β-keto-1,3-dithiane **24** in good yield. Then, the spiroacetalization reaction of this intermediate, performed under acidic condition, resulted in different stereoisomers of the spiroacetal derivatives **25** in specific isomeric ratios, due to the presence of the lone pair orbitals of the sulfur atoms [69]. Then, the dithiane group was removed by a chemoselective oxidation with $NaClO_4$ and the resulting ketone derivatives **26** were subsequently coupled with a variety of N-Fmoc-protected α-amino acids, thus obtaining a library of bicyclic spiroacetals **27** (Scheme 8.2).

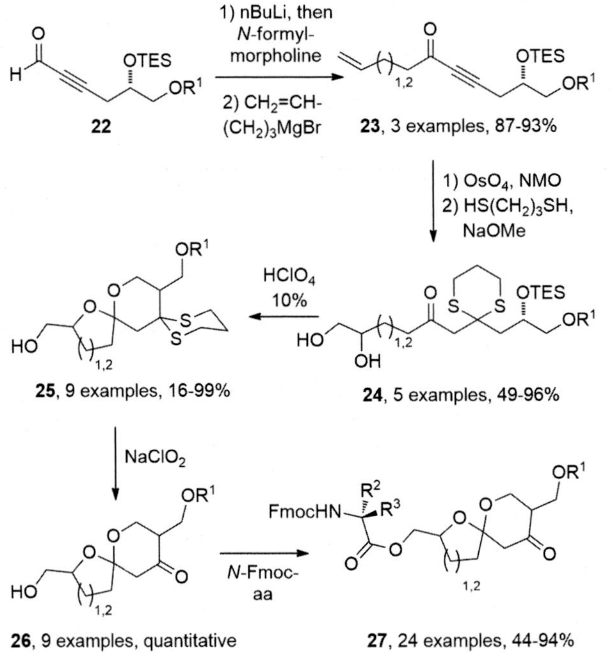

SCHEME 8.2

Synthesis of a library of [6,6]- and [6,5]-spiroacetals **26** and **27**.

A small collection of these spiroacetals was submitted to a cytotoxicity assay against CLL cells using flow cytometry [67]. Compound **22** was identified as the most active molecule (LD$_{50}$ = 1.6 μM) and used to develop a second-generation library around this structure, following SAR indications for the aromatic substitution, scaffold stereochemistry and carbamate derivatization (Fig. 8.7). The screening of this second-generation library showed an increase in the biological activity, although not so significant. In particular, the most potent spiroacetal compound was found to be **29**, with the ketone functionality replaced by the bulkier dithiane group that showed higher potency than fludarabine, the reference blockbuster drug for this type of cells. Preliminary biological investigation indicated that the mode of action of these spiroacetals follows the activation of an apoptotic pathway that was found to be selective for CLL cell lines, as these compounds were inactive in other tumoral cell lines, including those of breast cancer (MCF7) and adenocarcinoma (A549).

8.4.2 Spiroacetal derivatives as modulator of tubulin cytoskeleton integrity and phosphatase inhibitors

A wide range of naturally occurring spiroacetals, including spongistatin (**3**), are extremely potent inhibitors of the tubulin polymerization. Also, some of them, such as okadaic acid (**30**, Fig. 8.8) [70,71] or tautomycin (**31**, Fig. 8.8) [72,73],

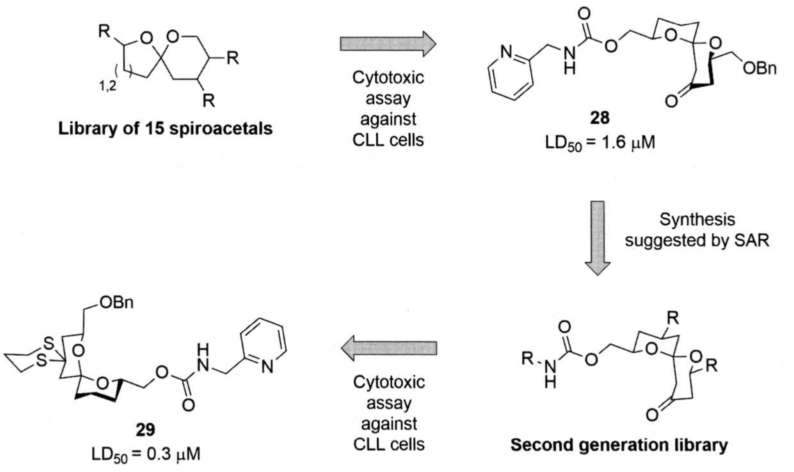

FIGURE 8.7

First-generation and second-generation library of spiroacetals and biological activity toward CLL cells.

are able to strongly inhibit phosphatase 1 and 2, thus inducing apoptosis. All these compounds contain one or two [6,6]-spiroacetal molecular frameworks. Starting from this observation, Waldmann's research group reasoned to develop the solid-phase synthesis of compounds with general formula 32 (Fig. 8.8) able to incorporating up to eight sites of possible diversification, in agreement with the principles of combinatorial chemistry [74].

More than 250 [6,6]-spiroacetals were prepared starting from commercial polystyrene resin 33, through two different synthetic strategies. In the first one (Scheme 8.3, path a), aldehyde 34 was converted into the alkynone 35 by addition of the corresponding alkyne and subsequent oxidation. Then, cleavage of the substrate from the resin and intramolecular Michael addition of the free hydroxyl groups allowed to obtain spiroacetals 37 in an overall yield of 5%–45% over 7 steps [75]. In the second strategy (Scheme 8.3, path b), the aldehyde 34 was transformed into the aldol adduct 36 in five steps. The treatment with dichlorodicyanobenzoquinone (DDQ) allowed to simultaneously cyclize this immobilized intermediate into released spiroacetals 38 [76]. Then, further chemical diversification was obtained from ketone-containing spiroacetals 37 by using Grignard addition, reductive amination, and oxime formation in the solution phase (see compounds 39, 40, and 41 in Scheme 8.3).

Some selected components of this [6,6]-spiroacetal collection were tested for their biological activity as phosphatase inhibitors and as tubulin cytoskeleton formation modulators, in order to see if these structurally simplified spiroacetals were able to retain the same biological activities of their parent natural products (Fig. 8.8). In

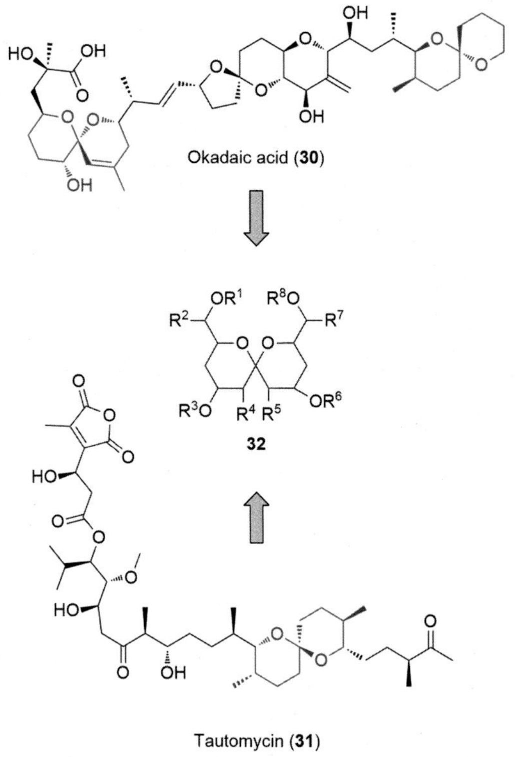

FIGURE 8.8

Structure of naturally occurring okadaic acid (**30**) and tautomycin (**31**) and general formula (**32**) of the spiroacetal collection developed by Waldmann and coworkers.

particular, protein tyrosine phosphatase PTP1b, the vaccinia virus—related phosphatase VHR and the cell regulator Cdcd25a and PP1 were chosen as representative enzymes, as they are involved in a variety of pathophysiological signaling pathways. The results of these biochemical assays showed that none of the synthesized spiroacetals were able to inhibit Cdc25a or PP1, whereas a significant inhibition of the VHR enzyme was observed by the action of spiroacetal **42**, but not by its stereoisomer **43**. Interestingly, the opposite behavior was observed when these compounds were assessed for their ability to affect the tubulin and actin networks in MDA-MB-231 cells [77]. While this observation alone is not sufficient to show a relationship between these two phenomena, it can open the way to further studies in this direction, able to potentially identify more biological targets that can be affected by spiroacetals (Fig. 8.9).

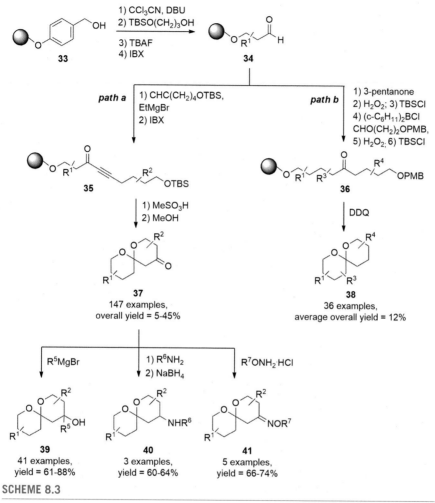

SCHEME 8.3

Solid-phase synthesis of [6,6]-spiroacetal collections, as reported by Waldmann and coworkers.

8.4.3 Spiroacetal derivatives as potential therapeutic agents for the treatment of oxidative stress—related pathologies

The pyrrolomorpholine spiroacetal of the family of acortatarin, shensongine, and pollenopyrroside (Scheme 8.4) have been used in Traditional Chinese Medicine for the treatment of different diseases [78,79]. These natural products show the ability to inhibit the production of reactive oxygen species (ROS) in different cell models, thus avoiding the oxidative stress that plays a critical role in many diseases. Just to give an example, acortatarins can significantly attenuate the hyperglycemia-

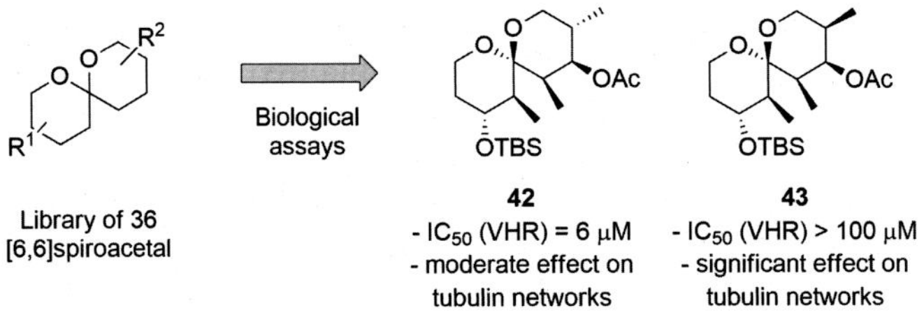

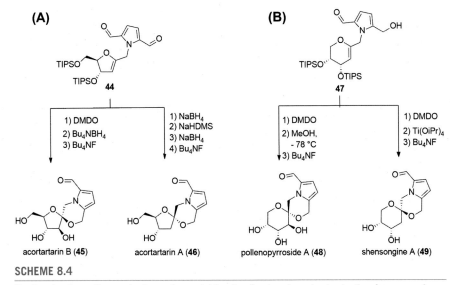

FIGURE 8.9

Phosphatase inhibition and tubulin modulation activity of [6,6]-spiroacetals **42** and **43**.

SCHEME 8.4

Synthesis of pyrrolomorpholine spiroacetal family of natural products starting from pyrrole functionalized glycal substrates **44** and **47**.

induced activation of NADPH oxidase, reducing the extracellular matrix production of ROS typical of diabetic nephropathy [80].

In this context, Tan and coworkers reported the synthesis of these natural products, as well as novel structurally related analogues, in order to have a direct comparison of the antioxidant activities of the entire D-enantiomeric series of furanose and pyranose pyrrolomorpholine spiroacetal isomers [81]. In particular, they were able to access pyrrolomorpholine natural products acortatarins (**45**–**46**) [82,83], pollenopyrroside (**48**) [84], and shensongine A (**49**) [85], starting from pyrrole-functionalized D-furanose glycal substrates **44** (Scheme 8.4A) [86] and D-pyranose glycal substrates **47** (Scheme 8.4B) [81]. To synthetize these compounds, as well as other structurally related small molecules, they exploited kinetically controlled spirocyclization that are independent from the thermodynamic stability of the resulting products. In particular, inversion of the configuration was obtained when the spirocyclization was performed under methanol-mediated conditions, whereas retention of configuration was obtained by using titanium(IV)-isopropoxide control [87].

In the same way, the authors prepared other five novel C2-hydroxy analogues that were assayed, together with the naturally occurring ones, in a glucose-induced oxidative stress test performed in rat mesangial cells. This biological assay revealed that compounds **50** and **51** were more active than their parent natural products, pollenopyrroside A (**48**) and shensongine A (**49**) (Fig. 8.10), although with incomplete ROS inhibition, possibly due to undetermined and nonspecific mechanisms.

8.5 Conclusion and future directions

Initially studied only within the field of total synthesis, spiroacetal molecular frameworks are now attracting the interest of the medicinal chemistry community. These moieties play in fact a key role in the biological outcome of the entire molecule, whether directly interacting with the target or acting as a scaffold to address side chains into specific directions. The distinct conformational flexibility of these frameworks assists the molecule in assuming the three-dimensional profile required for optimal binding to a biomacromolecule. More importantly, acetal chemistry is intrinsically able to give access to skeletally and stereochemically different arrangements of a spiroacetal architecture, thus providing a valuable synthetic tool for the development of high-quality compound collections. In fact, considering the great presence of spiroacetals in natural products, many synthetic methods have been reported in the literature for accessing these molecular frameworks. With this plethora of methods in hand, acetal chemistry can be easily applied to the production of large

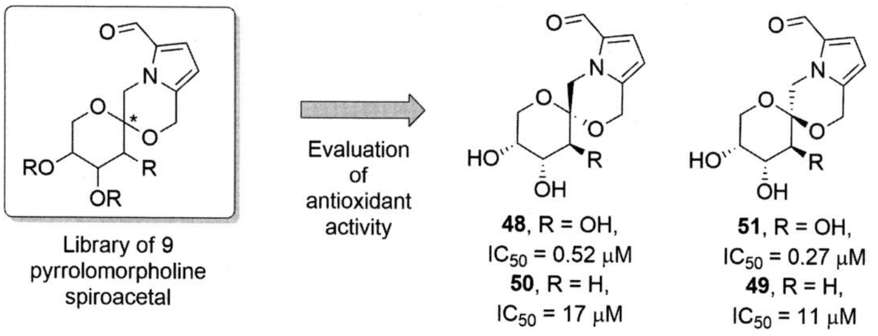

Library of 9 pyrrolomorpholine spiroacetal

Evaluation of antioxidant activity

48, R = OH, IC$_{50}$ = 0.52 μM
50, R = H, IC$_{50}$ = 17 μM

51, R = OH, IC$_{50}$ = 0.27 μM
49, R = H, IC$_{50}$ = 11 μM

FIGURE 8.10

Inhibition of high glucose-induced ROS production by pyrrolomorpholine spiroacetal natural products **48** and **49** and analogues **50** and **51**.

collections of spiroacetals, following diversity-oriented synthesis approaches, as reported along this chapter. Although there is still much room left in this field, their application in high-throughput screening and chemical genetics studies already allowed for the discovery of novel biologically active compounds and opened the way to further biological studies, directed to the identification of novel molecular targets potentially addressable by acetal-containing small molecules. We expect that in the future spiroacetals will be taken into account more often for the generation of high-value small molecules useful for drug discovery.

References

[1] F. Perron, K.F. Albizati, Chemistry of spiroketals, Chem. Rev. 89 (7) (1989) 1617–1661.

[2] W. Francke, W. Kitching, Spiroacetals in insects, Curr. Org. Chem. 5 (2001) 233–251.

[3] M.A. Brimble, D.P. Furkert, Chemistry of bis-spiroacetal systems: natural products, synthesis and stereochemistry, Curr. Org. Chem. 7 (2003) 1461–1484.

[4] J.E. Aho, P.M. Pihko, T.K. Rissa, Nonanomeric spiroketals in natural products: structures, sources, and synthetic strategies, Chem. Rev. 105 (12) (2005) 4406–4440.

[5] Y.K. Booth, W. Kitching, J.J. De Voss, Biosynthesis of insect spiroacetals, Nat. Prod. Rep. 26 (2009) 490–525.

[6] R. Quach, D.F. Chorleya, M.A. Brimble, Recent developments in transition metal-catalysed spiroketalisation, Org. Biomol. Chem. 12 (2014) 7423–7432.

[7] F.-M. Zhang, S.-Y. Zhang, Y.-Q. Tu, Recent progress in the isolation, bioactivity, biosynthesis, and total synthesis of natural spiroketals, Nat. Prod. Rep. 35 (2018) 75–104.

[8] S. Favre, P. Vogel, S. Gerber-Lemaire, Recent synthetic approaches toward non-anomeric spiroketals in natural products, Molecules 13 (13) (2008) 2570–2600.

[9] E. Lenci, G. Menchi, F.I. Saldívar-Gonzalez, J.L. Medina-Franco, A. Trabocchi, Bicyclic acetals: biological relevance, scaffold analysis, and applications in diversity-oriented synthesis, Org. Biomol. Chem. 17 (2019) 1037–1052.

[10] I. Čorič, B. List, Asymmetric spiroacetalization catalysed by confined Brønsted acids, Nature 483 (2012) 315–319.

[11] G.R. Pettit, Z.A. Cichacz, F. Gao, C.L. Herald, M.R. Boyd, J.M. Schmidt, J.N.A. Hooper, Isolation and structure of spongistatin 1, J. Org. Chem. 58 (6) (1993) 1302–1304.

[12] M. Kobayashi, S. Aoki, I. Kitagawa, Absolute stereostructure of altohyrtin A and its congeners, potent cytotoxic macrolides from the Okinawan marine sponge *Hyrtios altum*, Tetrahedron Lett. 35 (1994) 1243–1246.

[13] S. Hosseini Bai, S. Ogbourne, Eco-toxicological effects of the avermectin family with a focus on abamectin and ivermectin, Chemosphere 154 (2016) 204–214.

[14] G.R. Pettit, M. Inoue, Y. Kamano, D.L. Herald, C. Arm, C. Dufresne, N.D. Christie, J.M. Schmidt, D.L. Doubek, T.S. Krupa, Isolation and structure of the powerful cell growth inhibitor cephalostatin 1, J. Am. Chem. Soc. 120 (1998) 2006–2007.

[15] I.M. Muller, V.M. Dirch, A. Rudy, N. Lopez-Anton, G.R. Pettit, A.M. Vollmar, Cephalostatin 1 inactivates Bcl-2 by hyperphosphorylation independent of M-phase arrest and DNA damage, Mol. Pharmacol. 67 (2005) 1684−1689.

[16] T. Ueno, H. Takahashi, M. Oda, M. Mizunuma, A. Yokoyama, Y. Goto, Y. Mizushina, K. Sakaguchi, H. Hayashi, Inhibition of human telomerase by rubromycins: implication of spiroketal system of the compounds as an active moiety, Biochemistry 39 (2000) 5995−6002.

[17] Z. Amzil, M. Sibat, F. Royer, N. Masson, E. Abadie, Report on the first detection of pectenotoxin-2, spirolide-a and their derivatives in French shellfish, Mar. Drugs 23 (5) (2007) 168−179.

[18] H. Tingmo, I.W. Burton, A.D. Cembella, J.M. Curtis, M.A. Quilliam, J.A. Walter, J.L.C. Wright, Characterization of spirolides A, C, and 13-desmethyl C, new marine toxins isolated from toxic plankton and contaminated shellfish, J. Nat. Prod. 64 (3) (2001) 308−312.

[19] M.A. Brimble, F.A. Fares, Synthesis of bis-spiroacetal ring systems, Tetrahedron 55 (1999) 7661−7706.

[20] M.F. Jacobs, W. Kitching, A straightforward route to spiroketals, Curr. Org. Chem. 2 (1998) 395−436.

[21] J.A. Palmes, A. Aponick, Strategies for spiroketal synthesis based on transition-metal catalysis, Synthesis 44 (24) (2012) 3699−3721.

[22] R. Quach, D.P. Furkert, M.A. Brimble, Gold catalysis: synthesis of spiro, bridged, and fused ketal natural products, Org. Biomol. Chem. 15 (2017) 3098−3104.

[23] M.A. Brimble, L.A. Stubbing, Synthesis of saturated oxygenated heterocycles I, in: J. Cossy (Ed.), Topics in Heterocyclic Chemistry, vol. 35, Springer-Verlag, Berlin Heidelberg, 2014, pp. 189−267.

[24] R.W. Hoffmann, Flexible molecules with defined shape-conformational design, Angew. Chem. Int. Ed. 31 (9) (1992) 1124−1134.

[25] R.W. Hoffmann, Conformation design of open-chain compounds, Angew. Chem. Int. Ed. 39 (12) (2000) 2054−2070.

[26] E.M. Larsen, M.R. Wilson, R.E. Taylor, Conformation-activity relationships of polyketide natural products, Nat. Prod. Rep. 32 (8) (2015) 1183−1206.

[27] P.A. Wender, J. De Brabander, P.G. Harran, J.M. Jimenez, M.F.T. Koehler, B. Lippa, C.M. Park, C. Siedenbiedel, G.R. Pettit, The design, computer modeling, solution structure, and biological evaluation of synthetic analogs of bryostatin 1, Proc. Natl. Acad. Sci. USA 95 (12) (1998) 6624−6629.

[28] P. Sun, Q. Zhao, H. Zhang, J. Wu, W. Liu, Effect of stereochemistry of avermectin-like 6,6-spiroketals on biological activities and endogenous biotransformations in Streptomyces avermectinius, ChemBioChem 15 (5) (2014) 660−664.

[29] R.E. Hibbs, E. Gouaux, Principles of activation and permeation in an anion-selective Cys-loop receptor, Nature 474 (7349) (2011) 54−60.

[30] M.P. Sauviat, D. Gouiffes-Barbin, E. Ecault, J.F. Verbist, Blockade of sodium channels by Bistramide A in voltage-clamped frog skeletal muscle fibres, Biochim. Biophys. Acta 1103 (1) (1992) 109−114.

[31] D. Riou, C. Roussakis, N. Robillard, J.F. Biard, J.F. Verbist, Bistramide A-induced irreversible arrest of cell proliferation in a non-small-cell bronchopulmonary carcinoma is similar to induction of terminal maturation, Biol. Cell. 77 (3) (1993) 261−264.

[32] S.A. Rizvi, V. Tereshko, A.A. Kossiakoff, S.A. Kozmin, Structure of bistramide A-actin complex at a 1.35 angstroms resolution, J. Am. Chem. Soc. 128 (12) (2006) 3882−3883.

[33] T. de Lange, Activation of telomerase in a human tumor, Proc. Natl. Acad. Sci. USA 91 (8) (1994) 2882−2885.

[34] C.W. Greider, Telomerase activation. One step on the road to cancer? Trends Genet. 15 (3) (1999) 109−112.

[35] S. Sharma, E. Raymond, H. Soda, D. Sun, S.G. Hilsenbeck, A. Sharma, E. Izbicka, B. Windle, D.D. Von Hoff, Preclinical and clinical strategies for development of telomerase and telomere inhibitors, Ann. Oncol. 8 (11) (1997) 1063−1074.

[36] V. Urquidi, D. Tarin, S. Goodison, Telomerase in cancer: clinical applications, Ann. Med. 30 (5) (1998) 419−430.

[37] T.-Y. Yuen, Y.-P. Ng, F.C.F. Ip, J.L.-Y. Chen, D.J. Atkinson, J. Sperry, N.Y. Ip, M.A. Brimble, Telomerase inhibition studies of novel spiroketal containing rubromycin derivatives, Aus. J. Chem. 66 (5) (2013) 530−533.

[38] E.M. Seward, E. Carlson, T. Harrison, K.E. Haworth, R. Herbert, F.J. Kelleher, M.M. Kurtz, J. Moseley, S.N. Owen, A.P. Owens, S.J. Sadowski, C.J. Swain, B.J. Williams, Spirocyclic NK(1) antagonists I: [4.5] and [5.5]-spiroketals, Bioorg. Med. Chem. Lett. 12 (18) (2002) 2515−2518.

[39] Y. Ohtake, T. Sato, T. Kobayashi, M. Nishimoto, N. Taka, K. Takano, K. Yamamoto, M. Ohmori, M. Yamaguchi, K. Takami, S.-Y. Yeu, K.-H. Ahn, H. Matsuoka, K. Morikawa, M. Suzuki, H. Hagita, K. Ozawa, K. Yamaguchi, M. Kato, S. Ikeda, Discovery of tofogliflozin, a novel C-arylglucoside with an O-spiroketal ring system, as a highly selective sodium glucose cotransporter 2 (SGLT2) inhibitor for the treatment of type 2 diabetes, J. Med. Chem. 55 (17) (2012) 7828−7840.

[40] M. Scheepstra, S.A. Andrei, M.Y. Unver, A.K.H. Hirsch, S. Leysen, C. Ottmann, L. Brunsveld, L.-G. Milroy, Designed spiroketal protein modulation, Angew. Chem. Int. Ed. 56 (20) (2017) 5480−5484.

[41] T.T. Mai, A. Hamaï, A. Hienzsch, T. Cañeque, S. Müller, J. Wicinski, O. Cabaud, C. Leroy, A. David, V. Acevedo, A. Ryo, C. Ginestier, D. Birnbaum, E. Charafe-Jauffret, P. Codogno, M. Mehrpour, R. Rodriguez, Salinomycin kills cancer stem cells by sequestering iron in lysosomes, Nat. Chem. 9 (10) (2017) 1025−1033.

[42] A. Markowska, J. Kaysiewicz, J. Markowska, A. Huczyński, Doxycycline, salinomycin, monensin and ivermectin repositioned as cancer drugs, Bioorg. Med. Chem. Lett. 29 (13) (2019) 1549−1554.

[43] J. Dewangan, S. Srivastava, S. Mishra, A. Divakar, S. Kumar, S.K. Rath, Salinomycin inhibits breast cancer progression via targeting HIF-1α/VEGF mediated tumor angiogenesis in vitro and in vivo, Biochem. Pharmacol. 24 (164) (2019) 326−335.

[44] S.J. Dixon, K.M. Lemberg, M.R. Lamprecht, R. Skouta, E.M. Zaitsev, C.E. Gleason, D.N. Patel, A.J. Bauer, A.M. Cantley, W.S. Yang, B. Morrison, B.R. Stockwell, Ferroptosis: an iron-dependent form of nonapoptotic cell death, Cell 149 (5) (2012) 1060−1072.

[45] R. Liu, X. Li, K.S. Lam, Combinatorial chemistry in drug discovery, Curr. Opin. Chem. Biol. 38 (2017) 117−126.

[46] H. Maher, Combinatorial chemistry in drug research from a new vantage point, Bioorg. Med. Chem. 5 (3) (1997) 473−491.

[47] S.L. Crooks, L.J. Charles, Overview of combinatorial chemistry, Curr. Protoc. Pharmacol. 10 (1) (2001), 9.3.1-9.3.16.

[48] S.L. Schreiber, Target-oriented and diversity-oriented organic synthesis in drug discovery, Science 287 (5460) (2000) 1964–1969.

[49] A. Trabocchi (Ed.), Diversity-Oriented Synthesis: Basics and Applications in Organic Synthesis, Drug Discovery, and Chemical Biology, Wiley and Sons, Hoboken, 2013.

[50] M. Feher, J.M. Schmidt, Property distributions: differences between drugs, natural products, and molecules from combinatorial chemistry, J. Chem. Inf. Comput. Sci. 43 (1) (2003) 218–227.

[51] D.J. Faulkner, Marine natural products, Nat. Prod. Rep. 18 (2001) 1–49.

[52] T. Zarganes-Tzitzikas, A. Dömling A, Modern multicomponent reactions for better drug syntheses, Org. Chem. Front. 1 (7) (2014) 834–883.

[53] D.B. Ramachary, R. Mondal, C. Venkaiah, Rapid two-step synthesis of drug-like polycyclic substances by sequential multi-catalysis cascade reactions, Org. Biomol. Chem. 21 (8) (2010) 321–325.

[54] M.O. Sydnes, One-pot reactions: a step towards greener chemistry, Curr. Green Chem. 1 (3) (2014) 216–226.

[55] K.T. Mead, B.N. Brewer, Strategies in spiroketal synthesis revisited: recent applications and advances, Curr. Org. Chem. 7 (3) (2003) 227–256.

[56] P. Deslongchamps, Stereoelectronic effects in organic chemistry, in: J.E. Baldwin (Ed.), Organic Chemistry Series, vol. 1, Pergamon Press, Oxford, 1983, pp. 5–20.

[57] P. Nagorny, Z. Sun, G.A. Winschel, Chiral phosphoric acid catalyzed stereoselective spiroketalizations, Synlett 24 (6) (2013) 661–665.

[58] B. Liu, J.K. De Brabander, Metal-catalyzed regioselective oxy-functionalization of internal Alkynes: an entry into ketones, acetals, and spiroketals, Org. Lett. 8 (21) (2006) 4907–4910.

[59] S. Elgafi, L. D Field, B.A. Messerle, Cyclisation of acetylenic carboxylic acids and acetylenic alcohols to oxygen-containing heterocycles using cationic rhodium(I) complexes, J. Organomet. Chem. 607 (1) (2000) 97–104.

[60] A. Aponick, C.Y. Li, J.A. Palmes, Au-catalyzed cyclization of monopropargylic triols: an expedient synthesis of monounsaturated spiroketals, Org. Lett. 11 (1) (2009) 121–124.

[61] B.D. Sherry, L. Maus, B.N. Laforteza, F.D. Toste, Gold(I)-catalyzed synthesis of dihydropyrans, J. Am. Chem. Soc. 128 (25) (2006) 8132–8133.

[62] M.A. Rizzacasa, A. Pollex, The hetero-Diels–Alder approach to spiroketals, Org. Biomol. Chem. 7 (2009) 1053–1059.

[63] S.K. Ghosh, R.P. Hsung, J. Wang, Ketal-tethered ring-closing metathesis. An unconventional approach to constructing spiroketals and total synthesis of an insect pheromone, Tetrahedron Lett. 45 (28) (2004) 5505–5510.

[64] P.A.V. van Hooft, F. El Oualid, H.S. Overkleeft, G.A. van der Marel, J.H. van Boom, M.A. Leeuwenburgh, Synthesis and elaboration of functionalised carbohydrate-derived spiroketals, Org. Biomol. Chem. 2 (2004) 1395–1403.

[65] B.A. Kulkarni, G.P. Roth, E. Lobkovsky, J.A. Porco Jr., Combinatorial synthesis of natural product-like molecules using a first-generation spiroketal scaffold, J. Comb. Chem. 4 (1) (2002) 56–72.

[66] B.D. Sherry, L. Maus, B.N. Laforteza, F.D. Toste, Gold(I)-Catalyzed synthesis of dihydropyrans, J. Am. Chem. Soc. 128 (25) (2006) 8132–8133.

[67] L.-G. Milroy, G. Zinzalla, F. Loiseau, Z. Qian, G. Prencipe, C. Pepper, C. Fegan, S.V. Ley, Natural-product-like spiroketals and fused bicyclic acetals as potential

therapeutic agents for B-cell chronic lymphocytic leukaemia, ChemMedChem 3 (12) (2008) 1922–1935.

[68] G. Zinzalla, L.-G. Milroy, S.V. Ley, Chemical variation of natural product-like scaffolds: design and synthesis of spiroketal derivatives, Org. Biomol. Chem. 4 (10) (2016) 1977–2002.

[69] M.J. Gaunt, D.F. Hook, H.R. Tanner, S.V. Ley, A practical and efficient synthesis of the C-16–C-28 spiroketal fragment (CD) of the spongistatins, Org. Lett. 5 (25) (2003) 4815–4818.

[70] P.J. Ferron, K. Hogeveen, V. Fessard, L. Le Hégarat, Comparative analysis of the cytotoxic effects of okadaic acid-group toxins on human intestinal cell lines, Mar. Drugs 12 (8) (2014) 4616–4634.

[71] J.M. Oliveira, C.B. da Cruz E Silva, T. Müller, T.S. Martins, M. Cova, O.A.B. da Cruz E Silva, A.G. Henriques, Toward neuroproteomics in biological psychiatry: a systems approach unravels okadaic acid-induced alterations in the neuronal phosphoproteome, Omics 21 (9) (2017) 550–563.

[72] X. Chen, X. Zhu, Y. Ding, Y. Shen, Antifungal activity of tautomycin and related compounds against Sclerotinia sclerotiorum, J. Antibiot. 64 (8) (2011) 563–569.

[73] T. Kawamura, S. Matsuzawa, Y. Mizuno, K. Kikuchi, H. Oikawa, M. Oikawa, M. Ubukata, A. Ichihara, Different moieties of tautomycin involved in protein phosphatase inhibition and induction of apoptosis, Biochem. Pharmacol. 55 (7) (1998) 995–1003.

[74] S. Sommer, H. Waldmann, Solid phase synthesis of a spiro[5.5]ketal library, Chem. Commun. (2005) 5684–5686.

[75] S. Sommer, M. Kuhn, H. Waldmann, Solid-phase synthesis of [5.5]-Spiroketals, Adv. Synth. Catal. 350 (11) (2008) 1736–1750.

[76] O. Barun, S. Sommer, H. Waldmann, Asymmetric solid-phase synthesis of 6,6-spiroketals, Angew. Chem. Int. Ed. 43 (24) (2004) 3195–3199.

[77] O. Barun, K. Kumar, S. Sommer, A. Langerak, T.U. Mayer, O. Muller, H. Waldmann, Atural product-guided synthesis of a spiroacetal collection reveals modulators of tubulin cytoskeleton integrity, Eur. J. Org. Chem. 22 (2005) 4773–4788.

[78] B. Ding, Y. Dai, Y.-L. Hou, X.-S. Yao, Spiroalkaloids from shensong yangxin capsule, J. Asian Nat. Prod. Res. 17 (5) (2015) 559–566.

[79] M. Li, J. Xiong, Y. Huang, L.-J. Wang, Y. Tang, G.-X. Yang, X.-H. Liu, B.-G. Wei, H. Fan, Y. Zhao, W.-Z. Zhai, J.-F. Hu, A. and B. Xylapyrrosides, Two rare sugar-morpholine spiroketal pyrrole-derived alkaloids from Xylaria nigripes: isolation, complete structure elucidation, and total syntheses, Tetrahedron 71 (2015) 5285–5295.

[80] J.M. Forbes, M.T. Coughlan, M.E. Cooper, Oxidative stress as a major culprit in kidney disease in diabetes, Diabetes 57 (6) (2008) 1446–1454.

[81] A.L. Verano, D.S. Tan, Family-level stereoselective synthesis and biological evaluation of pyrrolomorpholine spiroketal natural product antioxidants, Chem. Sci. 8 (5) (2017) 3687–3693.

[82] Z.F. Zhao, L.L. Zhou, X. Chen, Y.X. Cheng, F.F. Hou, J. Nie, Acortatarin A inhibits high glucose-induced extracellular matrix production in mesangial cells, Chin. Med. J. 126 (7) (2013) 1230–1235.

[83] T. Teranishi, M. Kageyama, S. Kuwahara, Concise total synthesis of acortatarin A, Biosci. Biotechnol. Biochem. 77 (3) (2013) 676–678.

[84] M. Faisal, D. Shahzad, F.A. Larik, P. Dar, Synthetic approaches to access acortatarins, shensongines and pollenopyrroside; potent antioxidative spiro-alkaloids with a naturally rare morpholine moiety, Fitoterapia 129 (2018) 366–382.

[85] J.M. Wood, D.P. Furkert, M.A. Brimble, Synthesis of the 2-formylpyrrole spiroketal pollenopyrroside A and structural elucidation of xylapyrroside A, shensongine A and capparisine B, Org. Biomol. Chem. 14 (32) (2016) 7659–7664.

[86] J.M. Wurst, A.L. Verano, D.S. Tan, Stereoselective synthesis of acortatarins a and B, Org. Lett. 14 (17) (2012) 4442–4445.

[87] J.S. Potuzak, S.B. Moilanen, D.S. Tan, Stereocontrolled synthesis of spiroketals via a remarkable methanol-induced kinetic spirocyclization reaction, J. Am. Chem. Soc. 127 (40) (2005) 13796–13797.

Centrocountins— synthesis and chemical biology of nature inspired indoloquinolizines

Kamal Kumar[1,2]

[1]*Max Planck Institute of Molecular Physiology, Dortmund, Germany;* [2]*Aicuris Antiinfective Cures GmbH, Wuppertal, Germany*

9.1 Introduction

The advent of high-throughput screening (HTS) and genomics applications at the dawn of this century built an anticipation that small molecule collections generated by combinatorial kind of synthesis would offer a number of drug candidates. However, when such combichem collections failed [1] to deliver hit and lead molecules in HTS efforts, the importance of structural complexity and diversity in screening collections was realized and chemists had to resort back to complex natural products in discovery research. In fact, natural products remain a significant source of active ingredients of medicines with more than 50% of drug substances derived from natural products or inspired by a natural compound [2]. They cover a range of therapeutic indications from the expected anti-infective to anticancer, antidiabetic, etc. Despite their advantages and enormous track record in medicine, pharmaceutical industry had largely abandoned natural product research mainly because of the supply issues of natural products from natural or synthetic means as well as due to legal difficulties in acquiring intellectual property rights for natural products [3–5]. On the other side, the redundancy of the chemotypes is a major obstacle in the discovery of novel hit and lead molecules in HTS campaigns [6,7] which is often driven by library synthesis exploiting inexpensive substrates or following similar synthesis routes and thereby heavily compromising structural features of molecules in a screening collection [8,9]. The structural and molecular properties of natural products are quite distinct from combichem-based small molecules and bequeath them interesting biological activities [10,11]. Intriguing biological activities of highly complex natural products have inspired organic chemists to take up more challenging total synthesis studies of natural products [12–14] as well as strategies leading to a wider natural product-like chemical space for the discovery of bioactive small molecules [15–20]. Molecular scaffolds existing in natural product structures encode evolutionary selected properties for interacting with proteins and therefore represent biologically prevalidated scaffolds [15,21–23]. Natural product-inspired synthesis tends to exploit these core molecular frameworks of natural products as key chemotypes or scaffolds of compound collections [19,24]. The advantage of this approach is that not only the relative positions and nature of substituents and decoration of functional groups can be varied but also different relative stereochemistry patterns that nature may have missed or was not able to create in biosynthesis planning can be generated and therefore a rather wider chemical space around a biologically intriguing scaffold can be covered.

Polycyclic indoles represent core structures of a large number of natural products displaying interesting, potent, and a wide range of biological activities (Fig. 9.1) [25—29]. A surge in the development of new synthetic routes to this class of molecules has happened in recent years in order to tap their potential in medicinal and biological applications [30—33]. These nature inspired small molecules have proven to be important probes to elucidate deeper insights of different biological functions and thereby offering important information in chemical biology and drug discovery research. In this chapter, the development of the synthesis of centrocountins - the mitotic inhibitors and the unraveling of their cellular targets is described. Centrocountins, though structurally similar to some natural indole alkaloid, displayed very distinct biological activities and target centrosomal proteins in affecting the mitotic cell division of different cancer cell lines [34—37].

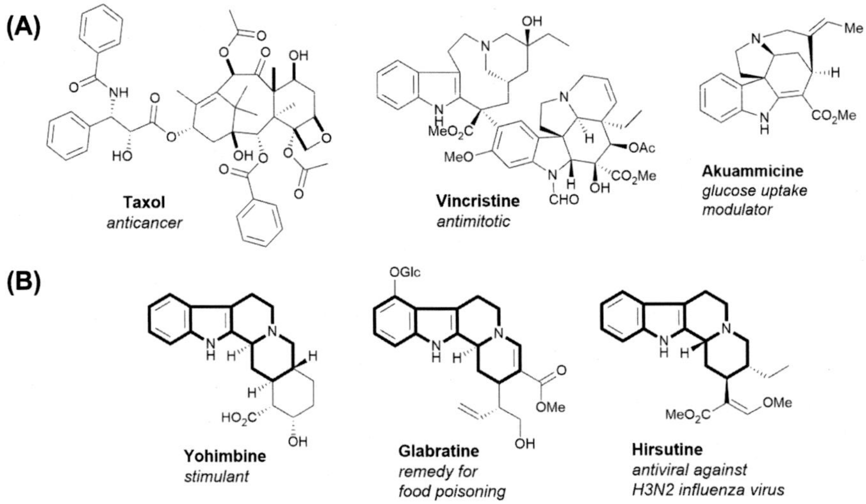

FIGURE 9.1

(A) Biologically active complex natural products and (B) indoloquinolizine-based natural products.

9.2 Synthesis of natural product-inspired Tetrahydroindolo [2,3-*a*]quinolizines

Structurally complex natural products like paclitaxel and vinblastine are successfully applied as anticancer agents [38]. The core molecular-scaffolds of natural products are considered as prevalidated frameworks because they have experienced binding to proteins of the biosynthetic machinery [9]. Therefore, small molecules embodying natural product scaffolds remain intriguing source of novel biologically active compounds. However, organic synthesis remains a demanding factor when it comes to generate structurally complex small molecules and in particular at higher scales (gram scale onwards). Even simple desired stereochemical permutations on a given chemotype can pose a huge synthetic challenge, in particular, on a complex natural product structure. A viable alternative to structurally complex nature's pharmacy is natural product-inspired compounds [39]. Nevertheless, generation of molecular complexity of the level of natural products in a compound collection calls for development of efficient synthesis methods. In this regard, cascade or domino reactions (see Box 9.1) wherein more than one chemical reaction happens consecutively in a reaction sequence in one-pot have gained great attention from chemists in the last decade.

These reactions can rapidly build up the molecular complexity [40] and therefore can offer efficient routes to nature inspired small molecule collections.

Indoloquinolizine-based tetracyclic indoles represent a range of synthetic as well as natural molecules which are often endowed with interesting biological activities (Fig. 9.1B). Biosynthesis of a number of monoterpene indole alkaloids makes use of a reaction of tryptamine with secologanin to form an intermediate strictosidine (Scheme 9.1A). The latter is a common intermediate to a diverse set of indole alkaloids. Inspired by this natural design, Waldmann, Kumar, and coworkers planned a similar synthesis involving tryptamine and benzopyrone (Scheme 9.1B) to gain access to novel indoloquinolizine-based small molecules.

Synthesis of the desired benzopyrone (**5**) was already established by [4 + 2] annulation of commercially available 3-formylchromones (**1**) with acetylene dicarboxylates (**2**)-derived zwitterions (**3**) via addition of catalytic phosphines (Scheme 9.2) [41]. Exploiting this methodology, a concise and efficient cascade synthesis of tetrahydroindolo[2,3-*a*]quinolizines involving a long 12-step sequence of reactions in a one-pot strategy was developed (Scheme 9.2) [42]. In this synthetic method, a solution of 3-formyl-chromones (**1**), acetylene dicarboxylates (**2**) and triphenylphosphine in toluene at 80°C was slowly treated with tryptamine derivatives (**6**) and followed by addition of camphorsulphonic acid (CSA). After 5−30 min, the tetrahydroindolo[2,3-a]quinolizines (**18**) were purified by flash column chromatography. The synthesis offered a very simple and practical method to build complex indoloquinolizines. Also, the commercial availability of differently substituted tryptamines (**6**), 3-formylchromones (**1**) as well as acetylene dicarboxylates (**2**) made it an ideal case for generating a focused small molecule collection based on

Box 9.1 Cascade reactions

Traditionally, a complex molecule is synthesized in a multistep approach by assembling simpler building blocks in a stepwise manner and that means generally performing one reaction at a time, doing a workup and the purification of the product which is then again employed in the next reaction step. Thus, a substrate A is transformed into first product P1 which is purified after the reaction workup and in another reaction transformed into second product P2. Another reaction, workup and purification then affords the final product P. In contrast to this tedious and time-consuming process, in cascade reactions multiple reactions happen one-after another in a sequence to deliver a final product while passing through a number of intermediary products (IP-1, IP-2). The intermediary products are not to be isolated and therefore one needs only single workup and a purification step. For example, Sharp et al. [59] reported a cascade synthesis of medicinally important pyrimido[1,6-*a*]indol-1(2*H*)-ones (**2**) involving two consecutive hydroamination reactions with anilinic diyne substrates (**1**) under the influence of gold catalysis. The reaction went through isolable intermediate **3** that was formed in the first intramolecular 5-*endo*-dig cyclization. A 6-*endo*-dig cyclization then followed to offer the final product **2**. Thus, synthesis involving cascade reactions is efficient, easy, and practical.

(A) Traditional Multi-step Synthesis

$$A \xrightarrow[\substack{\text{work up \&,}\\\text{purification}}]{Reaction\ 1} P1 \xrightarrow[\substack{\text{work up \&}\\\text{purification}}]{Reaction\ 2} P2 \xrightarrow[\substack{\text{work up \&}\\\text{purification}}]{Reaction\ 3} P \quad \substack{\text{Final}\\\text{Product}}$$

Cascade reactions

More than one reactions in a sequence

$$A \longrightarrow \left[\ \to IP\text{-}1 \longrightarrow IP\text{-}2\ \right] \to P \quad \substack{\text{Final}\\\text{Product}}$$

no isolation of intermediates requried   single work up & purification

(B)

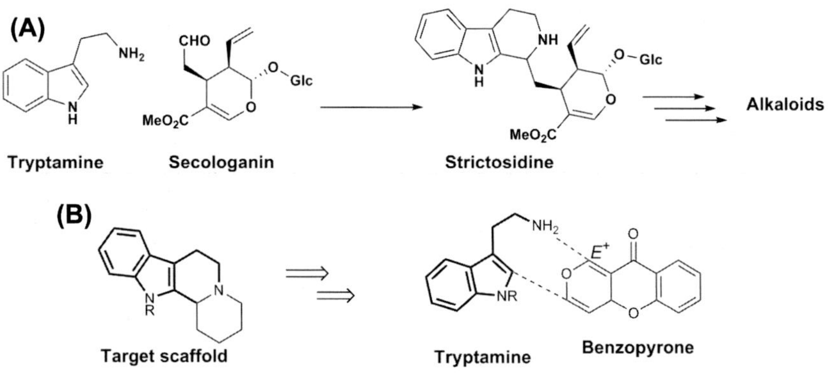

SCHEME 9.1

(A) Biosynthesis of monoterpene indole alkaloids; (B) a retrosynthetic design toward indoloquinolizine-based scaffold.

a natural product scaffold. The above cascade synthesis is the longest sequence of different individual steps and chemical reactions happening one-after another and leading to indoloquinolizines (**18**).

The mechanism of this long cascade reaction was established by isolating and characterizing some of the key intermediates depicted in Scheme 9.2 in dotted boxes. The reaction sequence began with a [4 + 2] annulation of 3-formylchromones **1** with acetylene carboxylates **2**. Addition of phosphine to acetylene dicarboxylates formed phospha-zwitterion (**3**) that underwent conjugated addition to C2-position of the chromone (**1**) and forming intermediate **4** which upon cyclization and phosphine elimination formed the first isolable intermediate— the tricyclic benzopyran (**5**) [41]. The addition of substrates (**1, 2** and PPh$_3$) was performed slowly at 80°C so that within minutes, 3-formylchromone and acetylene carboxylates were consumed (which could otherwise react with tryptamines added next to the reaction) and tryptamine was added slowly that followed a conjugate addition to ring C of the pyrones **5**. Pyrone ring opening was facilitated as phenol moiety served as a leaving group and affording intermediate **7**. The phenol moiety in **7** could add again to the newly generated α,β-unsaturated ester and led to another pyran ring opening to form intermediate **8** bearing an enamine and an α-keto ester in close proximity. A 6π−electrocyclization of triene **9**, which is the isomeric form of the **8** formed α-hydroxy-dihydropyridine **10** which can be isolated and characterized spectroscopically in the absence of acidic conditions. Addition of CSA at this stage (around 2−5 min after tryptamine addition) generated a mild acidic condition that catalyzed dehydration of **10** and leading to tricyclic dihydropyridines **11**. Acid-promoted chromone ring-opening led to form pyridinium salt **12** to which another phenolate addition yielded the tricyclic dienes **13** and thereby set the stage for a sigmatropic aza-Claisen rearrangement affording iminoesters **14**. However, the latter was possible only for intermediates **13** supporting diester function (R^4 = CO$_2$R). No reaction was observed for monoester substrates **13** (R^4 = H). Apparently, ester

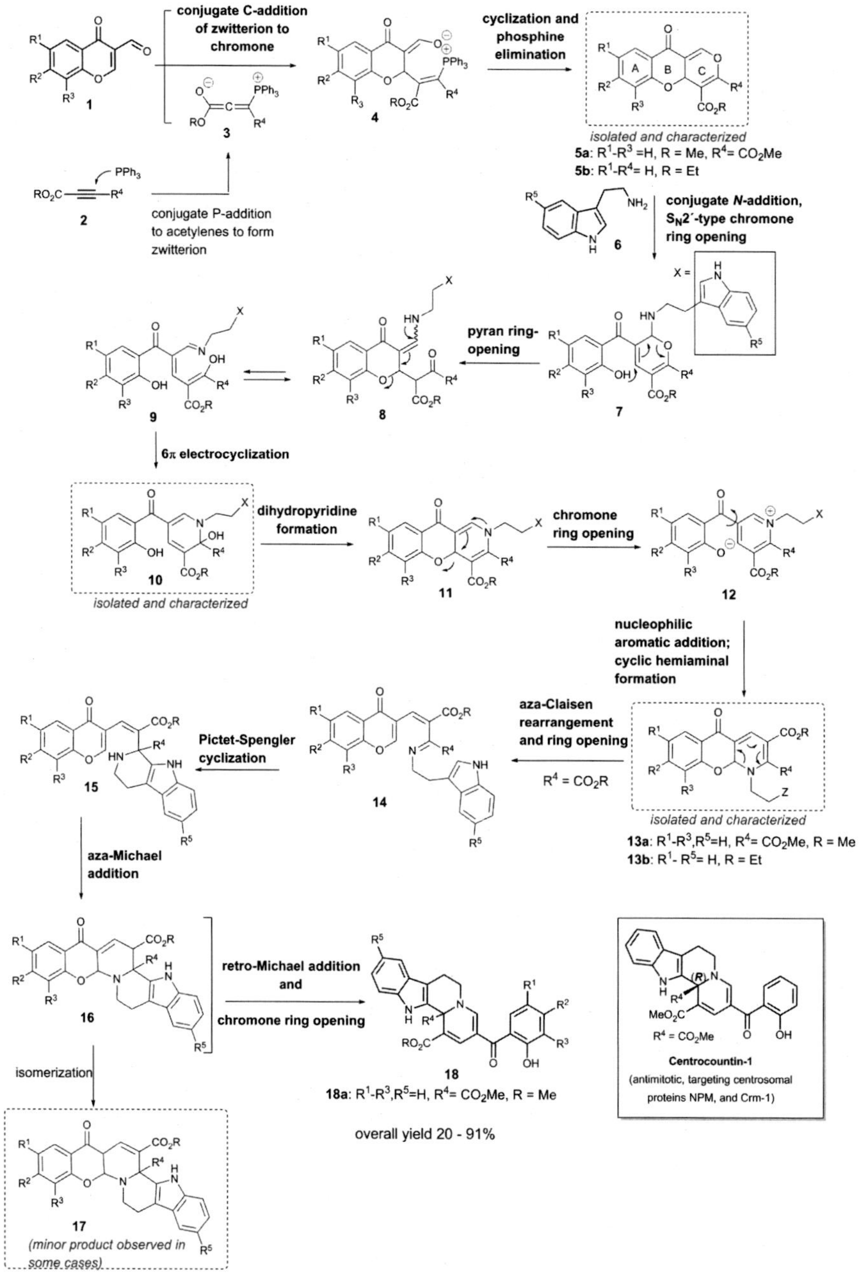

SCHEME 9.2

Cascade synthesis of tetrahydroindolo[2,3-a]quinolizines **(18)**.

function on α-position of dihydropyridine ring lowers the lowest occupied molecular orbital of the olefin and facilitates the sigmatropic rearrangement leading to iminoesters **14**.

Under mild acidic conditions, a Pictet–Spengler cyclization of indole ring on to imino ester moiety led to tetrahydro-β-carbolines **15** supporting a secondary amine in proximity to a highly conjugated ester. Among the final steps of the cascade sequence, a conjugated aza-Michael addition of secondary amines to the doubly vinylogous esters formed the hexacyclic benzopyrans **16**. Finally, an acid mediated pyran ring-opening with phenol serving again as a leaving group generated the indoloquinolizines **18**. Isolation and characterization of hexacyclic molecules **17** formed after a 1,3-*H* shift in **16** corroborated the proposed final steps of the cascade sequence. Although this was mechanistically a very complex cascade reaction, overall yields of products were very high (Scheme 9.2). Only for halogen-substituted 3-formylchromones, around 20% yield of the corresponding indoloquinolizines was observed. A closer look later revealed that for these chromones, the first step of the cascade reaction sequence, *i.e.* [4 + 2] annulation reaction with acetylene carboxylates was itself around 20% yielding. Therefore, this infers that all steps in the cascade reaction sequence after the phosphine-mediated annulation reaction of acetylene dicarboxylates were quantitatively yielding. The cascade synthesis was used to develop a focused library of more than 60 indoloquinolizines [43].

9.3 Phenotypic screening and discovery of centrocountins as novel mitotic inhibitors

Cell-based high-content or phenotypic screens monitor the effect of a compound on a complex living system in its entirety [44]. In recent years, phenotypic screenings have gained a great attention from both academic and industrial research labs, and the small molecules identified in these screens have proven their ability to influence and modulate complex systems in a desired manner [39,45–50]. This is a great advantage as the bioactive molecules obtained from a target-based assay do not necessarily offer the same phenotype due to a number of causes including sometimes their inability to permeate cell membranes, etc [51]. Although, an initial screening may define some structure-activity relationship (SAR) for a compound collection, often few rounds of synthesis and screening followed to acquire more useful information (Fig. 9.2).

Growth of an organism and propagation of life is warranted in a proper cell division. In eukaryotes, cell division is regulated by the cell cycle that consists of an interphase and M phase. The interphase prepares cells for the division process, and M phase ensures the equal segregation of the cytoplasmic and genetic contents of parental cells that were replicated and synthesized during the preceding interphase. The M phase is further divided into mitosis and cytokinesis, and the mitotic phase by itself has five subphases. Thus, M phase in a fully grown cell segregates the replicated chromosomes—the genetic material to opposite poles of a cell, as well as

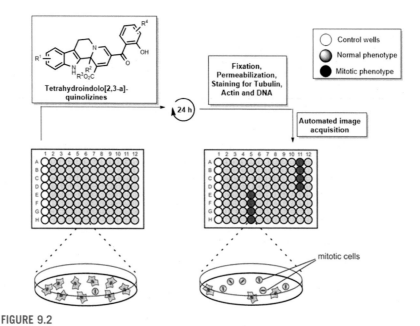

FIGURE 9.2

Phenotypic assay to identify mitotic inhibitors. Cells are seeded in a 96 multi-well plate prior to treatment with indoloquinoilzines for 24 h. Cells are then fixed and stained for tubulin, actin, and DNA. Mitotic inhibitors refer to the compounds which cause the accumulation of round-shaped cells using automated high-content imaging.

cytoplasm in cytokinesis, wherein the cell cleaves between the separated chromosomes to produce two daughter cells. The process of cell cycle is unidirectional and irreversible and is strictly regulated to ensure error-free segregation of chromosomes. This regulation is achieved by activation and inactivation of cyclin-dependent kinases (CDKs) in complex with defined cyclins whereby the presence of cyclins oscillates during the cell cycle. Highly regulated control mechanisms called checkpoints allow for entry into the next cell cycle phase only after a proper completion of the previous one is ensured.

Cancer remains one of the leading causes of death worldwide, and unfortunately, our understanding of these causes and progression of cancer remains limited. There are several hallmarks of cancer, for instance, sustained cellular proliferation and signaling that deregulates cell division [52]. Many of the current anticancer drugs target mitosis via different means, for example, taxol and vinblastine disturb microtubule dynamics [53]. However, a major concern with anticancer drugs is their severe adverse effects that calls for identification of novel antimitotic compounds targeting selectively cancer cells [54].

Cell-based phenotypic screenings have emerged as viable screening strategies to identify mitotic modulators. In order to identify novel small molecule modulators of mitosis, a high-content cell-based assay was established to monitor changes in the

Box 9.2 Immunostaining and Immunoblotting

Immunostaining is a method used to visualize cellular components in cells (immunocytochemistry) or tissues (immunohistochemistry) by using specific antibodies or dyes. Tissue preparation or fixation is essential for the preservation of cell morphology, tissue architecture, and localization of cellular components. This is achieved by seeding cells on glass bottom surfaces (for instance, glass coverslips) and cross-linking with formaldehyde. Cells are permeabilized with a detergent (for example, Triton X-100) to allow detection of intracellular targets using antibodies. Cells can also be fixed and permeabilized with organic solvents like methanol and acetone. The type of fixing method is determined by cellular components that need to be detected as well as the employed antibodies. Cellular components are then detected either directly by means of dyes (e.g., fluorescent dyes that bind to and stain tubulin or DNA) or antibodies that are labeled with fluorophores. In an indirect detection method, a primary antibody specifically binds to a protein of interest and a secondary antibody usually labeled with a fluorophore is then used to recognizes the primary antibody and for detection. The detection of cellular component is usually performed by means of fluorescence microscopy.

 Immunoblotting or Western blotting is an analytical technique to transfer proteins from a gel to a solid support matrix (also called membrane). The proteins are then detected by means of specific antibodies. In this experiment, protein samples are first separated according to their molecular weight by means of sodium dodecyl sulfate polyacrylamide gel electrophoresis (SDS-PAGE). Proteins are then transferred from the SDS gel to a polyvinylidene difluoride (PVDF) or nitrocellulose membrane using electroblotting. The employed membranes generally have high affinity for proteins and free sites on the membrane may bind antibodies. This is prevented by using a blocking solution that contains skimmed milk or bovine serum albumin (BSA) and covers the free sites on membrane. In the next step, the membrane is incubated with primary antibody that specifically binds to the protein of interest. The primary antibody generally lacks a label and therefore is not detectable. After some rinsing to remove unbound primary antibody, the membrane is incubated with a secondary antibody that specifically binds to the primary antibody and enables the indirect detection of the protein of interest by various means of detection. For instance, secondary antibody can be coupled to an enzyme for detection by means of an enzymatic reaction (for example, horseradish peroxidase [HRP] and a chemiluminescence detection or be labeled with a fluorophore to follow fluorescence detection.

cytoskeleton, *i.e.* tubulin and actin. To this aim, African green monkey BSC-1 cells were treated for 24 h with indoloquinolizines (**18**), and the cells were immunostained (see Box 9.2) for tubulin, actin, and DNA. Compounds inducing morphological changes to cells that were related to mitosis were considered as hits. Typically, inhibition of mitosis leads to a disturbed or misalignment of chromosomes in the equatorial plane during metaphase, i.e., metaphase plate, and this is often associated with activation of the spindle assembly checkpoint proteins. Cells arrested in mitosis by small molecules can remain in this phase for long period of time and thereby leading to an accumulation of round-shaped cells. These cells arrested in mitosis are marked with intense staining of the DNA and microtubules which can be detected by fluorescence staining using antibodies or dyes. The spindle assembly checkpoint in cells that are arrested in mitosis is inactivated only after proper error correction. Alternatively, cells often commit suicide by apoptosis triggered by the checkpoint controls of the cells.

The small compound collection of around 60 indoloquinolizines was subjected to the phenotypic assay to identify mitosis modulators at a concentration of 30 μM, and this led to the identification of **18a** (Fig. 9.3) as a mitotic inhibitor. When HeLa cells overexpressing GFP-tubulin and U2OS cells overexpressing

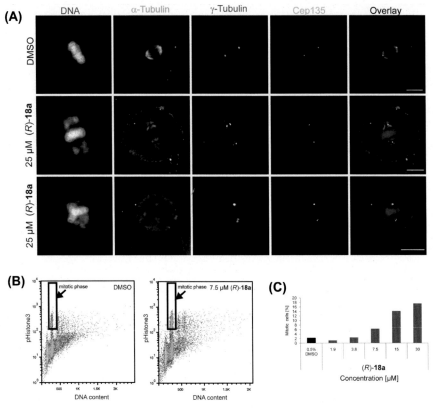

FIGURE 9.3

Influence of (R)-18a on chromosome congression, spindle pole formation and progression of mitosis in HeLa cells. (A) Multiple defects including chromosome congression defects (ii) and aberrant spindle structures (iii) induced by (R)-**18a**. HeLa cells were treated with 25 μM (R)-**18a** for 18 h prior to staining with antibodies for α-tubulin, γ-tubulin, cep135 and for DNA. Images are Z-projections of image stacks done with ImageJ software. Representative images of mitotic cells are shown for cells treated with DMSO (i) and cells treated with 25 μM (R)-**18a** (ii and iii). γ-Tubulin staining is indicative of centrosomes, cep135-staining is indicative of centrioles; Scale bar: 10 μm. (b and c) FACS analysis of HeLa cells treated with (R)-**18a**. Cells were incubated for 18 h with different concentrations of (R)-**18a** and DMSO as control prior to staining with anti-phospho-histone three antibody, Alexa Fluor® 488 secondary antibody, and propidium iodide. (B) Representative histograms for DMSO and 7.5 μM (R)-**18a** and (C) dose-dependent accumulation of mitotic cells are shown.

*Reproduced from Dückert et al., Nature Chem Biology, **2012**, 8, 179–184; https://doi.org/10.1038/NChemBio.758.*

mCherry—α-tubulin were treated with **(R)-18a**, multipolar cell division and abundant formation of three daughter cells during mitosis was observed (Fig. 9.3A). Intriguingly, **(R)-18a** enantiomer (only one chiral center is there in **18**) was significantly more active than **(S)-18a**. **(R)-18a** (centrocountin-1, Scheme 9.2) increased the number of cells with misaligned chromosomes in a dose-dependent manner (from 1 to 25 µM, Fig. 9.3). Increasing concentrations of centrocountin-1 (25—50 µM), further led to increase in the number of mitotic cells with multipolar mitotic spindles (Fig. 9.3A—C). There was a clear mitotic delay observed when cells were treated with **18a**. In fact, centrocountin-1 caused a fivefold prolonged mitosis as compared to the control which clearly indicates for an activated spindle assembly checkpoint and mitotic arrest. The mitotic delay expectedly led to an increase in cell apoptosis. Exposure of a number of different cancer cell lines to centrocountin-1 led to impairment of mitosis and induction of apoptosis and thereby exhibited the anticancer potential of centrocountins.

9.4 Identification and confirmation of cellular targets of Centrocountin-1

In contrast to the target-based reverse chemical genetic approach, a major hurdle for the development of compounds from a phenotypic screening is the identification and validation of their cellular targets related to the phenotype. Centrocountins displayed anticancer potential and a strong and interesting phenotype suggesting many potential targets related to mitosis. However, several potential proteins could be excluded as targets of centrocountin-1. Evaluation of active molecules revealed no inhibition of various kinases involved as checkpoint regulators in mitosis. A number of mitotic modulators target tubulin polymerization process. However, centrocountin-1 did not inhibit tubulin dynamics in tubulin polymerization studies in vitro and in cells. Some other related targets for instance, Eg5, Cdc25a as well as modulation of Wnt signaling etc. were also ruled out for indoloquinolizine **18a**. In order to unravel the mode of action of centrocountins, affinity-based approach that is also known as pull-down assays was thus followed.

An important prerequisite for pull-down approach is to have an SAR to develop chemical probes. Fortunately, the phenotypic screening had delivered a moderate SAR. (R)-enantiomer of **18a** and other active centrocountins were found to exhibit significantly more activity than (S)-enantiomers. Variations in the substitution pattern on the right side of molecule, *i.e.* the aromatic ketone substructure as well as on the central quinolizine core were not tolerated and led to substantial loss of activity. However, variations on the indole part of centrocountins were tolerated without loss of activity and hence provided a site for the attachment of linker in developing probe molecules. Based on this SAR, pull-down probes **19-21** were prepared wherein **19** was active probe with (R)-enantiomer and **20**—the (S)-enantiomer-based molecule was used as negative control (Fig. 9.4). A fluorescent probe **21** supporting Cy3 dye was also prepared for biophysical studies using

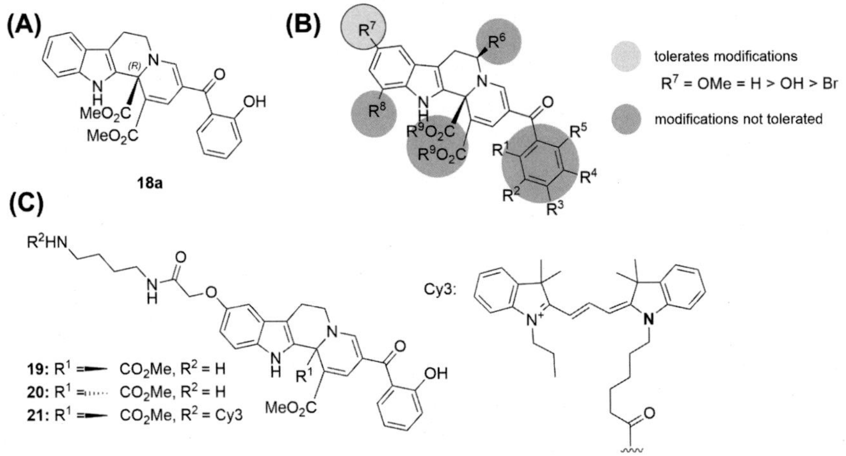

FIGURE 9.4

(A) Chemical structure of centrocountin-1 **(18a)**; (B) structure-activity relationship for centrocountin derivatives; (C) chemical structures of the employed probes **(19—21)**.

fluorescence polarization and FLIM (fluorescence lifetime imaging microscopy) experiments (see Box 9.3).

Based on the SAR, a pull-down probe **19** and the corresponding inactive enantiomer **20** were synthesized and immobilized on NHS-activated sepharose and incubated with the cell lysate. Later, the unspecific bound components were removed by stringent washing. Thus, proteins that remained bound to the matrix were eluted with 10-fold excess of centrocountin-1 (R)-**18a**. Samples of eluted proteins were subjected to SDS-PAGE. The complex protein mixtures was first digested with a trypsin protease that cleaves peptide bonds C-terminally of lysine and arginine and the resulting peptides were partially separated using reversed phase nano-HPLC and identified using MS—MS analysis. In the latter case, HPLC is coupled online with a mass spectrometer via a nanoelectrospray ion source and mass to charge ratio and the charge state of the peptides are determined. The peptides are further fragmented in the mass spectrometer to get partial sequence information of the peptides, i.e., peptide sequence tags. The combined peptide fragmentation fingerprint information is used for a database search in silico, and the experimental results are compared to these theoretical data. Comparison of proteins isolated with active probe **19** and the inactive enantiomer **20** revealed the nucleolar and centrosome-associated protein nucleophosmin (NPM) as a promising target candidate of centrocountin-1. The result was interestingly, further supported by the reported knock-down experiment of NPM by siRNA that in fact phenocopied the effect of centrocountin-1 in HeLa cells [55]. Moreover, exportin-1 or Crm-1 is a binding partner to NPM, and a similar phenotype was also observed for Crm-1 [56,57].

Box 9.3 Fluorescence Polarization (FP) and Fluorescence Lifetime Imaging Microscopy (FLIM)

FP or fluorescence anisotropy is a property of fluorescent molecules and provides information on the orientation and mobility of a fluorophore when exposed to plane polarized light. If a fluorescent labeled molecule is excited by a linearly polarized light, the degree of polarization of the emitted light depends on the rate at which a molecule rotates which is turn indicates the size of the fluorophore. A small fluorescent molecule usually has high motility (rotational and translational) and, when exposed to polarized light, will reduce the degree of polarization of the emitted light. However, a large-sized fluorescent molecule, for instance, a protein bound to a fluorophore, will have a reduced motility and will eventually increase the degree of polarization of the emitted light. FP is often used for the determination of binding constants of two interacting molecules, which cause change in motility of fluorescent molecules, for example, the binding of a compound that is labeled with a fluorophore to a protein. An advantage of this approach is that it is independent of the concentration of the fluorophore.

FLIM is a technique for producing an image based on the differences in the exponential decay rate of the fluorescence donor fluorophore when located in close proximity to an acceptor fluorophore through a fluorescence resonance energy transfer (FRET). In FRET, a radiation-less (without emission of radiation) energy transfer happens from a donor fluorophore to a suitable acceptor fluorophore. The efficiency of the energy transfer is inversely proportional to the distance between the two fluorophores. The fluorescence lifetime (τ) determines how long a fluorophore remains in an excited state upon excitation which depends on the microenvironment around a fluorophore as well as its concentration. FLIM finds applications in detecting protein-protein interaction or protein-compound interaction when they are labeled with suitable fluorophores that enable FRET to happen. Binding of two molecules (proteins or a compound to the protein) brings the fluorophores closer and upon excitation an energy transfer from donor fluorophore to acceptor fluorophore happens that reduces the fluorescence lifetime of the donor.

Further confirmation of NPM and/or Crm-1 as cellular targets of centrocountin-1 was obtained by subjecting the proteins enriched during the pull-down experiment to immunoblotting (see Box 9.2). Pull-down probe **19** enriched both NPM and Crm-1. However, in a competition approach, treating both proteins bound to the probe with increasing concentrations of centrocountin-1, i.e., ((R)-**18a**) abolished the binding of NPM and Crm-1 to the probe (see Fig. 9.5A) which is a strong evidence for NPM and Crm-1 as target proteins of centrocountin-1.

The dissociation constants K_d for individual binding of centrocountin-1-derivative **21** that was labeled with the fluorophore Cy3, to NPM and Crm-1 were determined in an FP experiment (see Box 9.3 and Fig. 9.5B). Furthermore, the binding of **21** to tagged target proteins citrine-NPM or EYFP-Crm-1 was also analyzed with the help of fluorescence lifetime imaging microscopy (FLIM, see Box 9.3). In FLIM experiment, a reduction in the lifetime of the donor fluorophores citrine and EYFP was detected upon addition of **21** (see Fig. 9.5C–D) which corroborate close proximity of the donor and acceptor fluorophore and thus of direct interaction of NPM and Crm-1 with centrocountin-1-based probe **21**.

Interestingly, besides their importance in nuclear transport process, both NPM and Crm-1 play key role in regulating duplication of centrosomes which is a central function guiding a cell division. It seems that NPM associates with the centrosome

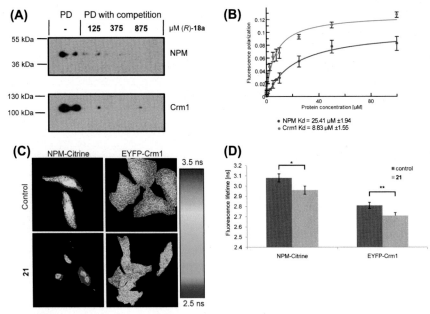

FIGURE 9.5

(A) Immunodetection of NPM and Crm-1 binding to immobilized probe-**19**. After pull-down using immobilized probe-**19**, proteins released by heating were resolved by SDS-PAGE, transferred to a polyvinylidene fluoride membrane and subsequently detected with NPM- and Crm-1-specific antibodies. For competition experiments, HeLa cell lysates were preincubated with different concentrations of (R)-**18a**; (B) binding of **21** to NPM and Crm-1 as determined by means of fluorescence polarization. The Cy3-labeled probe **21** was titrated with increasing concentrations of His-tagged NPM or Crm-1 until saturation was reached. K_d values were determined from the fit of fluorescence polarization vs. His-NPM or Crm-1 concentration; (C) fluorescence lifetime imaging microscopy showing specific binding of NPM-citrine and EYFP-Crm-1 (donor) to Cy3-labeled **21** (acceptor). HeLa cells were transfected with NPM-citrine or EYFP-Crm-1. Twenty-four hours later, 2.5 μM **21** was added. The decrease in the lifetime of the donor after addition of **21** is shown in the lifetime maps of cells. Scale bars, 30 μm; (D) graph showing the decrease in the donor lifetime as compared to the control. (**Citrine** is a member of green fluorescent proteins [GFP] and is a type of yellow fluorescent protein [YFP] which is particularly stable to acidic environment and also possess higher photostability. **EYFP:** *Enhanced yellow fluorescent protein* is one of the brightest fluorescent proteins.)

Reproduced from Dückert et al., Nature Chem Biology, 2012, 8, 179–184; https://doi.org/10.1038/NChemBio.758.

through binding to Crm-1 and blocks centrosomal duplication. Dissociation of NPM (via phosphorylation by CDK2) from the centrosome is a green signal for the duplication of the centrosomes during S-G2 phase of cell cycle. Before entry into mitosis, NPM reassociates with the two centrosomes to prevent their further duplication and thereby it ensures the proper formation of a bipolar mitotic spindle and a proper mitosis [58]. However, cells treated with centrocountin-1 led to an accumulation of mitotic cells with defective mitotic spindles observed by employing specific markers for centrosomes (γ-tubulin) and centrioles (cep135). The results clearly demonstrate that centrocountin-1 impairs the process of centrosome duplication resulting in overduplication as well as fragmentation of centrosomes and formation of acentrosomal spindle poles. These modulations eventually result in uncontrolled mitotic division and finally triggering the apoptotic cell death.

9.5 Conclusion

In the search for novel biologically and medicinally relevant small molecules which not only have therapeutic potential but can also be used to elucidate the complexities of various biological functions in chemical biology research, natural products can still play an important inspirational role. Although structure and to some extent synthesis of centrocountins was based on monoterpene indole alkaloids, in particular, yohimbine and hirsutine, none of these and related indole alkaloids have displayed antimitotic activity. Therefore, natural product-based small molecules although represent natural product-like chemical space, they can potentially reach out to novel biological space. This is an important goal for both chemical biology and drug discovery research. The synthesis of indoloquinolizine compound collection followed by a high-content screening for mitotic inhibitors and the elucidation of mode of action of centrocountin-1 is an exemplary case of a forward chemical genetics approach. In particular, the work demonstrated the strength of organic synthesis, biology, and proteomics as well as biophysical techniques in identification of biologically active small molecules in cells as well as their target deconvolution and validation. A long cascade reaction sequence was used to quickly and efficiently build a focused compound collection of tetrahydroindolo[2,3-*a*]quinolizines which led to the discovery of centrocountins as mitotic inhibitors. In the whole collection, Centrocountin-1 was the most potent mitotic inhibitor and impaired the proper chromosome congression, induced chromosomal misalignment and mitotic spindle defects. The SAR guided to design probes for target identification and validation. NPM and Crm-1 were identified as cellular targets of centrocountins by means of affinity-based proteomics approach coupled to LC-MS/MS analysis and later confirmed as direct targets with the help of immunodetection of the proteins after target enrichment, fluorescence polarization, and fluorescence lifetime imaging microscopy. Most likely, centrocountin-1 binds to NPM and Crm-1 and disturbs the cell cycle by inducing mitotic spindle defects, centrosomal fragmentation, and overduplication and leading to mitotic arrest and finally cell death. Centrocountin-1 remains the only probe molecule binding reversibly to NPM and that can be used as probe for further investigation of these centrosomal proteins to shed more insights into their role in basic biology and applications in drug discovery research.

References

[1] P. Ertl, S. Jelfs, J. Muhlbacher, A. Schuffenhauer, P. Selzer, Quest for the rings. In silico exploration of ring universe to identify novel bioactive heteroaromatic scaffolds, J. Med. Chem. 49 (15) (2006) 4568–4573.

[2] M.S. Butler, Natural products to drugs: natural product-derived compounds in clinical trials, Nat. Prod. Rep. 25 (3) (2008) 475–516.

[3] K.S. Lam, New aspects of natural products in drug discovery, Trends Microbiol. 15 (6) (2007) 279–289.

[4] S.B. Singh, J.F. Barrett, Empirical antibacterial drug discovery — foundation in natural products, Biochem. Pharmacol. 71 (7) (2006) 1006–1015.

[5] D.D. Baker, M. Chu, U. Oza, V. Rajgarhia, The value of natural products to future pharmaceutical discovery, Nat. Prod. Rep. 24 (6) (2007) 1225–1244.

[6] H.O. Villar, M.R. Hansen, Design of chemical libraries for screening, Expert Opin. Drug Discov. 4 (12) (2009) 1215–1220.

[7] S.R. Langdon, N. Brown, J. Blagg, Scaffold diversity of exemplified medicinal chemistry space, J. Chem. Inf. Model. 51 (9) (2011) 2174–2185.

[8] M.L. Lee, G. Schneider, Scaffold architecture and pharmacophoric properties of natural products and trade drugs: application in the design of natural product-based combinatorial libraries, J. Comb. Chem. 3 (3) (2001) 284–289.

[9] A.A. Shelat, R.K. Guy, Scaffold composition and biological relevance of screening libraries, Nat. Chem. Biol. 3 (8) (2007) 442–446.

[10] A. Nören-Muller, I. Reis-Correa, H. Prinz, C. Rosenbaum, K. Saxena, H.J. Schwalbe, D. Vestweber, G. Cagna, S. Schunk, O. Schwarz, H. Schiewe, H. Waldmann, Discovery of protein phosphatase inhibitor classes by biology-oriented synthesis, Proc. Natl. Acad. Sci. USA 103 (28) (2006) 10606–10611.

[11] W. Wilk, T.J. Zimmermann, M. Kaiser, H. Waldmann, Principles, implementation, and application of biology-oriented synthesis (BIOS), Biol. Chem. 391 (5) (2009) 491–497.

[12] K.C. Nicolaou, T. Montagnon (Eds.), Molecules that Changed the World, Wiley-VCH Verlag GmbH & Co. KGaA, Weinheim, Germany, 2008.

[13] K.C. Nicolaou, S.A. Snyder (Eds.), Classics in Total Synthesis II: More Targets, Strategies, Methods, Wiley-VCH Verlag GmbH & Co. KGaA, Weinheim, Germany, 2003.

[14] K.C. Nicolaou, E.J. Sorensen (Eds.), Classics in Total Synthesis: Targets, Strategies, Methods, Wiley-VCH Verlag GmbH & Co. KGaA, Weinheim, Germany, 1996.

[15] R.S. Bon, H. Waldmann, Bioactivity-guided navigation of chemical space, Acc. Chem. Res. 43 (8) (2010) 1103–1114.

[16] R. Breinbauer, I.R. Vetter, H. Waldmann, From protein domains to drug candidates — natural products as guiding principles in the design and synthesis of compound libraries, Angew. Chem. Int. Ed. 41 (16) (2002) 2879–2890.

[17] S. Rizzo, H. Waldmann, Development of a natural-product-derived chemical toolbox for modulation of protein function, Chem. Rev. 114 (9) (2014) 4621–4639.

[18] K. Hubel, T. Lessmann, H. Waldmann, Chemical biology — identification of small molecule modulators of cellular activity by natural product inspired synthesis, Chem. Soc. Rev. 37 (7) (2008) 1361–1374.

[19] K. Kumar, H. Waldmann, Synthesis of natural product inspired compound collections, Angew. Chem. Int. Ed. 48 (18) (2009) 3224–3242.

[20] A.M. Walji, D.W.C. MacMillan, Strategies to bypass the taxol problem. Enantioselective cascade catalysis, a new approach for the efficient construction of molecular complexity, Synlett 10 (2007) 1477–1489.

[21] H. van Hattum, H. Waldmann, Biology-Oriented synthesis: harnessing the power of evolution, J. Am. Chem. Soc. 136 (34) (2014) 11853–11859.

[22] S. Wetzel, A. Schuffenhauer, S. Roggo, P. Ertl, H. Waldmann, Cheminformatic analysis of natural products and their chemical space, Chimia 61 (6) (2007) 355–360.

[23] S. Wetzel, R.S. Bon, K. Kumar, H. Waldmann, Biology oriented synthesis, Angew. Chem. Int. Ed. 50 (2011) 10800–10826.

[24] J.P. Nandy, M. Prakesch, S. Khadem, P.T. Reddy, U. Sharma, P. Arya, Advances in solution- and solid-phase synthesis toward the generation of natural product-like libraries, Chem. Rev. 109 (5) (2009) 1999–2060.

[25] P.J. Praveen, P.S. Parameswaran, M.S. Majik, Bis(indolyl)methane alkaloids: isolation, bioactivity, and syntheses, Synthesis 47 (13) (2015) 1827–1837.

[26] J. Zhang, J.B. Wu, B.K. Hong, W.Y. Ai, X.M. Wang, H.H. Li, X. Lei, Diversity-oriented synthesis of Lycopodium alkaloids inspired by the hidden functional group pairing pattern, Nat. Commun. 5 (2014) 4614.

[27] S.B. Bharate, S. Manda, N. Mupparapu, N. Battini, R.A. Vishwakarma, Chemistry and biology of fascaplysin, a potent marine-derived CDK-4 inhibitor, Mini Rev. Med. Chem. 12 (7) (2012) 650–664.

[28] D. Bialonska, J.K. Zjawiony, Aplysinopsins – marine indole alkaloids: chemistry, bioactivity and ecological significance, Mar. Drugs 7 (2) (2009) 166–183.

[29] K. Higuchi, T. Kawasaki, Simple indole alkaloids and those with a nonrearranged monoterpenoid unit, Nat. Prod. Rep. 24 (4) (2007) 843–868.

[30] J.D. Podoll, Y.X. Liu, L. Chang, S. Walls, W. Wang, X. Wang, Bio-inspired synthesis yields a tricyclic indoline that selectively resensitizes methicillin-resistant *Staphylococcus aureus* (MRSA) to beta-lactam antibiotics, Proc. Natl. Acad. Sci. USA 110 (39) (2013) 15573–15578.

[31] N. Chernyak, D. Tilly, Z. Li, V. Gevorgyan, Cascade carbopalladation-annulation approach toward polycyclic derivatives of indole and indolizine, ARKIVOC (5) (2011) 76–91.

[32] W.W. Zi, Z.W. Zuo, D.W. Ma, Intramolecular dearomative oxidative coupling of indoles: a unified strategy for the total synthesis of indoline alkaloids, Acc. Chem. Res. 48 (3) (2015) 702–711.

[33] Z. Amara, J. Caron, D. Joseph, Recent contributions from the asymmetric aza-Michael reaction to alkaloids total synthesis, Nat. Prod. Rep. 30 (9) (2013) 1211–1225.

[34] M. Garcia-Castro, S. Zimmermann, M.G. Sankar, K. Kumar, Scaffold diversity synthesis and its application in probe and drug discovery, Angew. Chem. Int. Ed. 55 (27) (2016) 7586–7605.

[35] A. Pahl, H. Waldmann, K. Kumar, Exploring natural product fragments for drug and probe discovery, Chimia 71 (10) (2017) 653–660.

[36] D.A. Horton, G.T. Bourne, M.L. Smythe, The combinatorial synthesis of bicyclic privileged structures or privileged substructures, Chem. Rev. 103 (3) (2003) 893–930.

[37] M. Shiri, Indoles in multicomponent processes (MCPs), Chem. Rev. 112 (6) (2012) 3508–3549.

[38] D.J. Newman, G.M. Cragg, Natural products as sources of new drugs over the last 25 years, J. Nat. Prod. 70 (3) (2007) 461–477.

[39] G.T. Carter, Natural products and Pharma 2011: strategic changes spur new opportunities, Nat. Prod. Rep. 28 (11) (2011) 1783−1789.

[40] L.F. Tietze, Domino reactions in organic synthesis, Chem. Rev. 96 (1) (1996) 115−136.

[41] H. Waldmann, V. Khedkar, H. Duckert, M. Schumann, I.M. Oppel, K. Kumar, Asymmetric synthesis of natural product inspired tricyclic benzopyrones by an organocatalyzed annulation reaction, Angew. Chem. Int. Ed. 47 (36) (2008) 6869−6872.

[42] H. Duckert, V. Pries, V. Khedkar, S. Menninger, H. Bruss, A.W. Bird, Z. Maliga, A. Brockmeyer, P. Janning, A. Hyman, S. Grimme, M. Schürmann, H. Preut, K. Hübel, S. Ziegler, K. Kumar, H. Waldmann, Natural product-inspired cascade synthesis yields modulators of centrosome integrity, Nat. Chem. Biol. 8 (2) (2012) 179−184.

[43] V. Eschenbrenner-Lux, H. Duckert, V. Khedkar, H. Bruss, H. Waldmann, K. Kumar, Cascade syntheses routes to the centrocountins, Chem. Eur. J. 19 (7) (2013) 2294−2304.

[44] J.C. Yarrow, Y. Feng, Z.E. Perlman, T. Kirchhausen, T.J. Mitchison, Phenotypic screening of small molecule libraries by high throughput cell imaging, Comb. Chem. High Throughput Screen. 6 (4) (2003) 279−286.

[45] M.A. Bray, S. Singh, H. Han, C.T. Davis, B. Borgeson, C. Hartland, M. Kost-Alimova, S.M. Gustafsdottir, C. C Gibson, A.E. Carpenter, Cell Painting, a high-content image-based assay for morphological profiling using multiplexed fluorescent dyes, Nat. Protoc. 11 (9) (2016) 1757−1774.

[46] I.H. Gilbert, D. Leroy, J.A. Frearson, Finding new hits in neglected disease projects: target or phenotypic based screening? Curr. Top. Med. Chem. 11 (10) (2011) 1284−1291.

[47] K.A. Kelly, S.Y. Shaw, M. Nahrendorf, K. Kristoff, E. Aikawa, S.L. Schreiber, P.A. Clemons, R. Weissleder, Unbiased discovery of in vivo imaging probes through in vitro profiling of nanoparticle libraries, Integr. Biol. 1 (4) (2009) 311−317.

[48] S.J. Swamidass, C.N. Schillebeeckx, M. Matlock, M.R. Hurle, P. Agarwal, Combined analysis of phenotypic and target-based screening in assay networks, J. Biomol. Screen 19 (5) (2014) 782−790.

[49] A. Ursu, H.R. Scholer, H. Waldmann, Small-molecule phenotypic screening with stem cells, Nat. Chem. Biol. 13 (6) (2017) 560−563.

[50] F. Vincent, P. Loria, M. Pregel, R. Stanton, L. Kitching, K. Nocka, R. Doyonnas, C. Steppan, A. Gilbert, T. Schroeter, M.-C. Peakman, Developing predictive assays: the phenotypic screening "rule of 3, Sci. Transl. Med. 7 (293) (2015) 293ps15.

[51] S. Ziegler, V. Pries, C. Hedberg, H. Waldmann, Target identification for small bioactive molecules: finding the needle in the haystack, Angew. Chem. Int. Ed. Engl. 52 (10) (2013) 2744−2792.

[52] D. Hanahan, R.A. Weinberg, Hallmarks of cancer: the next generation, Cell 144 (5) (2011) 646−674.

[53] C. Dumontet, M.A. Jordan, Microtubule-binding agents: a dynamic field of cancer therapeutics, Nat. Rev. Drug Discov. 9 (10) (2010) 790−803.

[54] E. Manchado, M. Guillamot, M. Malumbres, Killing cells by targeting mitosis, Cell Death Differ. 19 (3) (2012) 369−377.

[55] M.A. Amin, S. Matsunaga, S. Uchiyama, K. Fukui, Nucleophosmin is required for chromosome congression, proper mitotic spindle formation, and kinetochore-microtubule attachment in HeLa cells, FEBS Lett. 582 (27) (2008) 3839−3844.

[56] Q. Liu, Q. Jiang, C. Zhang, A fraction of Crm1 locates at centrosomes by its CRIME domain and regulates the centrosomal localization of pericentrin, Biochem. Biophys. Res. Commun. 384 (3) (2009) 383–388.

[57] W. Wang, A. Budhu, M. Forgues, X.W. Wang, Temporal and spatial control of nucleophosmin by the Ran-Crm1 complex in centrosome duplication, Nat. Cell Biol. 7 (8) (2005) 823–830.

[58] M. Okuda, H.F. Horn, P. Tarapore, Y. Tokuyama, A.G. Smulian, P.K. Chan, E.S. Knudsen, I.A. Hofmann, J.D. Snyder, K.E. Bove, K. Fukasawa, Nucleophosmin/B23 is a target of CDK2/Cyclin E in centrosome duplication, Cell 103 (1) (2000) 127–140.

[59] P.P. Sharp, M.G. Banwell, J. Renner, K. Lohmann, A.C. Willis, Consecutive gold(I)-Catalyzed cyclization reactions of o-(Buta-1,3-diyn-1-yl-)-Substituted N-aryl ureas: a one-pot synthesis of pyrimido[1,6-a]indol-1(2H)-ones and related systems, Org. Lett. 15 (11) (2013) 2616–2619.

PPIs as therapeutic targets for anticancer drug discovery: the case study of MDM2 and BET bromodomain inhibitors

Margarida Espadinha[1], Stuart J. Conway[2], Maria M.M. Santos[1]

Research Institute for Medicines (iMed.ULisboa), Faculty of Pharmacy, Universidade de Lisboa, Lisbon, Portugal[1]; Department of Chemistry, Chemistry Research Laboratory, University of Oxford, Oxford, United Kingdom[2]

10.1 Introduction

Protein-protein interactions (PPIs) are essential in the regulation of many biological pathways including apoptosis and cell cycle arrest. When there is an alteration in the balance of PPIs certain pathologies may appear. So, there is significant interest in developing small molecules that can modulate the activity of PPIs.

Most protein-protein interactions (PPIs) have large interfaces and are consequently more difficult to target than enzymes or receptors that typically have well-defined catalytic sites. However, there are already a number of small molecules in clinical trials that can target PPIs, some of which bind to a deep hydrophobic cleft on the protein interface. The rational to develop these ligands is to design compounds that will bind at a "hot spot" disrupting key binding interactions made by the partner (Fig. 10.1).

Using this strategy, it has been possible to identify novel small molecule PPI modulators, several of which have entered clinical trials for the treatment of cancer [1–3]. In particular, potent and highly specific PPI inhibitors have been discovered in the last decade that inhibit antiapoptotic BCL-2 proteins [4–6], the BET bromodomain family [7,8], and p53-MDMs interactions [9–11]. In this chapter, we present two case studies of potent and selective PPIs inhibitors and an overview of some representative anticancer PPI small molecule inhibitors that have reached clinical trials.

10.2 The case study of the p53-MDM2 PPI inhibitor APG-115

p53 is a transcription factor that is responsible for the control of signaling pathways related with, for example, cell cycle, cell differentiation, senescence, angiogenesis,

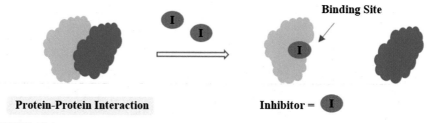

FIGURE 10.1

Orthosteric inhibition of a protein-protein interaction by a small molecule.

and DNA repair [12–14]. In almost half of human cancers, the p53 function is inactivated by negative regulators, which include the proteins MDM2 and MDMX [15–17]. Medicinal chemistry approaches to reactivate the p53 pathway have been mainly focused on inhibiting the p53-MDM2 interaction. However, it is now considered that for an efficient reactivation of p53 it is also important to develop potent and selective p53-MDMX interaction inhibitors (Fig. 10.2) [11]. In fact, in cancer cell lines with wild-type (wt) p53, the use of p53-MDM2/X dual inhibitors for activation of p53 was more effective than the use of only MDM2 inhibitors [18–20]. Crystallographic studies revealed that the interaction surface between p53-MDM2 is surprisingly small for a PPI, at approximately 700 Å^2, presenting an opportunity to use small inhibitors to disrupt this hydrophobic interaction [21]. MDMX also has a deep and hydrophobic pocket and the cleft at the surface is similar

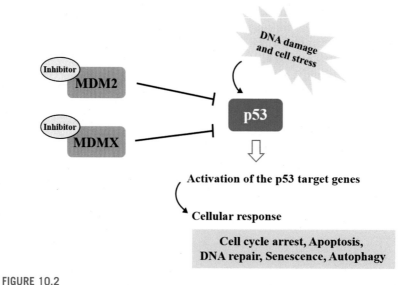

FIGURE 10.2

Activation of the p53 pathway by inhibition of p53-MDM2/X interactions.

to that found in MDM2. However, of the 14 residues responsible to form the ligand-binding cavities of MDM2 and MDMX, four of them are different in the two proteins. This change is responsible for alterations in shape and size of the MDMX pocket and makes it difficult to identify potent p53-MDM2/X dual inhibitors [20].

The first potent p53-MDM2 inhibitors reported were based on the cis-imidazoline scaffold. Nutlin-3a, a cis-imidazoline derivative optimized by Hoffman-La Roche, was the first small molecule to reach clinical trials. However, these compounds did not progress to the clinic because of poor bioavailability and high toxicity. Even so, nutlin-3a is still the tool of choice to study p53 biology. In addition, the first crystal structure of a small molecule in complex with MDM2 was obtained with nutlin-3a, representing an important mark for the development of new classes of MDM2 inhibitors [22,23]. Initially, most of the inhibitors were designed to mimic the three key hydrophobic residues in p53 involved in the interaction with MDM2: Phe19, Trp23, and Leu26 (Fig. 10.3) [24,25]. Later, the p53 Leu22 was also considered a key residue in the design of potent MDM2 inhibitors [26].

One example of a successful hit-to-lead optimization of a MDM2 inhibitor is the drug candidate APG-115. This pyrrolidine spirooxindole is a potent p53-MDM2 interaction inhibitor developed by the research group of Professor Shaomeng Wang. The optimization process took 11 years from the initial structure-based design until obtaining a molecule with adequate pharmacological properties to enter clinical trials [27]. The structural proximity between the p53 Trp23 residue and the spirooxindole core was the rationale behind the discovery of the spirooxindole scaffold (Fig. 10.4).

Substructure search techniques to identify spirooxindole-based natural products, followed by modeling studies, showed that although these alkaloids (e.g., *Spirotryprostatin A* and *Alstonisine*) fit poorly into the MDM2 cleft due to steric hindrance, the spiro(oxindole-3,3′-pyrrolidine) was a good starting point for the design of new class of MDM2 inhibitors. The optimization involved the use of the spirooxindole

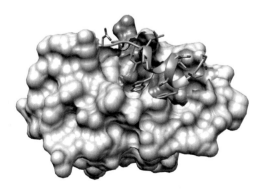

FIGURE 10.3

The three critical p53 residues (Phe19, Trp23, and Leu26) in the p53 binding pocket of MDM2 (PDB entry 1YCR).

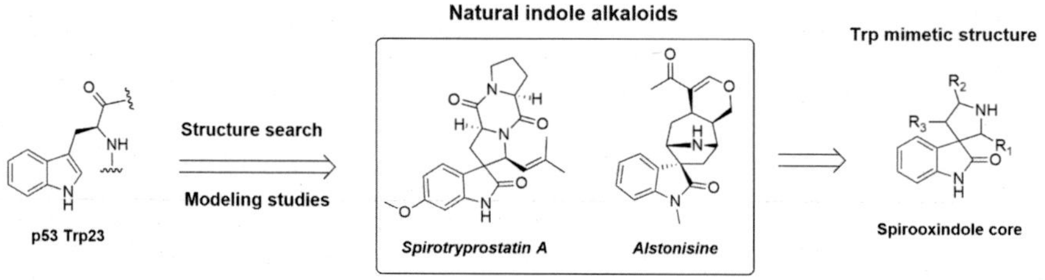

FIGURE 10.4

In silico identification of spirooxindole scaffold for the development of promising p53-MDM2 inhibitors [28].

skeleton as a rigid chemical structure from which two other hydrophobic groups could be projected, as to mimic the Phe19 and p53 Leu26 residues of p53. Libraries of compounds were generated and docked resulting in hit compound **1** (Fig. 10.5) with a K_i value of 8.46 µM. Compound **1** contains a 6-chloro substituent on the oxindole ring that occupies a smaller hydrophobic cavity in MDM2 and that improves the binding to MDM2 [28].

To further improve the binding affinity to MDM2 the other two substituents were changed, and the new analogues were evaluated against MDM2 using a fluorescence polarization-based (FP-based) binding assay.

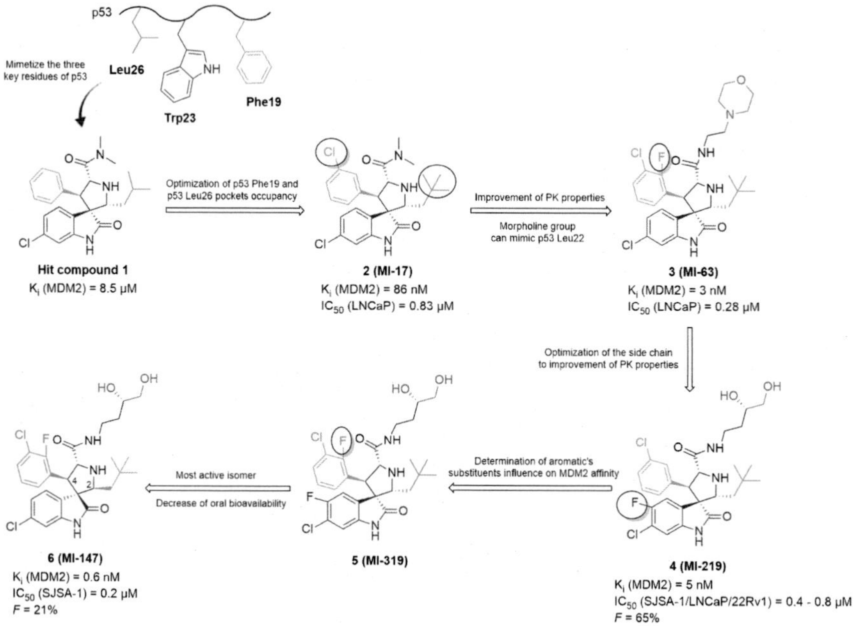

FIGURE 10.5

Structure-based optimization from compound 1 to compound 6 (MI-147).

MI-17 (**2**) (Fig. 10.5), with 2,2-dimethylpropyl and *m*-chlorophenyl substituents, revealed to be 98 times more potent ($K_i = 86$ nM, FP assay) than the initial hit compound **1** ($K_i = 8.46$ μM, FP assay) for MDM2. Moreover, MI-17 (**2**) has an IC_{50} value of 0.83 μM in wt p53 LNCaP human prostate cancer cells, has 27-fold selectivity over deleted p53 cancer cell lines, and is 13 times less toxic in normal human prostate epithelial cells with wild-type p53 [28]. This first set of compounds allowed the identification of spiropyrrolidine oxindoles as a promising chemical family for the development of potent p53-MDM2 interaction inhibitors. Although MI-17 (**2**) had good affinity for MDM2, there was still a significant difference between the activity of this compound and most of the peptide-based MDM2 inhibitors. This result suggested that there could be other important interactions responsible for the higher binding affinity of the small peptides to MDM2 [28]. *In silico* studies of MI-17-MDM2 and p53-MDM2 interactions allowed identification of a fourth p53 residue (Leu22) that should be mimicked to obtain more potent MDM2 inhibitors. As this residue is partially solvent-exposed, the additional substituent to mimic Leu22 could contain some polar moieties improving not only the binding to MDM2 but also the physiochemical properties of the resulting compounds. *In silico* studies of compound MI-17 (**2**) suggested that the replacement of the *N,N*-dimethylamine group by a 2-morpholin-4-yl-ethylamine moiety would allow the establishment of an additional hydrogen bond with MDM2 Lys90 and mimic p53 Leu22. The introduction of a fluorine atom in *ortho* position of the *meta*-chlorophenyl ring that mimics p53 Phe19 was investigated in order to increase metabolic stability. The new derivative MI-63 (**3**) has a K_i value of 3 nM (FP assay) against MDM2 (Fig. 10.5). Interestingly, the enantiomer of MI-63 (**3**) showed a K_i value of only 4.0 μM against MDM2 (FP assay). The results in LNCaP human prostate cancer cell lines were consistent with higher binding affinities to MDM2, with MI-63 (**3**) showing the lowest IC_{50} value (0.28 μM) (Fig. 10.5) [26]. To find an equally potent derivative with better pharmacokinetic (PK) properties an extensive hit-to-lead optimization led to MI-219 (**4**, Fig. 10.5), with a butyl-1,2-diol group mimicking p53 Leu22. This molecule was a potent and selective MDM2 inhibitor with improved oral bioavailability properties ($F = 65\%$) (K_i value of 5.0 nM (FP assay) for MDM2 and IC_{50} values between 0.4 and 0.8 μM in SJSA-1, LNCaP, and 22Rv1 cancer cells). Compound MI-219 (**4**) activated the p53 pathway, inducing cell cycle arrest and apoptosis, in wt p53 cancer cell lines but not in normal cells. Additionally, *in vivo* studies in xenograft mouse models showed complete tumor growth inhibition by compound MI-219 (**4**) [29,30]. Based on the chemical structure of MI-63 (**3**), two other compounds, MI-319 (**5**) and MI-147 (**6**), were designed and synthesized (Fig. 10.5) [29,31]. MI-147 (**6**), with an opposite stereochemistry at positions 2 and 3 of the pyrrolidine ring, was the most potent, specific and cell permeable spiropyrrolidine oxindole, but had lower oral bioavailability (K_i value of 0.6 nM to MDM2, $IC_{50} = 0.2$ μM in SJSA-1 and $F = 21\%$) (Fig. 10.5). Some of these compounds, such as compounds MI-219 (**4**) and MI-319 (**5**), were tested in combination with other chemotherapeutic agents and showed promising results [32–34].

Despite the excellent results obtained with compounds MI-219 (**4**) and MI-147 (**6**), it was observed that in protic solvents some of the spirooxindole MDM2 inhibitors can afford four diastereomers by a reversible ring-opening-cyclization reaction of the pyrrolidine ring. After a meticulous isomerization study, new MDM2 inhibitors with a different stereochemistry were identified. The cis-2,3-cis-3,4 isomer was the most stable and active configuration. An improvement of PK properties led to MI-888 (**7**, K_i [MDM2] = 0.44 nM, IC_{50} [SJSA-1] = 0.08 μM) (Fig. 10.6), able to induce a complete tumor regression in a SJSA-1 tumor xenograft model [35,36]. Compound MI-77301/SAR405838 [37] (**8**, K_i [MDM2] = 0.88 nM, IC_{50} [SJSA-1] = 0.09 μM) (Fig. 10.6), with a 4-hydroxycyclohexyl group, revealed similar potency to compound MI-888 (**7**) but improved PK properties [37−39]. This small molecule was selected for clinical development and completed phase I clinical trial in patients with advanced cancer.

In 2014, a second generation of spiropyrrolidine oxindoles was developed in order to avoid the reversible ring-opening and cyclization of the pyrrolidine ring and consequent isomerization, leading to MI-1061 (**9**, K_i [MDM2] = 0.16 nM, IC_{50} [SJSA-1] = 0.10 μM) (Fig. 10.6). The presence of a symmetrical pyrrolidine C2 position allowed a rapid and irreversible conversion to the preferred diastereoisomer for MDM2 binding [40]. Based on this second generation of spirooxindoles, an extensive structure−activity relationship (SAR) study was performed to improve

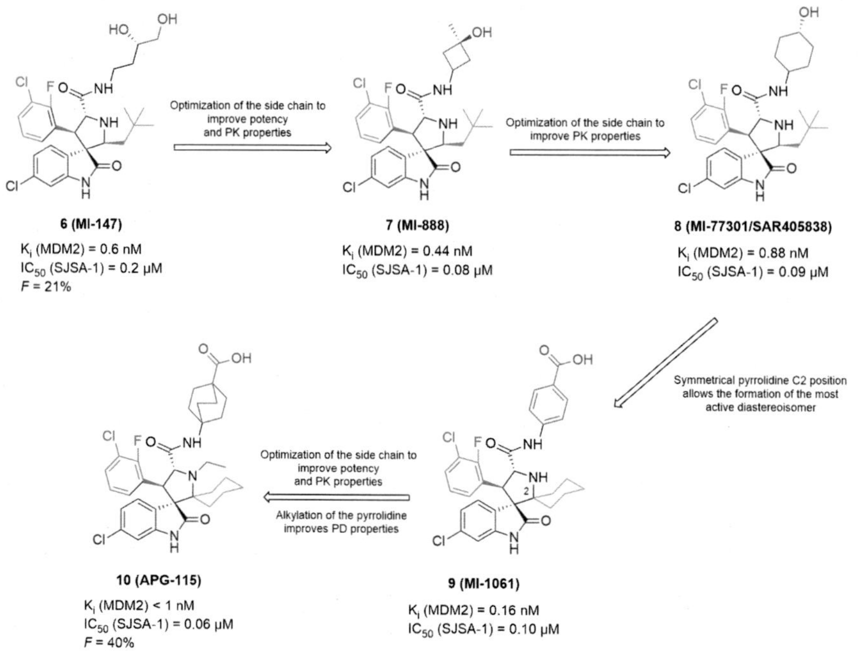

6 (MI-147)
K_i (MDM2) = 0.6 nM
IC_{50} (SJSA-1) = 0.2 μM
F = 21%

Optimization of the side chain to improve potency and PK properties

7 (MI-888)
K_i (MDM2) = 0.44 nM
IC_{50} (SJSA-1) = 0.08 μM

Optimization of the side chain to improve PK properties

8 (MI-77301/SAR405838)
K_i (MDM2) = 0.88 nM
IC_{50} (SJSA-1) = 0.09 μM

10 (APG-115)
K_i (MDM2) < 1 nM
IC_{50} (SJSA-1) = 0.06 μM
F = 40%

Optimization of the side chain to improve potency and PK properties
Alkylation of the pyrrolidine improves PD properties

9 (MI-1061)
K_i (MDM2) = 0.16 nM
IC_{50} (SJSA-1) = 0.10 μM

Symmetrical pyrrolidine C2 position allows the formation of the most active diastereoisomer

FIGURE 10.6

Structure-based optimization from compound 1 to APG-115 (10).

potency and PK and PD properties. The substitution of a 4-hydroxycyclohexyl group by a bicyclo[2.2.2]octane-4-carboxylic acid group and, also, the presence of an ethyl group on the pyrrolidine moiety led to APG-115 (**10**) (Fig. 10.6). This drug candidate is currently in phase I/II clinical trials (sponsored by *Ascentage Pharma*) [41−43].

APG-115 (**10**) showed tight binding to MDM2 ($K_i < 1$ nM) and IC_{50} values lower than 100 nM in several human cancer cell lines with wild-type p53 of different tumor types (SJSA-1, Saos2, RS4;11, LNCaP, PC3, HCT116). p53-mediated apoptosis and increase of p53 and p21 levels were observed in human tumor cells with wt p53 status. In xenograft mice studies, APG-115 (**10**) achieved completed regression of SJSA-1 xenograft tumors and showed to be well tolerated [27].

APG-115 (**10**), designed to mimic the short helix of p53, showed interaction with MDM2 and competes directly with p53 in the p53-MDM2 interaction. APG-115 (**10**) is a clear example that the p53-MDM2 interaction is druggable and can be targeted using small chemical ligands, combining potency, cell permeability, and oral activity.

10.3 Development of the BET bromodomain ligand I-BET762

Epigenetic phenomena are defined as "a stably heritable phenotype resulting from changes in a chromosome without alterations in the DNA sequence" [44]. These changes include chemical modification of DNA [45], for example, methylation, modification of RNA [46], and the posttranslational modification of histone proteins [47,48]. A wide range of PTMs are found on histone proteins [47]; of these, lysine acetylation is one of the most studied.

Lysine residues are acetylated by lysine acetyl transferases (KATs) and deacetylated by lysine deacetylases (KDACs) [49]. Bromodomains are protein modules that bind to acetylated lysine residues (KAc) and consequently mediate PPIs, or protein-chromatin interactions (PCIs) (Fig. 10.7). These modules are viewed as "readers" of the epigenetic code [50,51], of which KAc is a key component [49,52,53].

While 3600 lysine acetylation sites have been identified in 1750 proteins throughout the cell [54], the role of bromodomains binding to histones has been most heavily studied. This has led to much work on the role of bromodomain-containing proteins (BCPs) in transcription [52]. Humans possess 61 bromodomains found in 46 BCPs. Of these, the bromodomain and extraterminal domain (BET) family of BCPs has been the focus of most research. This family comprises BRD2, BRD3, BRD4, and the testis-specific BRDT, all of which contain two adjacent bromodomains located toward the N-terminus. There has been substantial effort invested in developing small molecule ligands and tools to aid the study of the BET bromodomains [55−58], which has resulted in 10 BET bromodomain ligands in >30 clinical trials (see below).

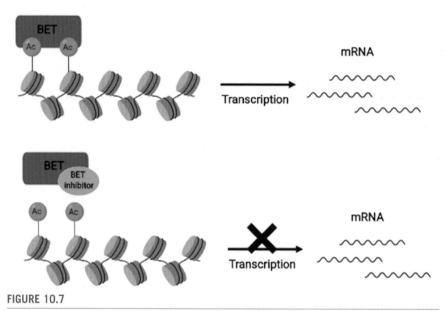

FIGURE 10.7

The effect of a BET inhibitor on posttranslation modifications.

In 2010, two high-affinity BET bromodomain ligands, I-BET762 (originally I-BET) and (+)-JQ1, were published and used to demonstrate that inhibition of BET bromodomain function can cause beneficial anticancer and anti-inflammatory effects [59,60]. This work demonstrated that a PPI, which had previously been considered "undruggable" could be targeted with small molecules [55]. Since then, inhibition of bromodomain function has shown promising results in a number of different cancer types, including in combination with other drugs [61]. I-BET762 (**20**, Fig. 10.13), which was developed by GlaxoSmithKline, was one of the first potent and selective BET bromodomain ligands reported [60,62]. I-BET762 (**20**) was discovered through phenotypic screening to identify molecules with the ability to upregulate apolipoprotein A-1 (ApoA1). The triazolobenzodiazepine (BDZ) **11** was identified as a potent inducer of ApoA1 gene expression, with an EC_{50} value of 440 nM [62] (Fig. 10.8) in a luciferase reporter assay in HepG2 cells. The upregulation of this gene is associated with anti-inflammatory effects and is responsible for preventing the progression of atherosclerosis [63].

Compound **11** (Fig. 10.8) was viewed as a good starting point to find more potent ApoA1 upregulating compounds; however, the hit-to-lead optimization program was carried out with no knowledge of the cellular target. To identify the cellular target of compound **11** a set of structurally related compounds were evaluated against panel of known drug targets (including kinases, GPCRs, nuclear receptors, and ion channels). As this provided no hits, a chemoproteomics approach was adopted [62]. Affinity matrices based on active and inactive derivatives of I-BET762 were prepared. Affinity chromatography using these two matrices and HegG2 cells,

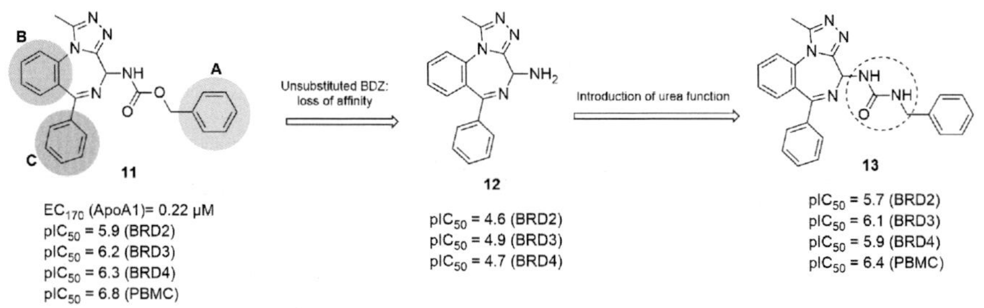

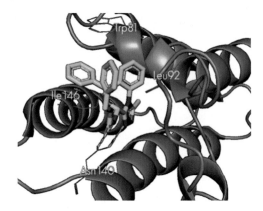

FIGURE 10.8

SAR study around the BZD ring [64].

followed by proteomics analysis, identified the cellular targets of I-BET762 as the bromodomains of BRD2, BRD3, and BRD4 [58,62]. siRNA knock-down of BRD4 resulted in modulation of ApoA1 levels, while siRNA knock-down of BRD2 or BRD3 had no effect, indicating that this was the primary cellular target responsible for the observed activity. BZD **11** revealed similar low micromolar affinity for the BRD2, 3, and 4 ($pIC_{50} = 5.9$ [BRD2], $pIC_{50} = 6.2$ [BRD3], $pIC_{50} = 6.3$ [BRD4]) (Fig. 10.8) determined using a fluorescence polarization assay.

The identification of the BET bromodomains as the cellular targets of compound **11** allowed X-ray structures (PDB entry 2YEL) to be obtained, which helped to rationalize the SAR that had been observed (Fig. 10.9) [62].

Unsubstituted BZD derivative **12** showed a loss of affinity, compared to compound **11**, to the three bromodomain-containing proteins (BRD2, BRD3, and BRD4) (Fig. 10.8). Based on the crystal structure of compound **11** with BRD4, the presence of the benzyl carbamate group is important for bromodomain binding, as the NH carbamate forms a hydrogen bond Asn140 (Fig. 10.9). Given this interaction, the chemical space around this group was explored by inclusion of

FIGURE 10.9

Crystal structure of BRD4 (1) in complex with compound 11 (PDB entry 2YEL).

functionalities including amides, ureas, and sulfonamides. The anti-inflammatory properties of these compounds were also evaluated based on the inhibition of IL-6 production in peripheral blood mononuclear cells (PBMC). The urea derivative **13** showed similar binding affinities to compound **11**, although, when their anti-inflammatory properties in cell lines (PBMC) was evaluated, a loss of activity was observed (Fig. 10.8). This difference of activities can be explained by the lower solubility of the urea derivative **13**. Therefore, the carbamate substituent seemed to be the most adequate choice as substituent in 3-position of the BZD ring.

The phenyl ring (B) fused to the benzodiazepine core is another important part feature of the molecule. This group occupies the area between Leu92 and Pro82, which form part of the ZA channel (Fig. 10.9). The impact of electron-withdrawing (8-nitro and 8-chloro) or donating (8-methoxy and 9-methyl) groups on the affinity to the three BRD subunits was evaluated (Fig. 10.10). Compounds with electron-donating groups showed to be more active than the ones with electron-withdrawing groups but with similar activity to compound **11**. Nevertheless, the BDZ with the 8-methoxy group (compound **14**, $pIC_{50} = 6.1$ [BRD2], $pIC_{50} = 6.1$ [BRD3], $pIC_{50} = 6.3$ [BRD4]) was the most active on the PBMC cell-based assay ($pIC_{50} = 6.9$), and it was selected for further optimization (Fig. 10.10).

The SAR of the pendant phenyl ring (C) binding to the WPF shelf region was explored (Fig. 10.11). Modification of the *ortho*-position led to complete loss of affinity toward BRD2, BRD3, and BRD4 (compound **15**, Fig. 10.11). However, substitutions on the *meta-* or *para*-positions only showed a slight decrease in activity against the three bromodomains, though *para* substitutions exhibit a wider tolerability than *meta* substitutions. For example, compound **16** (4-methoxy) and compound **17** (4-methyl) showed the same affinity to the three bromodomains, but when compound **16** was tested in PBMC cell line there was a loss of potency compared to compound **11** (Fig. 10.11). Fascinatingly, the SAR study around this region was crucial to solve selectivity issues regarding to this family of compounds. The substitution at the 3-position of the BZD ring was described by Filippakopoulos et al. as an effective way to abolish the GABA receptor activity of BZD derivatives, as example JQ1 [65]. GABA receptor is known to be the molecular target of benzodiazepines [66,67], and for example, compound **11** was found to inhibit diazepam to

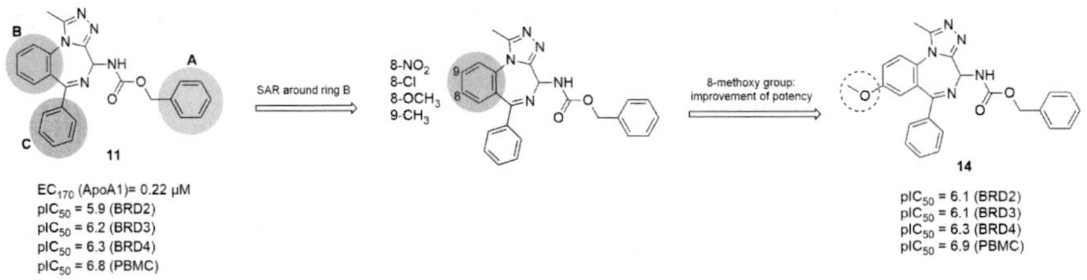

FIGURE 10.10

SAR study around the phenyl ring (2) fused to the BZD core [64].

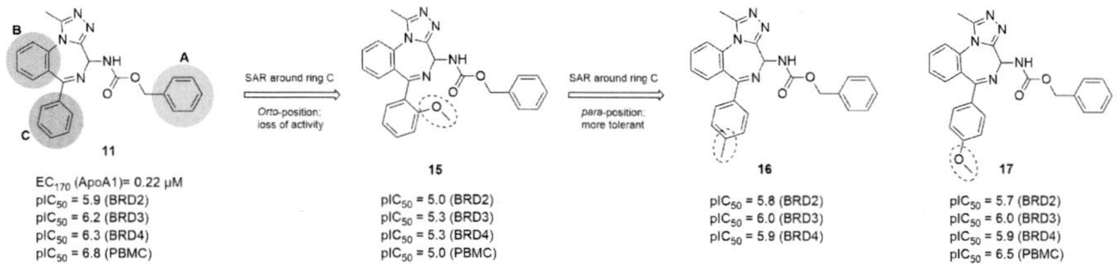

FIGURE 10.11

SAR study around the phenyl ring (C) [64].

bind to the central GABA receptor, revealing affinity to this receptor. Although, the presence of groups on *meta* or *para* positions on the pendant phenyl ring (C) helped to abolish affinity toward GABA receptor. Compounds **16** and **17**, for example, were found to be inactive on GABA receptors.

Based on the crystal structure of **11** bound to BRD4 (1) (PDB entry 2YEL, Fig. 10.9), the carbamate function (3-position of the BZD ring) points toward a solvent-exposed area. This limits the possibility of increasing the bromodomain affinity through modifications in this region. However, substitutions in this position did lead to improvement of the physicochemical properties of the BDZ derivatives. To reduce the molecular weight (MW) and lipophilicity of the compounds, the benzyl group (A) was replaced by alkyl groups, such an ethyl substituent (compound **18**) (Fig. 10.12). The affinity of compound **18** toward the three bromodomain-containing proteins was maintained ($pIC_{50} = 5.6$ [BRD2], $pIC_{50} = 6.0$ [BRD3], $pIC_{50} = 6.1$ [BRD4]), while there was a decrease in cellular potency ($pIC_{50} = 6.3$ [PBMC]) (Fig. 10.12). Due to the drop of potency in the cellular assay, the impact

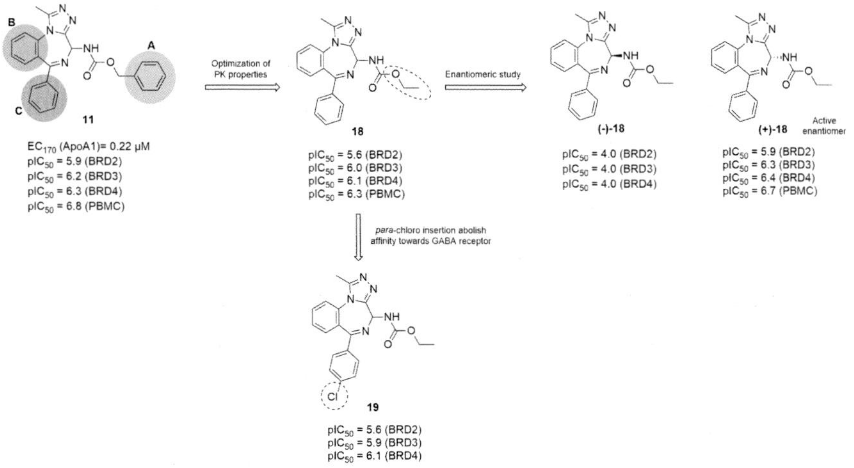

FIGURE 10.12

SAR study around the phenyl ring (C) [64].

of the compound stereochemistry on the affinity for the BET bromodomains was evaluated. Both enantiomers of compound **18** were tested, and surprisingly, the (S)-enantiomer of **18** was found to have lower affinity for BRD2-4 (Fig. 10.12). For example, (S)-enantiomer of **18** showed 251 times higher affinity for BRD4 (1) compared to the (R)-enantiomer of **18**. Combining the alkyl side chain in the CBZ ring and the introduction of *para*-chloro in the pendent ring (C) gave compound **19**, which show no affinity for GABA receptors, as expected. Compound **19** has similar activities to **18** ($pIC_{50} = 5.6$ [BRD2], $pIC_{50} = 5.9$ [BRD3], $pIC_{50} = 6.1$ [BRD4]) but showed improved physicochemical properties (Fig. 10.12).

Another concern regarding these compounds was their susceptibility to hydrolyze in acid, giving rise to ring-opening reaction and formation of the corresponding benzophenone. To mitigate this risk, the nitrogen atom at 3-position of the benzodiazepine ring was removed. The optimized compound was I-BET762 (**20**) which was viewed as lead compound, showing submicromolar affinity toward BRD4 (1) ($pIC_{50} = 6.2$) and potent in the cellular assay ($pIC_{50} = 6.5$ (PBMC)) (Fig. 10.13).

Additionally, I-BET762 (**20**) showed to have excellent preclinical development properties, including solubility, metabolic stability, permeability, and no mutagenic effects. Mutagenic studies with I-BET762 (**20**) were also performed, revealing that the compound is safe. I-BET762 (**20**) also revealed selectivity through other bromodomain-containing proteins, as CREBBP, PCAF, BAZ2B, SP140 and ATAD2. *In vivo* studies demonstrated the efficacy of I-BET762 (**20**) in four animal models (mouse, rat, dog, and primate). I-BET762 (**20**) showed higher clearance in mouse and rat models than dog and primate models. The volumes of distribution were moderate (rat, dog, and primate) and high in mouse. The opposite was observed in I-BET762 (**20**) half-life studies, which was short in rat and longer in the other three models. I-BET762 (**20**) has good oral bioavailability in mouse, dog, and primate models and good *in vitro* solubility and permeability. Discrepancies in the *in vitro* and *in vivo* PK studies for I-BET762 (**20**) led to the use of computational models to predict its PK profile. The results from this work were favorable, and so the compound was progressed to clinical studies in 2012.

I-BET762 (**20**) has been evaluated mainly for nuclear protein in testis (NUT) midline carcinoma (NMC) [68], a rare and lethal form of cancer in which an NUT gene can be found in oncogenic fusion products with bromodomain proteins (BRD4 or BRD3).

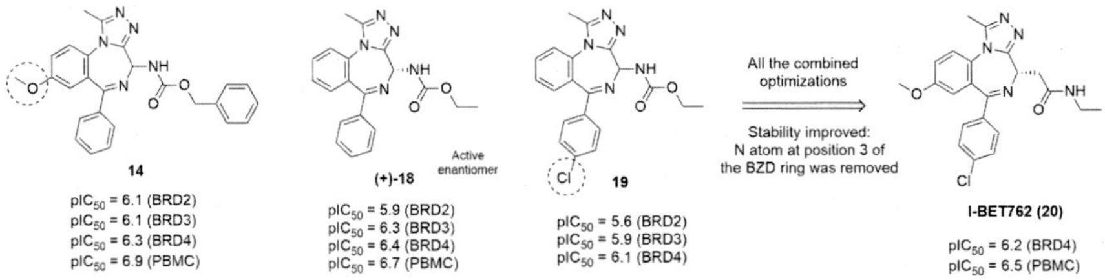

FIGURE 10.13

Lead optimization of I-BET762 [64].

10.4 Inhibitors of PPIs in clinical trials

The biggest success stories of PPIs inhibitors in anticancer drug discovery are the two small molecules that target the antiapoptotic BCL-2 protein, venetoclax (ABT-199) and navitoclax (ABT-263) (Table 10.1). These two drug candidates have completed phase II clinical trials against acute myelogenous leukemia (AML) [69] and platinum-resistant or refractory ovarian cancer [70], respectively. Moreover, venetoclax was the first BH3 mimetic (BCL-2 inhibitor) approved by the Food and Drug Administration (FDA) for cancer treatment of patients with chronic lymphocytic leukemia (CLL) or small lymphocytic lymphoma (SLL), with or without 17p deletion, who have received at least one prior therapy [71].

Two drug candidates that inhibit myeloid cell leukemia-1 (MCL-1), another member of antiapoptotic BCL-2 family, have also advanced to clinical trials (Table 10.1). AMG176, a potent and selective MCL-1 inhibitor, is currently in phase I clinical trials for relapsed or refractory multiple myeloma and for relapsed or refractory acute myeloid leukemia [72]. S64315 (also named MIK665) is in phase I clinical trials for acute myeloid leukemia or myelodysplastic syndrome [73].

Moreover, several small molecules that are pan-BCL-2 family protein inhibitors have advanced to clinical trials such as the natural product R-($-$)-gossypol acetic acid (also known as AT-101) [5].

Another success story in the development of PPI inhibitors is the case of p53-MDM2 interaction inhibitors. For this interaction, many different chemical scaffolds have been identified as MDM2 inhibitors, and some have entered clinical trials for the treatment of multiple cancer types (Table 10.2) [10,11]. Although some drug candidates did not advance to phase II trials (e.g., the nutlin-optimized compound RG7112 developed by Hoffman-La Roche [74], the spirooxindole MI-77301 (SAR405838) developed by Sanofi [75], and MK-8242 developed by Merck) [76], there are still at least seven p53-MDM2 interaction inhibitors in clinical evaluation (AMG-232, APG-115, RG-7112, NVP-CGM097, RG7388, DS-3032b and HDM201). Drug candidate RG7388 (also known as idasanutlin) [77], developed by Hoffman-La Roche, is in phase II clinical trials for treatment of patients with acute myeloid leukemia. APG-115 (**10**, Fig. 10.6) (discussed previously in more detail) [41], and drug candidates CGM097 [78] and HDM201 [79] (developed by Novartis) are currently in phase I clinical trials for advanced solid tumors, while compound AMG-232 (developed by Amgen) has recently completed phase I clinical trials in advanced solid tumors or multiple myeloma (Table 10.3) [80].

Other important therapeutic targets on the development of PPI inhibitors are the bromodomains [81]. Several bromodomain inhibitors have been developed and several have reached clinical trials [82]. Of these, MK-8628 (also known as OTX015), a small molecule that targets BET, has completed phase I clinical trials against advanced solid tumors [83] and hematologic malignancies [84], and

Table 10.1 Representative antiapoptotic BCL-2 family inhibitors in human clinical trials.

Compound	Target	Phase
 ABT-263	BCL-2	Phase II (completed)
 ABT-199	BCL-2	Phase II (completed)
 AMG-176	MCL-1	Phase I
 S64315	MCL-1	Phase I
 AT-101	BCL-2 BCL-XL MCL-1	Phase I/II

Table 10.2 - Representative MDM2 inhibitors in human clinical trials.

Compound	Target	Phase
RG-7388 (Idasanutlin)	MDM2	Phase II
AMG-232	MDM2	Phase I (completed)
APG-115	MDM2	Phase I
CGM097	MDM2	Phase I
HDM201	MDM2	Phase I

I-BET762 (also known as GSK525762, discussed previously in more detail) is in phase I clinical trials for cancer [85]; CPI-0610 [86] and BMS-986158 [87], two small molecule inhibitors of bromodomain and BET proteins, are in phase II clinical trials for myelofibrosis and advanced tumors, respectively, while ABBV-075 (also known as Mivebresib) is in phase I clinical trial in patients with advanced hematologic malignancies and solid tumors [88] (Table 10.3).

Table 10.3 - Representative bromodomain inhibitors in human clinical trials.

Compound	Target	Phase
I-BET-762	BET	Phase I
CPI-0610	BET	Phase II
OTX015	BET	Phase I (completed)
BMS-986158	BET	Phase II
ABBV-075	Pan-inhibitor of bromodomain and BET	Phase I

10.5 Conclusion

In conclusion, in the last years several PPI inhibitors have entered clinical trials for cancer treatment that have either been completed or are on-going. These trials have predominantly been phase I or II. The majority of these PPI inhibitors are being tested as single agents or in combination with other chemotherapeutic agents, and are leading to optimized treatments in cancer patients [10,11,82,89,90]. One successful example is venetoclax which was approved in May 2019 by the Food and Drug Administration (FDA) to treat two different types of blood cancer (chronic lymphocytic leukemia and small lymphocytic lymphoma) [91]. This milestone shows that targeting PPIs is a promising approach for the treatment of cancer.

Acknowledgments

FCT-Fundação para a Ciência e a Tecnologia, I.P., under the project PTDC/QUI-QOR/29664/2017, grant CEECIND/01772/2017 (M. M. M. Santos), and PhD fellowship SFRH/BD/117931/2016 (M. Espadinha). S. J. Conway thanks St Hugh's College, Oxford, for research support.

References

[1] X. Ran, J.E. Gestwicki, Inhibitors of protein—protein interactions (PPIs): an analysis of scaffold choices and buried surface area, Curr. Opin. Chem. Biol. 44 (2018) 75—86.

[2] M.R. Arkin, Y. Tang, J.A. Wells, Small-molecule inhibitors of protein-protein interactions: progressing toward the reality, Chem. Biol. 21 (9) (2014) 1102—1114.

[3] A.A. Ivanov, F.R. Khuri, H. Fu, Targeting protein—protein interactions as an anticancer strategy, Trends Pharmacol. Sci. 34 (7) (2013) 393—400.

[4] A.C. Timucin, H. Basaga, O. Kutuk, Selective targeting of antiapoptotic BCL-2 proteins in cancer, Med. Res. Rev. 39 (1) (2019) 146—175.

[5] W. Xiang, C.Y. Yang, L. Bai, MCL-1 inhibition in cancer treatment, Oncotargets Ther. 11 (2018) 7301—7314.

[6] A.N. Hata, J.A. Engelman, A.C. Faber, The BCL2 family: key mediators of the apoptotic response to targeted anticancer therapeutics, Cancer Discov. 5 (5) (2015) 475—487.

[7] P. Filippakopoulos, S. Knapp, Targeting bromodomains: epigenetic readers of lysine acetylation, Nat. Rev. Drug Discov. 13 (5) (2014) 337—356.

[8] Y. Duan, Y. Guan, W. Qin, X. Zhai, B. Yu, H. Liu, Targeting Brd4 for cancer therapy: inhibitors and degraders, Medchemcomm 9 (11) (2018) 1779—1802.

[9] C.J.A. Ribeiro, C.M.P. Rodrigues, R. Moreira, M.M.M. Santos, Chemical variations on the p53 reactivation theme, Pharmaceuticals 9 (25) (2016) 1—33.

[10] L. Skalniak, E. Surmiak, T.A. Holak, A therapeutic patent overview of MDM2/X-targeted therapies (2014—2018), Expert Opin. Ther. Pat. 29 (3) (2019) 151—170.

[11] M. Espadinha, V. Barcherini, E.A. Lopes, M.M.M. Santos, An update on MDMX and dual MDM2/X inhibitors, Curr. Top. Med. Chem. 18 (8) (2018) 647—660.

[12] L.T. Vassilev, p53 Activation by small molecules: application in oncology, J. Med. Chem. 48 (14) (2005) 4491—4499.

[13] D.F. Tschaharganeh, W. Xue, D.F. Calvisi, M. Evert, T.V. Michurina, L.E. Dow, A. Banito, S.F. Katz, E.R. Kastenhuber, S. Weissmueller, C.H. Huang, A. Lechel, J.B. Andersen, D. Capper, L. Zender, T. Longerich, G. Enikolopov, S.W. Lowe, p53-dependent Nestin regulation links tumor suppression to cellular plasticity in liver cancer, Cell 158 (3) (2014) 579—592.

[14] Y. Zhao, D. Bernard, S. Wang, Small molecule inhibitors of MDM2-p53 and MDMX-p53 interactions as new cancer therapeutics, Biodiscovery 8 (2013) 1—15.

[15] C.J. Brown, S. Lain, C.S. Verma, A.R. Fersht, D.P. Lane, Awakening guardian angels: drugging the p53 pathway, Nat. Rev. Cancer 9 (12) (2009) 862—873.

[16] M. Wade, Y.C. Li, G.M. Wahl, MDM2, MDMX and p53 in oncogenesis and cancer therapy, Nat. Rev. Cancer 13 (2) (2013) 83—96.

[17] K.H. Khoo, C.S. Verma, D.P. Lane, Drugging the p53 pathway: understanding the route to clinical efficacy, Nat. Rev. Drug Discov. 13 (3) (2014) 217—236.

[18] K. Khoury, G.M. Popowicz, T.A. Holak, A. Dömling, The p53-MDM2/MDMX axis - a chemotype perspective, Medchemcomm 2 (2011) 246—260.

[19] A. Gembarska, F. Luciani, C. Fedele, E.A. Russell, M. Dewaele, S. Villar, A. Zwolinska, S. Haupt, J. de Lange, D. Yip, J. Goydos, J.J. Haigh, Y. Haupt, L. Larue, A. Jochemsen, H. Shi, G. Moriceau, R.S. Lo, G. Ghanem, M. Shackleton, F. Bernal, J.C. Marine, MDM4 is a key therapeutic target in cutaneous melanoma, Nat. Med. 18 (8) (2012) 1239—1247.

[20] F. Toledo, G.M. Wahl, MDM2 and MDM4: p53 regulators as targets in anticancer therapy, Int. J. Biochem. Cell Biol. 39 (7—8) (2007) 1476—1482.

[21] P.H. Kussie, S. Gorina, V. Marechal, B. Elenbaas, J. Moreau, A.J. Levine, N.P. Pavletich, Structure of the MDM2 oncoprotein bound to the p53 tumor suppressor transactivation domain, Science 274 (5289) (1996) 948—953.

[22] P. Secchiero, R. Bosco, C. Celeghini, G. Zauli, Recent advances in the therapeutic perspectives of Nutlin-3 Curr, Pharm. Des. 17 (6) (2011) 569—577.

[23] L.T. Vassilev, B.T. Vu, B. Graves, D. Carvajal, F. Podlaski, Z. Filipovic, N. Kong, U. Kammlott, C. Lukacs, C. Klein, N. Fotouhi, E.A. Liu, In vivo activation of the p53 pathway by small-molecule antagonists of MDM2, Science 303 (5659) (2004) 844—848.

[24] J. Chen, V. Marechal, A.J. Levine, Mapping of the p53 and mdm-2 interaction domains, Mol. Cell. Biol. 13 (7) (1993) 4107—4114.

[25] A. Böttger, V. Böttger, C. Garcia-Echeverria, P. Chène, H.K. Hochkeppel, W. Sampson, K. Ang, S.F. Howard, S.M. Picksley, D.P. Lane, Molecular characterization of the hdm2-p53 interaction, J. Mol. Biol. 269 (5) (1997) 744—756.

[26] K. Ding, Y. Lu, Z. Nikolovska-Coleska, G. Wang, S. Qiu, S. Shangary, W. Gao, D. Qin, J. Stuckey, K. Krajewski, P.P. Roller, S. Wang, Structure-based design of spiro-oxindoles as potent, specific small-molecule inhibitors of the MDM2-p53 interaction, J. Med. Chem. 49 (12) (2006) 3432—3435.

[27] A. Aguilar, J. Lu, L. Liu, D. Du, D. Bernard, D. McEachern, S. Przybranowski, X. Li, R. Luo, B. Wen, D. Sun, H. Wang, J. Wen, G. Wang, Y. Zhai, M. Guo, D. Yang, S. Wang, Discovery of 4-((3'R,4'S,5'R)-6"-Chloro-4'-(3-chloro-2-fluorophenyl)-1'-ethyl-2"-oxodispiro[cyclohexane-1,2'-pyrrolidine-3',3"-indoline]-5'-carboxamido)bicyclo [2.2.2]octane-1-carboxylic acid (AA-115/APG-115): a potent and orally active murine double minute 2 (MDM2) inhibitor in clinical development, J. Med. Chem. 60 (7) (2017) 2819—2839.

[28] K. Ding, Y. Lu, Z. Nikolovska-Coleska, S. Qiu, Y. Ding, W. Gao, J. Stuckey, K. Krajewski, P.P. Roller, Y. Tomita, D.A. Parrish, J.R. Deschamps, S. Wang, Structure-based design of potent non-peptide MDM2 inhibitors, J. Am. Chem. Soc. 127 (29) (2005) 10130—10131.

[29] S. Yu, D. Qin, S. Shangary, J. Chen, G. Wang, K. Ding, D. McEachern, S. Qiu, Z. Nikolovska-Coleska, R. Miller, S. Kang, D. Yang, S. Wang, Potent and orally active

small-molecule inhibitors of the MDM2-p53 interaction, J. Med. Chem. 52 (24) (2009) 7970−7973.

[30] S. Shangary, D. Qin, D. McEachern, M. Liu, R.S. Miller, S. Qiu, Z. Nikolovska-Coleska, K. Ding, G. Wang, J. Chen, D. Bernard, J. Zhang, Y. Lu, Q. Gu, R.B. Shah, K.J. Pienta, X. Ling, S. Kang, M. Guo, Y. Sun, D. Yang, S. Wang, Temporal activation of p53 by a specific MDM2 inhibitor is selectively toxic to tumors and leads to complete tumor growth inhibition, Proc. Natl. Acad. Sci. USA 105 (10) (2008) 3933−3938.

[31] R.M. Mohammad, J. Wu, A.S. Azmi, A. Aboukameel, A. Sosin, S. Wu, D. Yang, S. Wang, A.M. Al-Katib, An MDM2 antagonist (MI-319) restores p53 functions and increases the life span of orally treated follicular lymphoma bearing animals, Mol. Cancer 8 (2009) 115.

[32] A.S. Azmi, A. Aboukameel, S. Banerjee, Z. Wang, M. Mohammad, J. Wu, S. Wang, D. Yang, P.A. Philip, F.H. Sarkar, R.M. Mohammad, MDM2 inhibitor MI-319 in combination with cisplatin is an effective treatment for pancreatic cancer independent of p53 function, Eur. J. Cancer 46 (6) (2010) 1122−1131.

[33] A.S. Azmi, P.A. Philip, F.W. Beck, Z. Wang, S. Banerjee, S. Wang, D. Yang, F.H. Sarkar, R.M. Mohammad, MI-219-zinc combination: a new paradigm in MDM2 inhibitor-based therapy, Oncogene 30 (1) (2011) 117−126.

[34] M. Zheng, J. Yang, X. Xu, J.T. Sebolt, S. Wang, Y. Sun, Efficacy of MDM2 inhibitor MI-219 against lung cancer cells alone or in combination with MDM2 knockdown, a XIAP inhibitor or etoposide, Anticancer Res. 30 (9) (2010) 3321−3331.

[35] Y. Zhao, L. Liu, W. Sun, J. Lu, D. McEachern, X. Li, S. Yu, D. Bernard, P. Ochsenbein, V. Ferey, J. Carry, J. Deschamps, D. Sun, S. Wang, Diastereomeric spirooxindoles as highly potent and efficacious MDM2 inhibitors, J. Am. Chem. Soc. 135 (19) (2013) 7223−7234.

[36] Y. Zhao, S. Yu, W. Sun, L. Liu, J. Lu, D. McEachern, S. Shargary, D. Bernard, X. Li, T. Zhao, P. Zou, D. Sun, S. Wang, A potent small-molecule inhibitor of the MDM2-p53 interaction (MI-888) achieved complete and durable tumor regression in mice, J. Med. Chem. 56 (13) (2013) 5553−5561.

[37] Phase 1 safety testing of SAR405838. Retrieved from: https://clinicaltrials.gov/ (Identification No. NCT01636479).

[38] S. Wang, W. Sun, Y. Zhao, D. McEachern, I. Meaux, C. Barrière, J.A. Stuckey, J.L. Meagher, L. Bai, L. Liu, C.G. Hoffman-Luca, J. Lu, S. Shangary, S. Yu, D. Bernard, A. Aguilar, O. Dos-Santos, L. Besret, S. Guerif, P. Pannier, D. Gorge-Bernat, L. Debussche, SAR405838: an optimized inhibitor of MDM2-p53 interaction that induces complete and durable tumor regression, Cancer Res. 74 (20) (2014) 5855−5865.

[39] C.G. Hoffman-Luca, C.Y. Yang, J. Lu, D. Ziazadeh, D. McEachern, L. Debussche, S. Wang, Significant differences in the development of acquired resistance to the MDM2 inhibitor SAR405838 between in vitro and in vivo drug treatment, PLoS One 10 (6) (2015) e0128807.

[40] A. Aguilar, W. Sun, L. Liu, J. Lu, D. McEachern, D. Bernard, J.R. Deschamps, S. Wang, Design of chemically stable, potent, and efficacious MDM2 inhibitors that exploit the retro-mannich ring-opening-cyclization reaction mechanism in spiro-oxindoles, J. Med. Chem. 57 (24) (2014) 10486−10498.

[41] APG-115 in patients with advanced solid tumors or lymphomas (APG-115). Retrieved from: https://clinicaltrials.gov/(Identification No. NCT02935907).

[42] A study of APG-115 in combination with pembrolizumab in patients with metastatic melanomas or advanced solid tumors. Retrieved from: https://clinicaltrials.gov/ (Identification No. NCT03611868).

[43] A phase I/II trial of APG-115 in patients with salivary gland carcinoma. Retrieved from: https://clinicaltrials.gov/(Identification No. NCT03781986).

[44] S.L. Berger, T. Kouzarides, R. Shiekhattar, A. Shilatifard, An operational definition of epigenetics, Genes Dev. 23 (7) (2009) 781–783.

[45] M.J. Booth, E.A. Raiber, S. Balasubramanian, Chemical methods for decoding cytosine modifications in DNA, Chem. Rev. 115 (6) (2015) 2240–2254.

[46] K. Meyer, R. Jaffrey, The dynamic epitranscriptome: N6-methyladenosine and gene expression control, Nat. Rev. Mol. Cell Biol. 15 (5) (2014) 313–326.

[47] M.M. Müller, T.W. Muir, Histones: at the crossroads of peptide and protein chemistry, Chem. Rev. 115 (6) (2015) 2296–2349.

[48] Z. Su, J.M. Denu, Reading the combinatorial histone language, ACS Chem. Biol. 11 (3) (2016) 564–574.

[49] M. Schiedel, S.J. Conway, Small molecules as tools to study the chemical epigenetics of lysine acetylation, Curr. Opin. Chem. Biol. 45 (2018) 166–178.

[50] K.E. Gardner, C.D. Allis, B.D. Strahl, Operating on chromatin, a colorful language where context matters, J. Mol. Biol. 409 (1) (2011) 36–46.

[51] B.D. Strahl, C.D. Allis, The language of covalent histone modifications, Nature 403 (6765) (2000) 41–45.

[52] T. Fujisawa, P. Filippakopoulos, Functions of bromodomain-containing proteins and their roles in homeostasis and cancer, Nat. Rev. Mol. Cell Biol. 18 (4) (2017) 246–262.

[53] S.J. Conway, Bromodomains: are readers right for epigenetic therapy? ACS Med. Chem. Lett. 3 (9) (2012) 691–694.

[54] C. Choudhary, C. Kumar, F. Gnad, M.L. Nielsen, M. Rehman, T.C. Walther, J.V. Olsen, M. Mann, Lysine acetylation targets protein complexes and co-regulates major cellular functions, Science 325 (5942) (2009) 834–840.

[55] D.S. Hewings, T. Rooney, l. Jennings, D. Hay, C. Schofield, P. Brennan, S. Knapp, S.J. Conway, Progress in the development and application of small molecule inhibitors of bromodomain–acetyl-lysine interactions, J. Med. Chem. 55 (22) (2012) 9393–9413.

[56] M. Brand, A.R. Measures, B.G. Wilson, W.A. Cortopassi, R. Alexander, M. Höss, D.S. Hewings, T.P. Rooney, R.S. Paton, S.J. Conway, Small molecule inhibitors of bromodomain-acetyl-lysine interactions, ACS Chem. Biol. 10 (1) (2015) 22–39.

[57] M. Schiedel, M. Moroglu, D. Ascough, A. Chamberlain, J. Kamps, A. Sekirnik, S.J. Conway, Chemical epigenetics: the impact of chemical- and chemical biology techniques on bromodomain target validation, Angew. Chem. Int. Ed. 58 (2019). Available only at: https://onlinelibrary.wiley.com/doi/epdf/10.1002/anie.201812164.

[58] L.E. Jennings, A.R. Measures, B.G. Wilson, S.J. Conway, Phenotypic screening and fragment-based approaches to the discovery of small-molecule bromodomain ligands, Future Med. Chem. 6 (2) (2014) 179–204.

[59] P. Filippakopoulos, J. Qi, S. Picaud, Y. Shen, W.B. Smith, O. Fedorov, E.M. Morse, T. Keates, T.T. Hickman, I. Felletar, M. Philpott, S. Munro, M.R. McKeown, Y. Wang, A.L. Christie, N. West, M.J. Cameron, B. Schwartz, T.D. Heightman, N. La Thangue, C.A. French, O. Wiest, A.L. Kung, S. Knapp, J.E. Bradner, Selective inhibition of BET bromodomains, Nature 468 (7327) (2010) 1067–1073.

[60] E. Nicodeme, K.L. Jeffrey, U. Schaefer, S. Beinke, S. Dewell, C.W. Chung, R. Chandwani, I. Marazzi, P. Wilson, H. Coste, J. White, J. Kirilovsky, C.M. Rice,

J.M. Lora, R.K. Prinjha, K. Lee, A. Tarakhovsky, Suppression of inflammation by a synthetic histone mimic, Nature 468 (7327) (2010) 1119–1123.

[61] A.C. Belkina, G.V. Denis, BET domain co-regulators in obesity, inflammation and cancer, Nat. Rev. Cancer 12 (7) (2012) 465–477.

[62] C.W. Chung, H. Coste, J.H. White, O. Mirguet, J. Wilde, R.L. Gosmini, C. Delves, S.M. Magny, R. Woodward, S.A. Hughes, E.V. Boursier, H. Flynn, A.M. Bouillot, P. Bamborough, J.M. Brusq, F.J. Gellibert, E.J. Jones, A.M. Riou, P. Homes, S.L. Martin, I.J. Uings, J. Toum, C.A. Clement, A.B. Boullay, R.L. Grimley, F.M. Blandel, R.K. Prinjha, K. Lee, J. Kirilovsky, E. Nicodeme, Discovery and characterization of small molecule inhibitors of the BET family bromodomains, J. Med. Chem. 54 (11) (2011) 3827–3838.

[63] P.J. Barter, S. Nicholls, K.A. Rye, G.M. Anantharamaiah, M. Navab, A.M. Fogelman, Antiinflammatory properties of HDL, Circ. Res. 95 (8) (2004) 764–872.

[64] O. Mirguet, R. Gosmini, J. Toum, C.A. Clément, M. Barnathan, J.M. Brusq, J.E. Mordaunt, R.M. Grimes, M. Crowe, O. Pineau, M. Ajakane, A. Daugan, P. Jeffrey, L. Cutler, A.C. Haynes, N.N. Smithers, C.W. Chung, P. Bamborough, I.J. Uings, A. Lewis, J. Witherington, N. Parr, R.K. Prinjha, E. Nicodème, Discovery of epigenetic regulator I-BET762: lead optimization to afford a clinical candidate inhibitor of the BET bromodomains, J. Med. Chem. 56 (19) (2013) 7501–7515.

[65] P. Filippakopoulos, S. Picaud, O. Fedorov, M. Keller, M. Wrobel, O. Morgenstern, F. Bracher, S. Knapp, Benzodiazepines and benzotriazepines as protein interaction inhibitors targeting bromodomains of the BET family, Bioorg. Med. Chem. 20 (6) (2012) 1878–1886.

[66] E. Sigel, M. Ernst, The benzodiazepine binding sites of GABA, Trends Pharmacol. Sci. 39 (7) (2018) 659–671.

[67] E. Sigel, B.P. Lüscher, A closer look at the high affinity benzodiazepine binding site on GABAA receptors, Curr. Top. Med. Chem. 11 (2) (2011) 241–246.

[68] Retrieved from: https://clinicaltrials.gov/(Identification Name: GSK525762).

[69] A phase 2 study of ABT-199 in subjects with acute myelogenous leukemia (AML). Retrieved from: https://clinicaltrials.gov/(Identification No. NCT01994837).

[70] A study of ABT-263 as single agent in women with platinum resistant/refractory recurrent ovarian cancer (MONAVI-1). Retrieved from: https://clinicaltrials.gov/(Identification No. NCT02591095).

[71] Available at: https://www.fda.gov/drugs/resources-information-approved-drugs/fda-approves-venetoclax-cll-or-sll-or-without-17-p-deletion-after-one-prior-therapy.

[72] AMG 176 first in human trial in subjects with relapsed or refractory multiple myeloma and subjects with relapsed or refractory acute myeloid leukemia. Retrieved from: https://clinicaltrials.gov/(Identification No. NCT02675452).

[73] Phase I study of S64315 administred intravenously in patients with acute myeloid leukaemia or myelodysplastic syndrome. Retrieved from: https://clinicaltrials.gov/(Identification No. NCT02979366).

[74] Retrieved from: https://www.roche.com/dam/jcr:c6e746ee-f1e9-4e83-9f2a-de228e73ee18/en/irp2q13e-annex.pdf.

[75] Retrieved from: https://www.sanofi.com/en/science-and-innovation/clinical-trials-and-results/our-disclosure-commitments/pharma/-/media/Project/One-Sanofi-Web/Websites/Global/Sanofi-COM/Home/common/docs/clinical-study-results/TCD13388_summary.pdf.

[76] Retrieved from: https://clinicaltrials.gov/(Identification Name: MK-8242). (accessed 19.07.2019).

[77] Retrieved from: https://clinicaltrials.gov/(Identification Name: RG7388).

[78] A phase I dose escalation study of CGM097 in adult patients with selected advanced solid tumors (CCGM097X2101). Retrieved from: https://clinicaltrials.gov/ (Identification No. NCT01760525).

[79] Study to determine and evaluate a safe and tolerated dose of HDM201 in patients with selected advanced tumors that are TP53wt. Retrieved from: https://clinicaltrials.gov/ (Identification No. NCT02143635).

[80] A phase 1 study evaluating AMG 232 in advanced solid tumors or multiple myeloma. Retrieved from: https://clinicaltrials.gov/(Identification No. NCT01723020).

[81] M. Pervaiz, P. Mishra, S. Günther, Bromodomain drug discovery — the past, the present, and the future, Chem. Rec. 18 (12) (2018) 1808—1817.

[82] A.G. Cochran, A.R. Conery, R.J. Sims, Bromodomains: a new target class for drug development, Nat. Rev. Drug Discov. (18) (2019), 609—628.

[83] A dose-finding study of MK-8628, a small molecule inhibitor of the Bromodomain and Extra-Terminal (BET) proteins, in adults with selected advanced solid tumors (MK-8628-003). Retrieved from: https://clinicaltrials.gov/(Identification No. NCT02259114).

[84] A dose-finding study of the bromodomain (Brd) inhibitor OTX015/MK-8628 in hematologic malignancies (MK-8628-001). Retrieved from: https://clinicaltrials.gov/ (Identification No. NCT01713582).

[85] A dose escalation study to investigate the safety, pharmacokinetics (PK), pharmacodynamics (PD) and clinical activity of GSK525762 in subjects with relapsed, refractory hematologic malignancies. Retrieved from: https://clinicaltrials.gov/(Identification No. NCT01943851).

[86] A phase 2 study of CPI-0610 with and without ruxolitinib in patients with myelofibrosis. Retrieved from: https://clinicaltrials.gov/(Identification No. NCT02158858).

[87] Study of BMS-986158 in subjects with select advanced cancers (BET). Retrieved from: https://clinicaltrials.gov/(Identification No. NCT02419417).

[88] A study evaluating the safety and pharmacokinetics of ABBV-075 in subjects with cancer. Retrieved from: https://clinicaltrials.gov/(Identification No. NCT02391480).

[89] B.L. Lampson, M.S. Davids, The development and current use of BCL-2 inhibitors for the treatment of chronic lymphocytic leukemia, Curr. Hematol. Malig. Rep. 12 (1) (2017) 11—19.

[90] G.F. Perini, G.N. Ribeiro, J.V. Pinto Neto, L.T. Campos, N. Hamerschlak, BCL-2 as therapeutic target for hematological malignancies, J. Hematol. Oncol. 11 (65) (2018) 1—15.

[91] Retrieved from: https://www.fda.gov/drugs/resources-information-approved-drugs/fda-approves-venetoclax-cll-and-sll.

Discovery of small molecules for the treatment of Alzheimer's disease

11

Praveen P.N. Rao, Amy Trinh Pham, Arash Shakeri

School of Pharmacy, Health Sciences Campus, University of Waterloo, Waterloo, ON, Canada

11.1 Introduction

Alzheimer's disease (AD) is a devastating brain disorder, which causes loss of memory, cognitive deficits and personality changes ultimately leading to dementia and death [1−3]. AD is the most common cause of dementia. With an aging population and increasing lifespan, the number of AD cases is on an upward swing worldwide that is affecting the health, economic and social landscapes of many countries [4−6]. A recent literature shows that currently there are about 40−50 million people suffering from dementia [7]. Recognizing this, the World Health Organization (WHO) endorsed and released *Global Action Plan on the Public Health Response to Dementia 2017−25* which calls for action to reduce the risk of dementia by establishing a number of initiatives across the globe [8,9]. The current standard of care for AD is therapy based on small molecules belonging to cholinesterase (ChE) inhibitors (Fig. 11.1) including donepezil (Aricept®), galantamine (Reminyl®), and rivastigmine (Exelon®). The N-methyl-D-Aspartate (NMDA) receptor antagonist memantine (Fig. 11.1) is also used in the pharmacotherapy [10,11]. Unfortunately, these drugs are able to provide only symptomatic relief and are not able to prevent the disease progression or offer any cure [12].

The last decade has seen tremendous advances in understanding the mechanisms of AD and disease pathways. It is now clear that the cholinergic dysfunction theory is only the tip of the iceberg and that AD pathophysiology is a lot more complex and is multifactorial. For example, amyloid cascade hypothesis has demonstrated the role of amyloid-beta (Aβ) in promoting neurotoxicity, neurodegeneration, and disease progression [13]. It gets accumulated and results in the formation of characteristic plaques in AD brain. This evidence also supported the development of therapeutics to reduce the accumulation of Aβ species in brain. Accordingly, misprocessing of the neuronal amyloid precursor protein (APP) by β- and γ-secretases results in the formation Aβ40 and Aβ42 peptides which undergo misfolding and aggregation to form toxic β-sheet structures with oligomers being more toxic as compared to fibrils. Furthermore, the imbalance in the formation and clearance of

Small Molecule Drug Discovery. https://doi.org/10.1016/B978-0-12-818349-6.00011-X

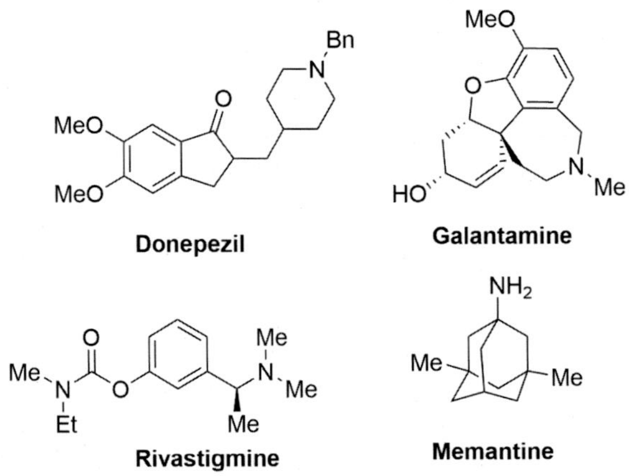

FIGURE 11.1

Chemical structures of currently used anti-AD drugs.

Aβ species can initiate AD early on. These findings provided the basis to target the amyloid cascade hypothesis to discover novel anti-AD drugs [12,14–17]. Many small molecule drug candidates were developed as inhibitors of β- and γ-secretases [18,19]. Furthermore, many natural and synthetic compounds including both small molecules and peptides were reported to prevent the aggregation of Aβ into toxic species [20–25]. Another promising approach to target the amyloid cascade was immunotherapy using anti-Aβ monoclonal antibodies solanezumab, bapineuzumab, and aducanumab [26–30]. Unfortunately, therapies based on preventing or reducing the accumulation of Aβ species have not lived up to the expectations. Many promising drug candidates went to clinical trials with plenty of hope in discovering novel anti-AD therapies. However, most of them failed at various stages of clinical trials due to lack of efficacy [31–36]. These negative outcomes have questioned the very basis of amyloid cascade as the main event in causing AD [37–39].

Neurofibrillary tangles (NFTs) are another characteristic feature of AD brain. The accumulation of NFTs forms the basis of tau cascade or tau aggregation pathology, yet another major hypothesis of AD, which posits that the microtubule binding protein tau can undergo self-assembly to form neurotoxic species that can lead to AD, a major form of tauopathy [40–45]. The tau protein undergoes hyperphosphorylation by protein kinases (e.g., cyclin-dependent kinase 5, Cdk-5; microtubule-associated regulatory kinase, MAPK; and glycogen synthase kinase 3β, GSK3β) [46,47]. Tau hyperphosphorylation can disrupt microtubule stability promoting self-aggregation of tau into dimers, oligomers, paired helical filaments (PHFs), and NFTs which are implicated in AD [48,49]. In addition, this process can impair microtubule function, synaptic function, causing neurodegeneration [50,51]. Therefore, preventing tau hyperphosphorylation and aggregation using small molecules is a valid approach to develop novel anti-AD therapies [48,49]. Due to clinical trial failures in anti-Aβ therapies, the research is shifting toward anti-tau therapies [52]. Many anti-tau therapies including modulators of tau phosphorylation, post-translational modifications, tau-aggregation, microtubule stabilizers, and immuno-therapies have been reported to date, and some are being taken through various stages of preclinical and clinical trials [53–55]. The research in this area is ongoing [12,48,56,57].

Other hypotheses of AD includes metal and reactive oxygen species (ROS)/oxidative stress–induced neurotoxicity, mitochondrial dysfunction, glutamate/NMDA receptor (NMDAR) excitotoxicity, neuroinflammation, role of monoamine oxidase (MAO), serotonin signaling, type 2 diabetes, role of pathogens, and calcium homeostasis [58–71]. Fig. 11.2 provides a summary on the multifactorial nature of AD. All these reports provide compelling evidence on the multifactorial nature of AD pathology and suggest that targeting a single pathway or molecular target will not provide effective therapies.

With the lack of any new treatment strategy for AD in the horizon, the current research focus is on developing multitarget-directed ligands (MTDLs) as therapeutic option for AD with the hope of discovering novel drugs (Fig. 11.2) that can provide symptomatic relief and more significantly can prevent disease progression and even cure AD [72–74]. This book chapter provides a summary on the recent developments in the design, synthesis, and evaluation of novel small molecules classified on the basis of their core ring templates. Promising lead small molecules with multi-targeting ability are highlighted.

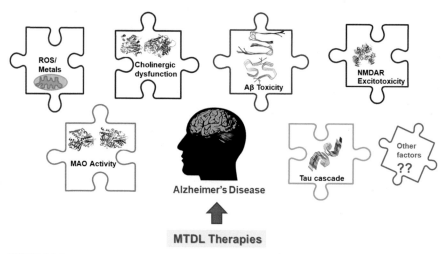

FIGURE 11.2

Multiple pathological factors of Alzheimer's disease and MTDL therapy.

11.2 Small molecules as multitargeting ligands multitarget-directed ligands

The last decade saw an incremental increase in the design and development of novel small molecules as MTDLs to treat AD. In their design, a wide variety of ring templates and scaffolds were used. A number of pharmacophores capable of binding to different molecular targets were discovered. This section will describe and highlight the key small molecules which were developed as MTDLs in recent years. The focus will be on their core ring templates, scaffolds/pharmacophores, synthetic methods used, in vitro and in vivo assay methodology, and biological activity data relevant to AD.

11.2.1 Quinone derivatives

The quinone derivative memoquin which possesses two methoxybenzyldiamino substituents (**1**, Fig. 11.3) was discovered in 2007 [75]. It exhibited multifunctional mechanism by inhibiting acetylcholinesterase (AChE), β-secretase (BACE-1), and Aβ aggregation. It was also a potent antioxidant which could reduce oxidative stress mediated neurodegeneration.

This discovery clearly showed the potential of targeting multiple pathological factors using a hybrid small molecule and heralded the beginning of MTDL discovery to treat AD. Unfortunately, the dimeric memoquin did not satisfy Lipinski rule as its molecular weight was more than 500 [76]. In order to address this, Bolognesi and coworkers designed derivatives of **1** which fall within Lipinski rule [77]. They synthesized a small library of eight memoquin derivatives. The lead compound **2** was synthesized by coupling 2-methoxynaphthalene-1,4-dione with the diamine as shown in Fig. 11.3. The monomeric memoquin derivative **2** was an MTDL with

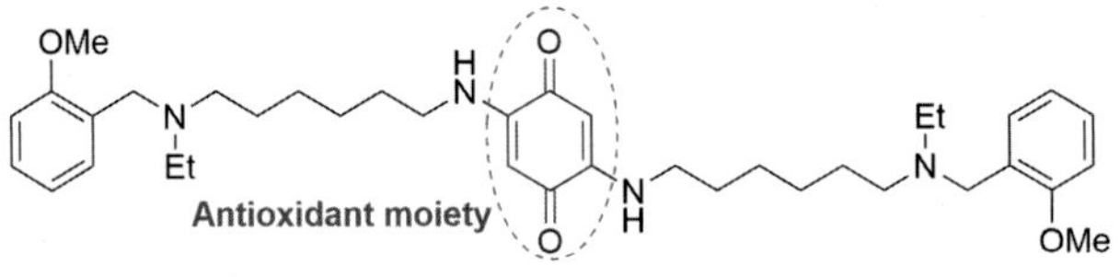

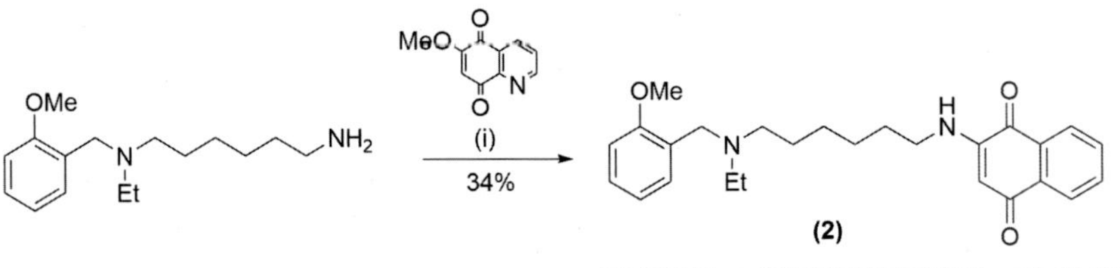

hAChE IC$_{50}$ = 9.7 nM; hBuChE IC$_{50}$ = 1.49 μM;
BACE-1 = 60.2% inhibition; Aβ42 aggregation
inhibition = 27%

FIGURE 11.3

Synthesis and activity of quinone MTDL 2. (i) EtOH, 4 h, reflux.

activity toward both human AChE (hAChE) and human butyrylcholinesterase (hBuChE), BACE-1 (hAChE $IC_{50} = 9.7$ nM; hBuChE $IC_{50} = 1.49$ µM; BACE-1 = 60.2% inhibition), AChE-induced Aβ40 aggregation (69% inhibition), and self-induced Aβ42 aggregation (27% inhibition). In addition, the presence of a benzoquinone moiety in **2** provides intrinsic antioxidant activity. Computational analysis was carried out by conducting molecular docking of **2** using X-ray structures of human AChE and BACE-1. These studies indicate favorable binding. All the experimental results were reported based on in vitro studies using human recombinant enzymes (ChE) and baculovirus-expressing BACE-1. Synthetic Aβ peptides were used in the assay [77]. This study demonstrated the advantages of targeting the cholinergic dysfunction, amyloid cascade, and oxidative stress pathways of AD as a strategy to discover novel anti-AD agents.

11.2.2 Selenazolone derivatives

Current pharmacotherapy of AD is dependent on cholinesterase inhibition. One of the commonly used drug is donepezil (Fig. 11.4) which is a known inhibitor of AChE and BuChE. In a novel study, Luo and coworkers designed and synthesized donepezil hybrids as MTDLs [78]. They replaced the five-membered cyclopentenone ring of donepezil with a bioisosteric selenazolone ring derived from the antioxidant and glutathione peroxidase (GPx) mimic ebselen (Fig. 11.4).

Selenazolone derivative **3** exhibited multiple activities. The synthesis was carried out by a diazotization reaction of aminobenzoic acids with sodium diselenide, followed by treatment with thionyl chloride and coupling with benzylpiperidine amine (Fig. 11.4). Ebselen has number of biological properties such as antioxidant, anti-inflammatory, and GPx mimic activity [79]. In this regard, literature evidence shows a reduction in the levels of several antioxidant enzymes including GPx, especially in the synaptosome and mitochondrial fractions of AD brain [80]. Furthermore, selenium is an essential micronutrient with a number of functions in maintaining the antioxidant defense mechanisms in the body [81]. Quite interestingly, the donepezil-ebselen hybrid **3** (selenpezil, Fig. 11.4) was able to retain its cholinesterase inhibition activity (hAChE $IC_{50} = 97$ nM; equine BuChE $IC_{50} = 1.58$ µM). It was also able to exhibit moderate inhibition of AChE-induced Aβ40 aggregation (21% inhibition) which was comparable to donepezil. Strikingly, compound **3** was able to scavenge hydrogen peroxide (H_2O_2) and another highly toxic species peroxynitrite ($ONOO^-$) unlike donepezil. In addition, compound **3** was acting as a substrate for the selenium containing enzyme thioredoxin reductase (TrxR) which suggests its ability to promote and maintain the activity of TrxR in countering oxidative stress. Authors also confirmed the blood-brain barrier (BBB) permeation of **3** in vitro using parallel artificial membrane permeability assay (PAMPA). In vivo studies were conducted to study the acute toxicity of **3** using single oral dose in mice in a 14-day study which demonstrated the safety profile of **3** as it was nontoxic up to 2000 mg/kg oral dose. This study demonstrates the modification of anti-AD drugs as MTDLs to enhance their efficacy in treating AD.

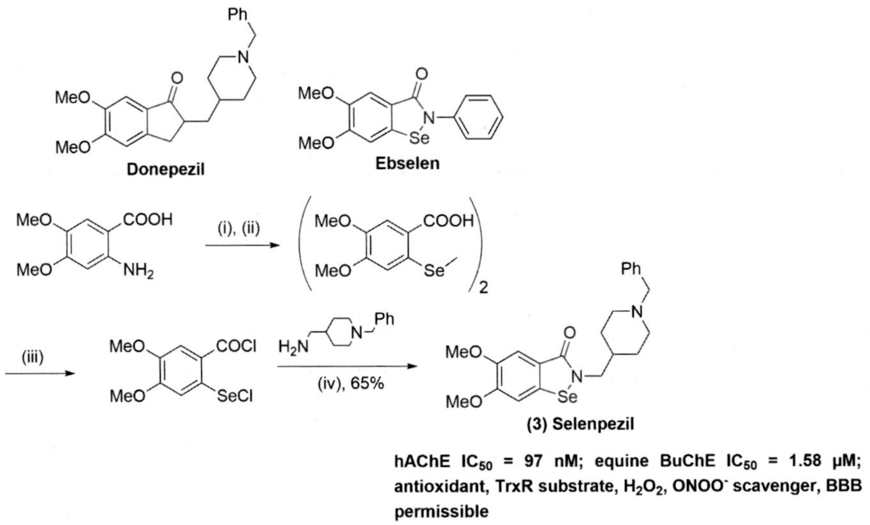

FIGURE 11.4

Synthesis and activity of donepezil-ebselen hybrid MTDL 3. (i) $NaNO_2$, HCl; (ii) Na_2Se_2; (iii) $SOCl_2$, reflux; (iv) TEA, DCM.

11.2.3 Stilbene derivatives

Resveratrol is a stilbene derivative present in red wine and is known to have antioxidant, anti-inflammatory, anti-Aβ, and number of other activities [82]. Despite these beneficial effects, resveratrol's pharmacokinetic profile is not favorable to be used as a drug [83]. In this regard, Lu and coworkers designed and developed a novel class of 20 resveratrol derivatives [84]. They fused stilbene template of resveratrol with a metal chelator and antioxidant compound clioquinol (Fig. 11.5) which afforded stilbene derivatives as MTDLs. They exhibited activity toward AChE, monoamine oxidases A (MAO-A) and B (MAO-B), Aβ42 aggregation, and metal chelating properties apart from antioxidant activity [84].

Compound **4** (Fig. 11.5) was identified from this library screen as a promising candidate. Its synthesis was accomplished by converting nitrobenzaldehyde to the Wittig reagent which was reacted with 3,5-dimethoxybenzaldehyde to give dimethoxy-nitrostyrylbenzene. In the next step, it was reduced to the corresponding amine and then coupled to hydroxybenzaldehyde under reducing conditions to afford resveratrol derivative **4** (Fig. 11.5). A number of in vitro assays were carried out. The antiaggregation property toward Aβ42 was determined by thioflavin T (ThT) and transmission electron microscopy (TEM) studies; antioxidant properties were evaluated using the oxygen radical absorbance capacity (ORAC-FL); and in SH-SY5Y cells, metal chelation assay toward copper, iron and zinc was carried out by UV spectroscopy, ChE inhibition was evaluated using electric eel AChE and equine BuChE enzymes, MAO inhibition in human MAO-A and B was

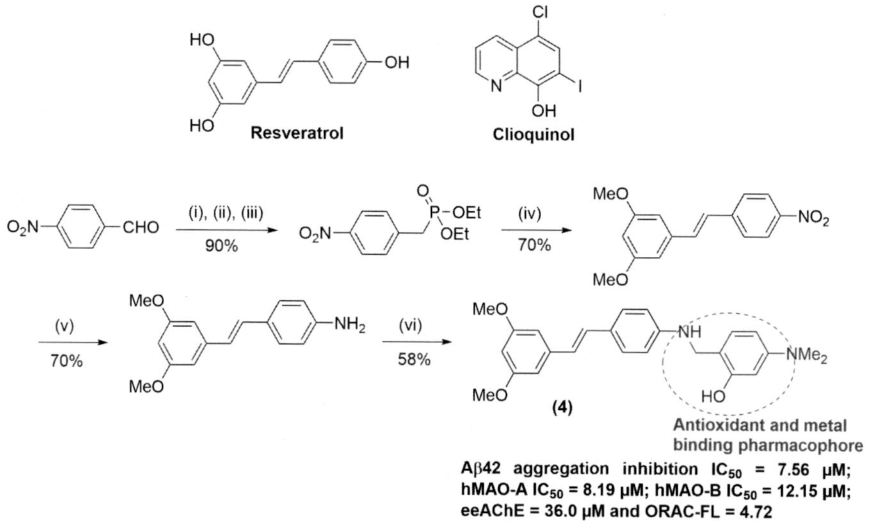

FIGURE 11.5

Synthesis and activity of resveratrol-clioquinol hybrid MTDL 4. (i) $NaBH_4$, MeOH, 0°C; (ii) PBr_3, pyridine; (iii) triethyl phosphite, reflux; (iv) 3,5-dimethoxybenzaldehyde, CH_3ONa; (v) $SnCl_2$, EtOAc; (vi) 2-hydroxybenzaldehyde, $NaBH_4$.

determined, BBB permeation was determined using PAMPA, and acute toxicity studies were evaluated in mice [84]. Compound **4** was able to exhibit a number of activities (Aβ42 aggregation inhibition $IC_{50} = 7.56$ μM; hMAO-A $IC_{50} = 8.19$ μM; hMAO-B $IC_{50} = 12.15$ μM; eeAChE $= 36.0$ μM, and ORAC-FL $= 4.72$, Fig. 11.5). Molecular docking studies of compound **4** were carried out using the NMR structure of Aβ42. Compound **4** exhibited excellent in vitro BBB permeation; it was capable of chelating to metals such as copper, iron, and zinc. In the acute toxicity studies in mice, it was well tolerated with no reported toxicity (up to 2000 mg/kg oral dose) which demonstrates its potential as an MTDL.

11.2.4 Hydroxyanthraquinone derivatives

Previous studies have shown that anthraquinone derivatives can inhibit tau aggregation, prevent the formation of paired helical filaments (PHFs) and Aβ aggregation [85,86]. They also exhibit antioxidant activity. Based on this observation Viayna and coworkers reported the discovery of hydroxyanthraquinone derivatives derived from the natural product rhein (Fig. 11.6).

They linked the rhein ring system with huprine Y which is known to exhibit AChE inhibition [87]. In vitro and in vivo studies were carried out for 10 rhein derivatives. Their inhibitory activity toward hAChE, hBuChE, hBACE-1, hAChE-induced Aβ40 aggregation, self-induced Aβ42 aggregation, Aβ42, and tau protein aggregation activity was measured in BL21 (DE3) competent *E. coli* cells, mice hippocampal slices were used to study the effect of rhein derivatives on long-term potentiation (LTP) and synaptic activity in the presence of Aβ oligomers, in vivo studies were carried out using APP-PS1 transgenic mice which mimics the Aβ pathology and molecular docking studies were performed using crystal structures of hAChE and hBACE-1 enzymes [87]. Among the 10 rhein-huprine hybrids synthesized and evaluated, two enantiopure compounds **5a** and **5b** (Fig. 11.6) exhibited very promising anti-AD activity. These two compounds were synthesized by coupling the hydroxyanthraquinone compound rhein with enantiopure huperine Y to afford enantiopure **5a** and **5b**. The general synthetic method is shown in Fig. 11.6. Both compounds were MTDL targeting the cholinergic, Aβ and tau cascade, which was confirmed by in vitro and in vivo experiments [87]. For example, compound **5b**'s activity data toward hAChE (IC$_{50}$ = 2.39 nM), hBuChE (IC$_{50}$ = 513 nM), BACE-1 (IC$_{50}$ = 80 nM), Aβ42 aggregation (47% inhibition at 10 μM), AChE-induced Aβ40 aggregation (38% inhibition at 10 μM) and tau aggregation (34% inhibition at 10 μM) demonstrated its multiple activity (Fig. 11.6). Furthermore, **5b** was able to protect synaptic failure induced by Aβ oligomers in mouse hippocampal slices, exhibited BBB permeability in vitro and was able to

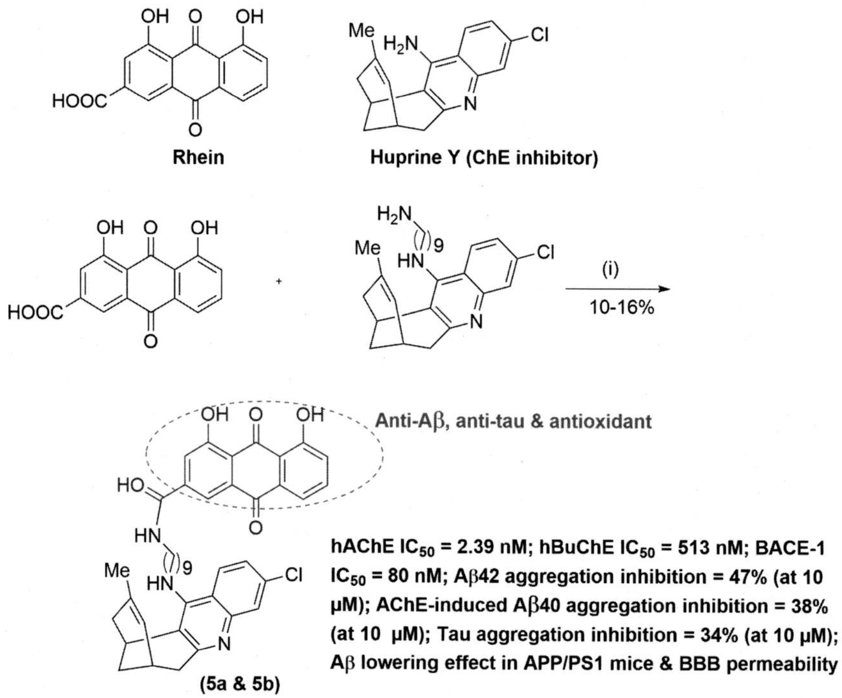

FIGURE 11.6

Synthesis and activity of rhein-huprine Y hybrid MTDLs **5a** and **5b**. (i) ClCO$_2$Et, TEA, DCM.

reduce Aβ aggregation in the APP/PS1 transgenic mice model confirming its multi-targeting effect [87].

11.2.5 Indole derivatives

The bicyclic indole ring scaffold has shown to be a desirable system to develop novel MTDLs as potential anti-AD drugs. Previously, indole-based derivatives were optimized to target the ChE and MAO pathway as potential MTDLs for AD [88]. It is known that propargylamine moiety is a pharmacophore for MAO inhibition and is capable of binding to flavin ring present in the catalytic site of MAO enzymes to exhibit irreversible inhibition [89]. In fact, increased MAO-B in AD brain is implicated in its pathology [66]. Based on this evidence, Bautista-Aguilera and coworkers designed indole derivatives linked to benzylpiperidines and propargylamine to target the ChE, Aβ, and MAO hypothesis of AD [88]. The chemical structures of MAO inhibitor selegiline and ladostigil, a dual MAO/ChE inhibitor, served as the starting point for SAR studies (Fig. 11.7). A systematic quantitative structure-activity relationship (QSAR) study led to the design of novel indole derivatives and their in vitro evaluation toward hAChE/BuChE, hMAO-A/B and PAMPA-BBB assays. These studies, revealed a novel indole derivative **6** (Fig. 11.7) which exhibited potent inhibition of hMAO-A/B (hMAO-A IC_{50} = 6.3 nM; hMAO-B IC_{50} = 183.6 nM) along with ChE inhibition (hAChE IC_{50} = 2.8 μM; BuChE IC_{50} = 4.9 μM). The synthesis was accomplished by reducing pyridine propanol to obtain Boc-protected piperidine, followed by chlorination, coupling with indole-N-propargylamine derivative, Boc-deprotection and alkylation to afford **6** (Fig. 11.7).

Another recent study described the development of a new class of MTDLs possessing an indole ring scaffold. The pathological process of AD is highly interconnected with the accumulation of the extracellular Aβ senile plaques within the brains of patient. As Aβ plaque load increases, it can promote ROS formation leading to its overproduction and consequent neuronal damage. In order to prevent this jeopardy, Bautista-Anguilera and coworkers reported a multitargeting so called tetratarget small molecule **7**, also known as Contilisant (Fig. 11.8) with inhibitory activity toward AChE/BuChE, MAO-A/B, the histamine H3 receptor (H3R), and the Sigma 1 receptor antagonism (S1R) [90,91]. Both H3R and S1R play a key role in learning and memory targets play a key role in learning and memory namely, H3R regulates the histamine and other neurotransmitters such as acetylcholine (ACh), serotonin, dopamine, and norepinephrine [90]; and the S1R activity can increase glutaminergic and cholinergic synapses [91].

Although compound **7** was structurally designed and synthesized based on the pharmacophore fragments of H3R antagonists (ciproxifan), the imidazole moiety in ciproxifan was replaced with cyclic aliphatic amine such as piperidine to reduce adverse effects [90]. The multiple biological activity of compound **7** was demonstrated by various in vitro assays [90,91]. The AChE and BuChE inhibitory activity was determined using human recombinant AChE and human plasma BuChE based on Ellman's method. MAO inhibition in human MAO-A and B was evaluated by

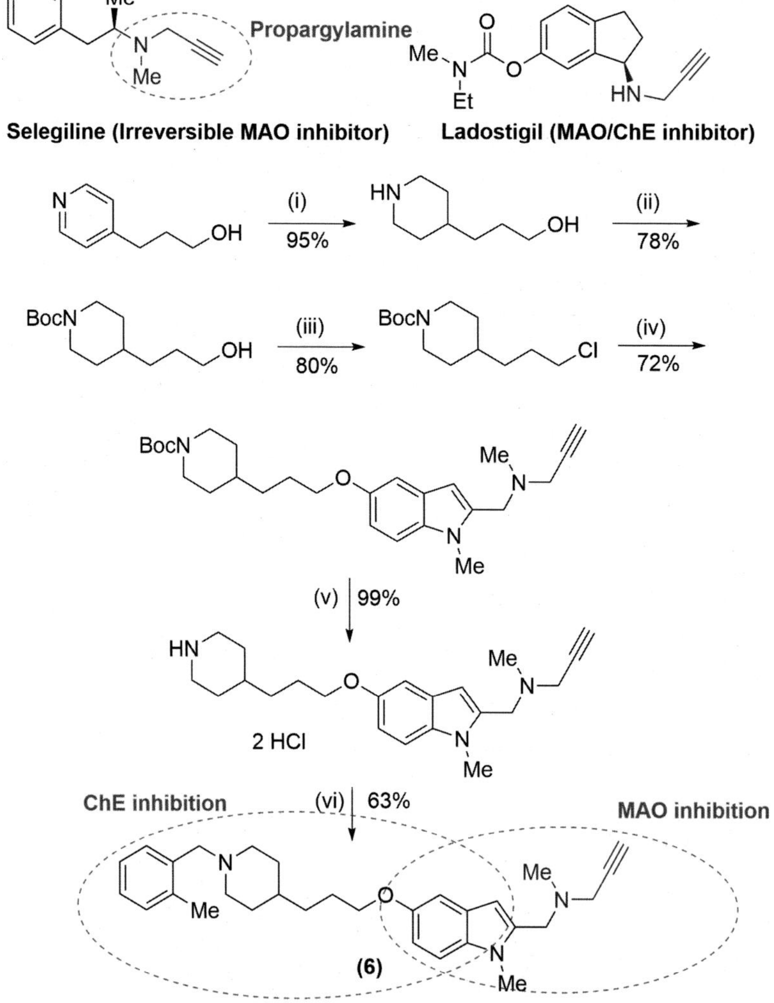

FIGURE 11.7

Synthesis and activity of indole hybrid MTDL 6. (i) H_2, PtO_2 20%, Pd/C 10%, HCl/dioxane, 45 psi; (ii) di-*tert*-butyl dicarbonate, NaOH, dioxane; (iii) CCl_4, PPh_3, DCM; (iv) indole-*N*-propargylamine derivative, NaH, DMF; (v) HCl/EtOAc; (vi) 1-(bromomethyl)-2-methylbenzene, DIPEA, CH_3CN, reflux.

determining the production of hydrogen peroxide. The binding affinity of compound **7** toward human H3R, human S1R, rat S2R, and rat vesicular acetylcholine transporter VAChT was determined under radioligand displacement experiments. The

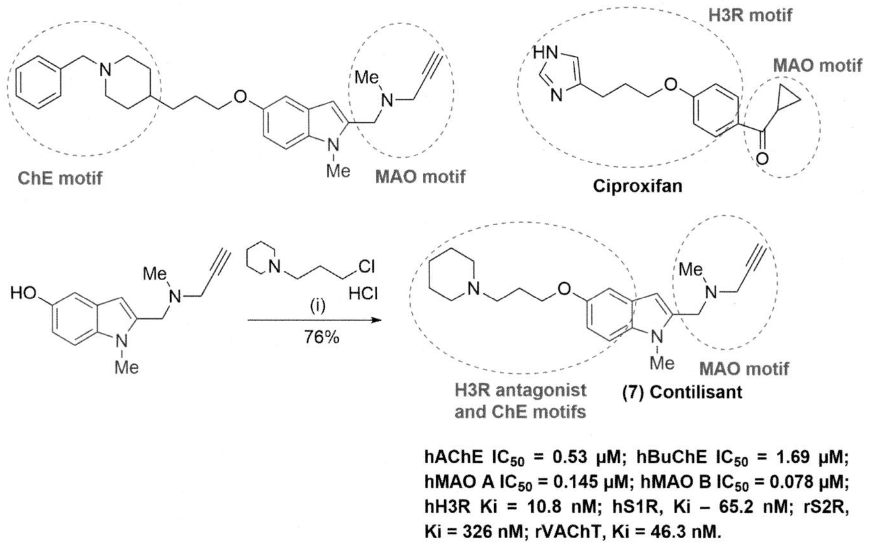

FIGURE 11.8

Synthesis and activity of indole MTDL **7**. (i) NaH, DMF, rt.

binding interactions of **7** in the cavities of AChE, MAO-A/B, H3R, and S1R were validated by molecular docking studies. The antioxidant activity of compound **7** was evaluated by the ORAC-FL method. The neuroprotection properties were studied using SH-SY5Y cells. Brain penetration was determined by using PAMPA. The recognition memory assessment was examined in vivo using novel object recognition (NOR) test in mice. It was synthesized by coupling 1-methyl-2-((methyl(prop-2-yn-1-yl)amino)methyl)-1*H*-indol-5-ol possessing an indole and an *N*-propargylamine group with 1-(3-chloropropyl)piperidine as shown in Fig. 11.8. In summary, compound **7** was identified as a nontoxic, brain permeable antioxidant with neuroprotective properties. It was able to display multifunctional activity (hAChE, $IC_{50} = 0.53$ μM; hBuChE, $IC_{50} = 1.69$ μM; hMAO A, $IC_{50} = 0.145$ μM; hMAO B, $IC_{50} = 0.078$ μM; hH3R, $K_i = 10.8$ nM; hS1R, $K_i = 65.2$ nM; rS2R, $K_i = 326$ nM; rVAChT, $K_i = 46.3$ nM). These studies highlight the usefulness of indole ring systems in designing MTDLs to treat AD.

11.2.6 Melatonin (*N*-acetyl-5-methoxytryptamine) derivatives

Melatonin is an indoleamine neurotransmitter secreted from the pineal gland in brain [92]. It has a wide range of beneficial effects in the brain including antioxidant activity, anti-inflammatory activity, anti-tau/anti-Aβ activity, regulation of mitochondrial function, and promoting neurogenesis which are all desirable in the treatment of AD [93,94]. Chemically, it can also be considered as an indole derivative as in the previous section (11.2.5 indole derivatives). In this regard, Lopez-Iglesias and coworkers designed melatonin hybrids as MTDLs. They linked the *N*-acetyl group of melatonin with an AChE binding pharmacophore *N,N*-dibenzyl(*N*-methyl)amine to provide multifunctional derivatives (Fig. 11.9) [94]. A library of 14 melatonin hybrids were synthesized by coupling 4-(bromomethyl)-benzonitrile with *N*-methyl-benzylamine derivatives followed by hydrolysis to obtain corresponding acids which were coupled with substituted-tryptamines to afford melatonin–*N,N*-dibenzyl(*N*-methyl)amine hybrids. A representative synthetic scheme to prepare the lead molecules **8** is shown in Fig. 11.9.

The in vitro assays conducted include hAChE/BuChE inhibition, propidium iodine displacement assay in the presence of AChE enzyme to determine the ability of compound **8** to bind to peripheral anionic site (PAS) of AChE, ORAC antioxidant activity, ROS-induced cell toxicity in SH-SY5Y cell line, neurogenesis potential in neural stem cell cultures derived from rat hippocampus, immunohistochemistry studies, and BBB permeation assay [94]. Compound **8** exhibited dual hAChE/BuChE inhibition (hAChE $IC_{50} = 6.1 \, \mu M$; BuChE $IC_{50} = 7.8 \, \mu M$), 67.3%

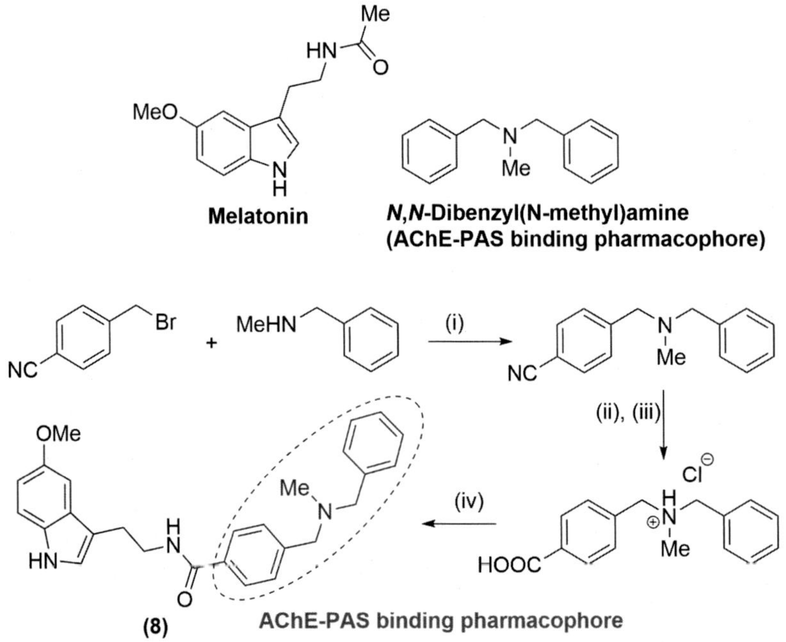

hAChE IC_{50} = 6.1 μM; hBuChE IC_{50} = 7.8 μM
67.3% AChE-PAS (at 3 μM); 36% neuroprotection
against ROS-induced cell toxicity, promotion of
neurogenesis

FIGURE 11.9

Synthesis and activity of melatonin hybrid MTDL **8**. (i) Diethyl ether, reflux; (ii) NaOH, reflux; (iii) HCl; (iv) 2-(5-methoxy-1*H*-indol-3-yl)ethanamine, PyBOP, TEA, DMF.

displacement of propidium iodide from AChE PAS (at 3 μM), 36% neuroprotection from ROS induced toxicity, good BBB permeation and ability to promote neurogenesis when incubated at 10 μM which all provide strong evidence on its potential to be developed as an anti-AD therapy.

11.2.7 Coumarin derivatives

Coumarins or 2*H*-chrome-2-ones represent an important class of phytochemicals present in many plants [95]. Farina and coworkers focused on designing MTDLs targeting the MAO and ChE pathology of AD using the coumarin template. Using computational docking studies, they designed coumarin derivatives linked to alkyl amines and benzylamines with the rationale that the bicyclic coumarin template (Fig. 11.10) would provide selective MAO-B inhibition whereas the alkyl amines/benzylamines would provide ChE inhibition especially by binding to the catalytic active site (CAS) of AChE [96]. Their previous studies demonstrated that the coumarin derivative MC1095 (Fig. 11.10) was a selective MAO-B inhibitor. To this coumarin template they incorporated the benzylamine moiety of the known ChE inhibitor donepezil (Fig. 11.10) to enhance AChE inhibition.

Computational studies suggested that the coumarin template can also bind to PAS of AChE and has the potential to prevent AChE-induced Aβ aggregation [96]. A large library of 44 coumarin hybrids were synthesized and evaluated as hChE/hMAO-B selective inhibitors, as antioxidants in H_2O_2-induced cytotoxicity, and their BBB permeability was assessed. This led to the identification of coumarin hybrid **9** (Fig. 11.10) as a promising MTDL. Compound **9** was synthesized by coupling 7-hydroxycoumarin derivative with dibromo-xylene to obtain the 4-(hydroxymethyl)-7-(4-(bromomethyl)benzyloxy)coumarin intermediate which was coupled to the *N*-benzylmethylamine to afford **9** as shown in Fig. 11.10. It exhibited a number of biological properties (hMAO-A $IC_{50} = 15.8$ μM; hMAO-B $IC_{50} = 10$ nM; hAChE $IC_{50} = 0.12$ μM; hBuChE $IC_{50} = 9.3$ μM; 25% neuroprotection in H_2O_2-induced cytotoxicity at 10 μM). The *N*-benzyl-*N*-methyl-1-phenylmethanamine substituent was a PAS binding pharmacophore in hAChE whereas the coumarin template itself provided selective MAO-B inhibition as well as exhibited CAS binding in hAChE.

11.2.8 Phenothiazine and phenoselenazine derivatives

The fused tricyclic phenothiazine is an important ring scaffold present in CNS agents such as chlorpromazine and fluphenazine, which are used in therapy to treat psychotic disorders [97]. As such, phenothiazines are known to exhibit ChE inhibition and antioxidant activity [98]. These properties make them a suitable template to design MTDLs for AD. Tin and coworkers developed novel phenothiazines as multifunctional agents capable of inhibiting hAChE, hBuChE, Aβ42 aggregation, and as antioxidants [99]. Furthermore, they developed novel phenoselenazines as bioisosteres of phenothiazines with multifunctional properties (Fig. 11.11).

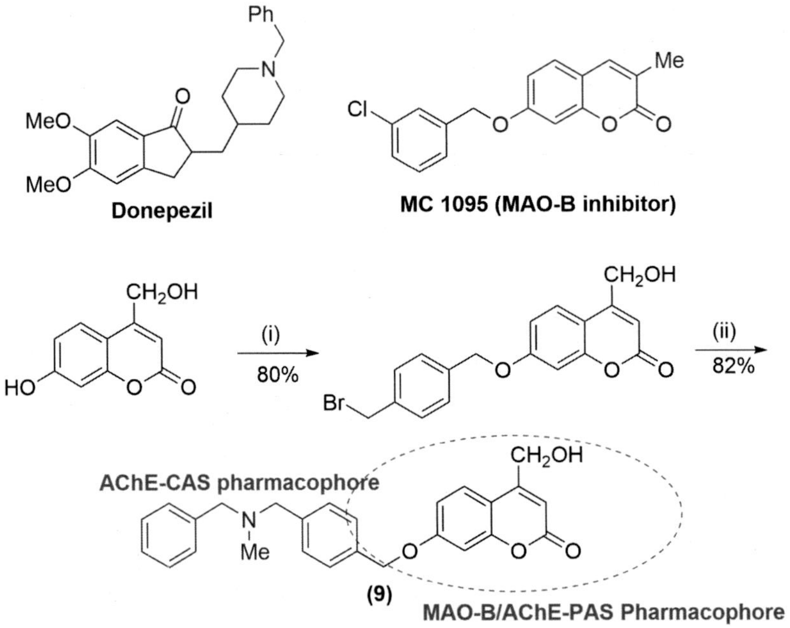

FIGURE 11.10

Synthesis and activity of coumarin hybrid MTDL **9**. (i) α,α'-Dibromo-p-xylene, K_2CO_3, CH_3CN, MW; (ii) N-benzylmethylamine, K_2CO_3, KI, CH_3CN, MW.

Selenium is an essential micronutrient and is present in the selenoproteins GPx which is part of the antioxidant defense system. Furthermore, AD patients exhibit lower levels of plasma selenium. These facts support the use of organoselenium-based therapies to treat AD [81]. A library of 28 phenothiazine and phenoselenazines were synthesized and evaluated using hAChE, hBuChE, Aβ42 peptide, diphenyl-1-picrylhydrazyl (DPPH) radical scavenging and SH-SY5Y neuroblastoma cell toxicity. Molecular docking studies in hAChE and hBuChE were carried out to study their binding orientation. The phenoselenazine derivatives were prepared by iodinating cyclohexenones followed by their reaction with substituted anilines under metal-free conditions to obtain diphenylamine derivatives which were treated with selenium and selenium dioxide in the presence of iodine under high temperature, high pressure conditions to afford phenoselenazine derivatives (Fig. 11.11). The phenothiazine derivatives **10a** (hAChE $IC_{50} = 7.3$ μM; hBuChE $IC_{50} = 5.8$ μM; 62% inhibition of Aβ42 aggregation at 25 μM and 92% scavenging of DPPH radical at 50 μM) and **10b** (hAChE $IC_{50} = 5.6$ μM; hBuChE $IC_{50} = 3.0$ μM; 45.6% inhibition of Aβ42 aggregation at 25 μM and 84.4%

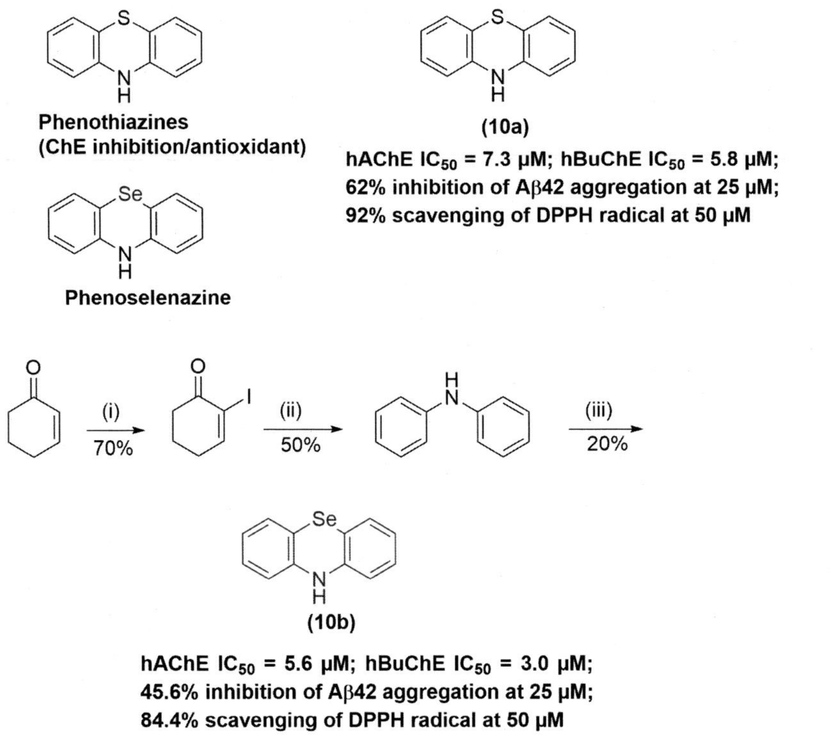

Phenothiazines
(ChE inhibition/antioxidant)

Phenoselenazine

(10a)

hAChE IC$_{50}$ = 7.3 μM; hBuChE IC$_{50}$ = 5.8 μM;
62% inhibition of Aβ42 aggregation at 25 μM;
92% scavenging of DPPH radical at 50 μM

(10b)

hAChE IC$_{50}$ = 5.6 μM; hBuChE IC$_{50}$ = 3.0 μM;
45.6% inhibition of Aβ42 aggregation at 25 μM;
84.4% scavenging of DPPH radical at 50 μM

FIGURE 11.11

Synthesis and activity of phenoselenazine MTDL **10b**. (i) I$_2$, DMAP, K$_2$CO$_3$, THF; (ii) aniline, EtOH, *p*-TsOH, reflux; (iii) Se, SeO$_2$, I$_2$, sulfolane, pressure vial, 150°C, 5 h.

scavenging of DPPH radical at 50 μM) were identified as lead compounds from this library (Fig. 11.11).

11.2.9 Pyrimidinylthiourea derivatives

Many small molecule MTDLs reported to date to treat AD, generally possess fused ring templates or bicyclic ring scaffolds or were derived from known ChE inhibitors in the market. In a deviation from this trend, Li and coworkers did a systematic SAR study to develop novel nonfused ring templates as MTDLs capable of targeting the ChE, Aβ, oxidative stress, and metal hypothesis of AD [100]. More than 50 derivatives were synthesized and evaluated. In vitro studies conducted include ChE inhibition using rabbit AChE and BuChE enzymes, antioxidant activity (ORAC-FL), metal binding activity toward Fe^{2+}, Cu^{2+}, Zn^{2+}, Aβ aggregation (immunoblot and TEM studies), cytotoxicity in human SH-SY5Y neuroblastoma cell lines, BBB permeability, and in vivo cognition assays in scopolamine-induced memory loss in mice [100]. Their library screening showed that a six-membered pyrimidine ring, linked to a 5-membered imidazole (4-(1*H*-imidazol-1-yl)pyrimidine, Fig. 11.12) was a suitable template which can be chemically modified to exhibit multifunctional properties.

They showed that incorporating a thiourea substituent at the pyrimidine C4-position of the 4-(1*H*-imidazol-1-yl)pyrimidine template, provided AChE-inhibition, antioxidant, and metal chelation properties, whereas addition of alkylamines and cycloalkylamines at the C4 of imidazole ring enhanced their lipophilicity and BBB permeability (Fig. 11.12). These studies demonstrate that pyrimidinylthiourea is a viable template to design and develop novel MTDLs to treat AD. The lead compound **11** was synthesized starting from 4-amino-6-chloropyrimidine which was reacted with imidazolecarboxaldehyde to obtain the amino-substituted 4-(1*H*-imidazol-1-yl)pyrimidine non-fused ring which was converted to a pyrimidinylthiourea by reacting with ethyl isothiocaynate, followed by reductive amination to afford **11** (Fig. 11.12). Compound **11** exhibited ChE inhibition (rAChE $IC_{50} = 0.204$ µM; rBuChE $IC_{50} = 32\%$ inhibition [at 40 µM]), was able to prevent metal promoted Aβ42 aggregation which was confirmed by dot blot/immunoblot assay using A11 oligomer specific Aβ antibody and 6E10 antibody to detect total

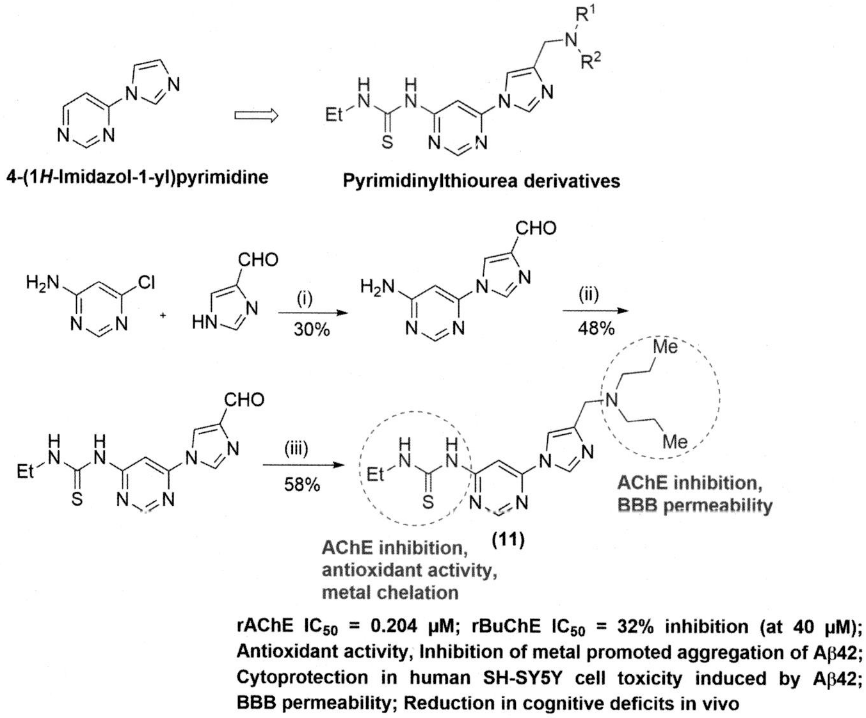

FIGURE 11.12

Synthesis and activity of pyrimidinylthiourea MTDL **11**. (i) Cs$_2$CO$_3$, DMF, reflux; (ii) EtNCS, DMF; (iii) dipropylamine, NaBH$_3$(CN), MeOH/DCE, AeOH.

Aβ which was further confirmed by TEM studies (at 50 μM). Compound **11** did not exhibit toxicity to human SH-SY5Y neuroblastoma cell lines at 10 or 30 μM and was able to protect these cells from Aβ42-mediated toxicity. It also exhibited excellent BBB permeation in the PAMPA assay. In the in vivo assay, at an oral dose of 200 mg/kg, compound **11** was able to reduce cognitive deficits in mice which demonstrates the effectiveness of compound **11** as a promising MTDL to treat AD.

11.2.10 Quinazoline and pyrido[3,2-*d*]pyrimidine derivatives

In an effort to develop novel ring scaffolds as MTDLs, Mohamed and coworkers reported the design and synthesis of 2,4-disubstituted-pyrimidines (*N*-benzylpyrimidin-4-amines, Fig. 11.13) as nonfused ring systems [101]. Their studies indicated that these small molecules exhibited dual ChE inhibition, hAChE-induced Aβ aggregation inhibition, BACE-1 inhibition, and Aβ40 aggregation (Fig. 11.13) [102].

In the next step of optimization, they modified the 2,4-disubstituted-pyrimidines by fusing a benzene ring to obtain a fused bicyclic quinazoline and varied substituents at the C2, C4, C6, C7 and C8 positions. More than 80-compound library was synthesized. These derivatives exhibited Aβ40 and Aβ42 aggregation inhibition (thioflavin T based fluorescence assay and TEM studies), hAChE and hBuChE inhibition (UV-based Ellman assay) and antioxidant properties (DPPH radical scavenging) [103,104]. The binding interactions of lead derivatives were studied by molecular docking studies using hAChE/BuChE enzymes and solid-state nuclear magnetic resonance (NMR) structures of Aβ40 using dimer and fibril models. They also replaced the quinazoline template with a pyrido[3,2-*d*]pyrimidine ring to incorporate metal chelating properties [104]. A number of derivatives with promising activity were identified. Some examples of small molecules (compounds **12a** and **12b**) and their synthetic routes are shown in Fig. 11.13. The quinazoline derivative **12a** exhibited dual ChE and Aβ inhibition (Aβ40 IC_{50} = 2.3 μM; hAChE IC_{50} = 2.1 μM; hBuChE IC_{50} = 8.3 μM) whereas the pyrido[3,2-*d*]pyrimidine derivative **12b** exhibited additional metal chelation property (Aβ40 IC_{50} = 1.1 μM; hAChE IC_{50} = 7.8 μM; hBuChE IC_{50} = 29.3 μM; 23% iron-chelation (at 50 μM)). These investigations led to the identification of quinazoline and pyrido[3,2-*d*]pyrimidine derivatives as a novel class of MTDLs.

11.2.11 Thiazole derivatives

Thiazole scaffold has been incorporated into various chemical entities with a broad spectrum of biological activities such as anticancer, antiviral, antimicrobial, and anti-inflammatory [105]. In a novel approach, Shidore and coworkers designed and synthesized a set of hybrid derivatives by fusing benzylpiperidine, a pharmacophoric feature of the anti-AD drug donepezil, with diarylthiazole as MTDLs [106]. Compound **13** (Fig. 11.14) was identified as the most potent derivative exhibiting excellent AChE (IC_{50} = 0.30 μM) and moderate BuChE (IC_{50} = 1.84 μM) inhibition. Furthermore, it showed significant in vivo antioxidant and neuroprotective

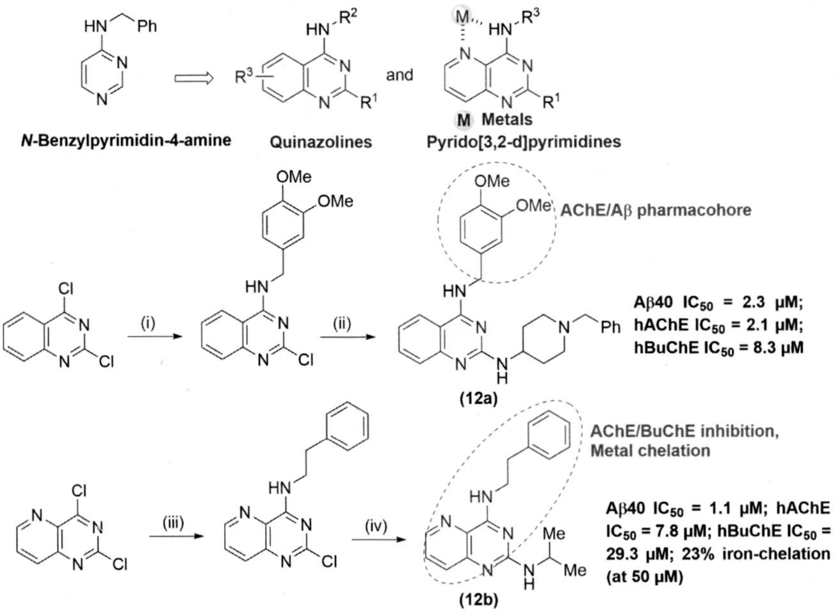

FIGURE 11.13

Synthesis and activity of quinazolines and pyrido[3,2-*d*]pyrimidines MTDL **12a** and **12b**. (i) 3,4-Dimethoxybenzylamine, DIPEA, EtOH, reflux; (ii) 4-amino-1-benzylpiperidine, DIPEA, 1,4-dioxane, pressure vial, 160–165°C; (iii) phenethylamine, DIPEA, EtOH, reflux; (iv) isopropylamine, DIPEA, 1,4-dioxane, pressure vial, 150–155°C.

properties and was well tolerated up to 2000 mg/kg oral dose without any adverse side effects [106]. Its synthesis involved four steps where in the first step *tert*-butyl 4-[4,5-bis(*p*-tolyl)thiazol-2-ylcarbamoyl]-piperidine-1-carboxylate was prepared by the reaction between 1-(*tert*-butoxycarbonyl)piperidine-4-carboxylic acid and (4,5-bis(*p*-tolyl)thiazol-2-ylamine)) followed by the deprotection of Boc-group using trifluoroacetic acid (TFA) to obtain an intermediate which was further treated with 3,5-diflourobenzyl bromide to yield the diarylthiazole derivative. This intermediate was reduced to afford the desired compound **13** (Fig. 11.14).

In molecular docking studies, compound **13** was able to form stable complex with AChE and showed similar binding mode to donepezil. A number of in vitro assays were also carried out where compound **13** exhibited moderate AChE-induced Aβ42 aggregation (27.65% inhibition at 10 μM) [106]. Compound **13** exhibited excellent BBB permeation via parallel artificial membrane permeability assay (PAMPA), no toxicity in SH-SY5Y cell line and neuroprotective effect in oxidative stress condition induced by the addition of H_2O_2 (59.5% at 20 μM). It exhibited 55% radical scavenging activity in the 2,2-diphenyl-1-picrylhydrazyl (DPPH) assay on par with the standard, ascorbic acid (at 10 μM). Pharmacokinetic analysis revealed that compound **13** can be absorbed orally and eliminated at good rate. This study

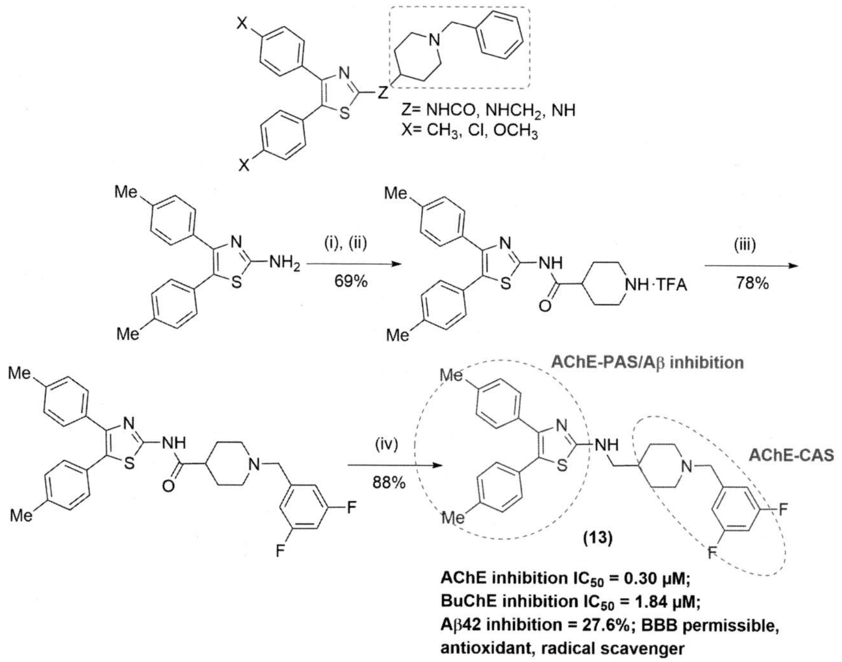

FIGURE 11.14

Synthesis and activity of thiazole MTDL **13**. (i) 1-(*tert*-Butoxycarbonyl)piperidine-4-carboxylic acid, DIPEA, BOP, dry ACN, 0°C to rt; (ii) TFA/DCM (70:30), rt; (iii) potassium carbonate, dry DMF, 3,5-difluorobenzyl bromide, rt to 60°C; (iv) BH₃-DMS, dry THF, 0°C to rt, 1N HCl, reflux 4 h, 5% sodium bicarbonate, rt to reflux, 2 h.

demonstrates the potential of thiazole based derivatives as potential candidates to develop MTDLs.

11.2.12 Tadalafil derivatives

Tadalafil, commercially available as Cialis, is a treatment for erectile dysfunction and is a potent inhibitor of phosphodiesterase 5 (PDE5), localized in the brain [107]. PDEs are essential in hydrolysis of cyclic guanosine monophosphate (cGMP) and cyclic adenosine monophosphate (cAMP) which are closely associated with neurotransmitter release, neuroplasticity, and neuroprotection [108]. PDE inhibitors have been able to effectively restore cognitive deficits in animal models. In this regard, Li and coworkers through drug repurposing approach designed and developed a set of novel tadalafil derivatives as dual AChE and PDE5 inhibitors [109]. According to their SAR studies, the *N*-methyl group of piperazine-2,5-dione was replaceable without significant loss of PDE5 inhibitory activity. A library of 40 derivatives were synthesized and compound **14a** (Fig. 11.15) was identified as the lead compound. In the first step of synthesis Pictet-Spengler reaction was carried out where the primary amine of L-tryptophan methyl ester hydrochloride and the aldehyde functional group of piperonal undergo condensation followed by cyclization to get the intermediate possessing the tadalafil core. In the next step, it was reacted with 2-chloroacetyl chloride, followed by coupling with 2-(1-benzylpiperidin-4-yl)ethane-1-amine to afford **14a** (Fig. 11.15).

Compound **14a** exhibited excellent AChE inhibition ($IC_{50} = 0.032$ μM), comparable to the potency of donepezil. It was significantly less potent for BuChE ($IC_{50} = 3.88$ μM) exhibiting strong selectivity toward AChE [109]. Using immobilized metal ion affinity-based fluorescence polarization (IMAP-FP) assay to evaluate PDE5 inhibitory activity, **14a** exhibited moderate inhibition ($IC_{50} = 1.53$ μM). It also showed good BBB permeation in PAMPA assay, and in the in vivo cognitive behavior assay, it was able to decrease the latency and the number of errors in the water maze experiment in scopolamine-induced cognitive deficit mouse model. Molecular docking studies of **14a** in crystal structures of hAChE and hPDE5A exhibited favorable interactions.

In their pursuit of second-generation AChE/PDE5 inhibitors, Li and coworkers further optimized **14a** [110]. A small library of 19 compounds were synthesized where derivative **14b** (Fig. 11.15) was identified as the most potent compound with improved water solubility. The ChE inhibitory activity of **14b** was evaluated at 40 μM and compared with known inhibitors huperzine A and donepezil. Compound **14b** exhibited 2.1-fold increase in inhibitory activity toward AChE ($IC_{50} = 0.015$ μM) in comparison to **14a**. However, the BuChE inhibition decreased by 1.95-fold ($IC_{50} = 7.61$ μM) with the a 507 selectivity index toward AChE. In the PDE5 inhibitory evaluation, **14b** showed 2.1-fold decrease in the inhibition ($IC_{50} = 3.23$ μM) compared to **14a** and exhibited no activity toward other isoforms of PDEs. Compound **14b** was BBB permeable and showed excellent safety profile in its corresponding citrate form at a dosage of 200 mg/kg, oral route in mice model [110]. The researchers were successful in improving the water solubility of **14a** by modifying its structure to **14b** (Fig. 11.15) as a citrate salt which showed good water solubility to enhance its oral absorption. These studies highlight the importance of drug repurposing concept in developing novel treatments in the field and the potential of dual AChE/PDE5 inhibitors as possible candidate in treatment of AD.

11.2.13 Pyridothiazole derivatives

Before tacrine was discontinued, its success in the treatment of memory and cognitive symptoms confirmed the cholinergic hypothesis of AD. Despite hepatotoxicity concerns with tacrine, it is the most commonly used template in the design of MTDLs owing to its high efficacy as a ChE inhibitor with a simple structure [111]. Aside from cholinergic hypothesis, tau is also a key factor in the pathogenesis of AD once it forms tangles via the hyperphosphorylation by protein kinases (e.g., GSK-3β). GSK-3β is a serine/threonine kinase mostly expressed in CNS that has been substantiated to be able to regulate tau phosphorylation (mainly at Ser396,

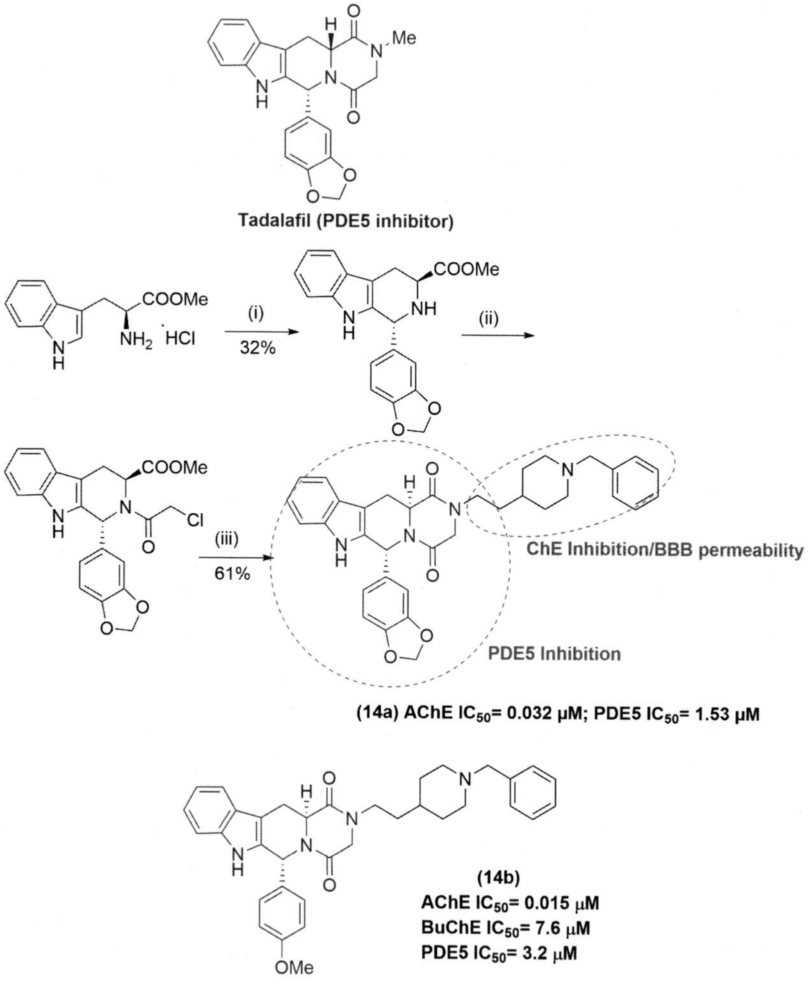

FIGURE 11.15

Synthesis and activity of AChE/PDE5 inhibitor MTDL **14a**. (i) Piperonal, i-PrOH, reflux, 24 h; (ii) 2-chloroacetyl chloride, Et₃N, DCM, −10°C to rt, 6 h; (iii) 2-(1-benzylpiperidin-4-yl)ethan-1-amine, Et₃N, MeOH, reflux, overnight.

Ser199, and Ser143) that is known to cause the dissociation of tau from the microtubules [111]. This finding was observed in transgenic mouse as GSK-3β was overexpressed, which also leads to the increased levels of Aβ accumulation [111]. In order to block the hyperphosphorylated tau and Aβ plaques as well as improve cognition, Jiang and coworkers demonstrated a new approach by developing a bifunctional inhibitor **15** (Fig. 11.16), based on the two crucial targets including hAChE and GSK-3β [111].

Pyridothiazole-tacrine hybrid compound **15** was synthesized through three step reaction. The pyridothiazole was obtained from 2-amino-4-cyanopyridine via acylation, addition and cyclization, followed by the hydrolysis to yield carboxyl intermediate, which was coupled with tacrine intermediate (Fig. 11.16) to afford the target compound **15** [111]. Molecular docking study was carried out using Discovery Studio software to predict the binding interactions of compound **15** with human AChE, BuChE, and GSK-3β. These studies demonstrate that compound **15** has the same binding mode as seen with the tacrine template. The inhibitory activity on the hyperphosphorylation of tau protein was determined by using Western blot assays in mouse neuroblastoma N2a-tau cells. Moreover, the cognition-improving potency and hepatotoxicity were investigated on scopolamine-induced memory impairment in institute of cancer research (ICR) male mice using the Morris water maze test and immunohistochemical study, respectively [111]. With a successful design, this novel pyridothiazole compound **15** (Fig. 11.16) exhibited potency toward both AChE and GSK-3β (hAChE $IC_{50} = 6.5$ nM; hGSK-3β $IC_{50} = 66.41$ μM), self-induced Aβ42 aggregation (46% inhibition) at 20 μM, exhibited cognition enhancement in scopolamine-induced mice model and reduced hepatotoxicity compared to tacrine.

11.2.14 Benzyloxyphenylpiperazine derivative

Recent study reported by Wieckowska and coworkers revealed a novel MTDL **16** (Fig. 11.17), which was designed to incorporate the cholinergic and serotonergic (5-HT) pharmacophore fragments [112].

They showed that benzyloxyphenylpiperazine moiety was able to exhibit 5-HT6 receptor antagonism. Multifunctional properties were incorporated by linking the benzyloxyphenylpiperazine 5-HT pharmacophore to tacrine, a known ChE inhibitor. Compound **16** exhibited inhibitory activity toward ChEs, self-induced Aβ42 aggregation, and serotonergic 5-HT6 receptor antagonism. In fact, the serotonergic 5-HT6 receptor [113] was found to be associated with the behavioral and psychological symptoms observed in AD [112]. Integrating 5-HT6 antagonistic property into novel anti-AD drugs would be beneficial, as they can regulate the synaptic remodeling and neuronal hyperexcitability, which are known to cause AD [112]. Compound **16** was synthesized by alkylating tacrine with 1,6-dibromohexane followed by another alkylation step to couple 1-(3-(benzyloxy)-2-methylphenyl)piperazine which afforded compound **16** (Fig. 11.17). The biological evaluation of **16** was carried out in a number of in vitro assays [112]. The binding affinity and functional activity of **16** on the recombinant human 5-HT6 receptors was determined using a radioligand displacement assay. The AChE and BuChE inhibitory activities of **16** was evaluated using AChE from electric eel (eeAChE) and BuChE from horse serum (eqBuChE) based on Ellman method. By using the molecular docking studies, the binding interactions of **16** were validated using serotonin 5-HT6 receptor and crystal structures of AChE. BBB penetration properties were determined by PAMPA. The metabolic stability of **16** were predicted using the in silico MetaSite 5.1.1 tool and evaluated in vitro using human liver microsomes (HLMs). The influence of

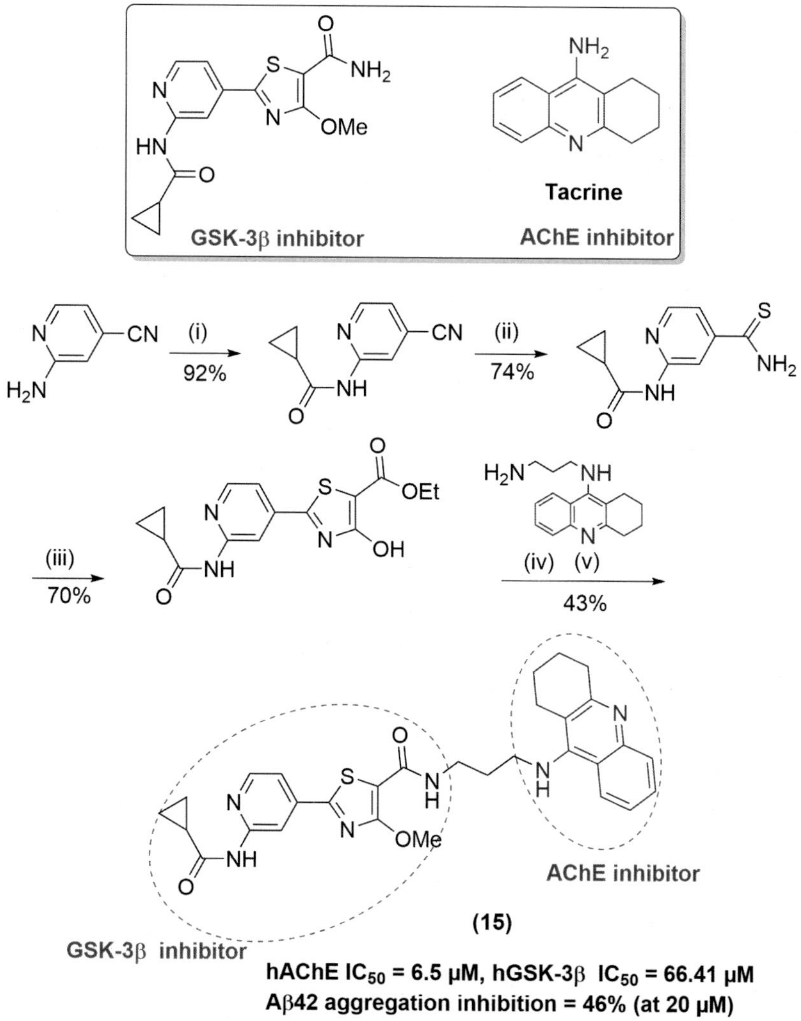

FIGURE 11.16

Synthesis and activity of pyridothiazole MTDL **15**. (i) Cyclopropanecarbonyl chloride, K_2CO_3, DCM, rt, 24 h; (ii) EtOH, $(NH_4)_2S$, 90°C, 3 h; (iii) EtOH, diethylbromomalonate, pyridine, 80°C, 20 h; (iv) anhydrous, MeOH, PPh_3, DIAD, rt, 18 h, 1 N LiOH, MeOH/THF, rt, 3 h; (v) HATU, DIPEA, DMF, rt, 2 h.

16 on CYP3A4 enzyme activity was determined using the luminescence method. Compound **16** with unique chemical features exhibited excellent permeability through BBB, and significant inhibitory activity on the selected biological targets with eeAChE, $IC_{50} = 14\ \mu M$; eqBuChE, $IC_{50} = 22\ \mu M$; 5-HT6, $K_i = 18\ nM$,

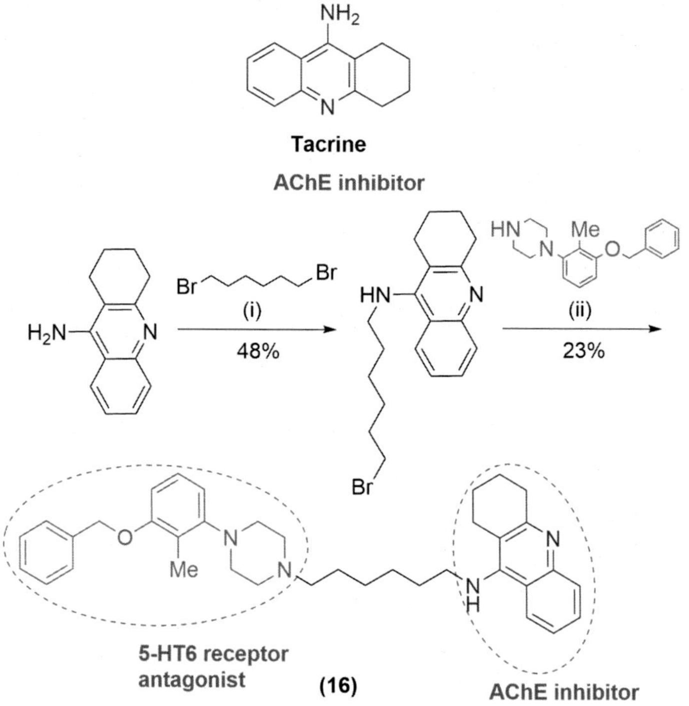

eeAChE IC_{50} = 14 μM; eqBuChE IC_{50} = 22 μM
5-HT6 Ki = 18 nM, K_b = 132 nM
Aβ42 aggregation inhibition IC_{50} = 1.27 μM

FIGURE 11.17

Synthesis and activity of benzyloxyphenylpiperazine MTDL **16**. (i) KOH, dry DMSO, rt, 24 h; (ii) K_2CO_3, KI, CH_3CN, reflux 24 h.

$K_b = 132$ nM; self-induced Aβ42 aggregation (94% inhibition, $IC_{50} = 1.27$ μM) at 10 μM (Fig. 11.17).

11.2.15 Quinoline-indole derivatives

AD is a consequence of the neuronal damage and loss that are caused by many factors, one of which is the intracellular oxidative stress, particularly mitochondrial ROS in the elderly. Wang and coworkers had taken this fact into consideration in the design of novel MTDLs by taking advantage of the melatonin-*N*-benzylamine structure, which has an indole framework that has the ability to promote the growth of neural stem cells into the neuronal phenotype [114]. They attempted to link the skeleton of a metal chelator and antioxidant clioquinol with an indole ring to generate a series of quinoline-indole scaffolds. Among them, compound **17** was identified as an excellent MTDL (Fig. 11.18). It was synthesized by coupling 5,7-dichloro-8-quinolinol with *o*-nitrobenzaldehyde, followed by the reductive cyclization of the o-nitrostyrene derivative, which was then converted by Pd catalyst under CO to produce the indole intermediate. Interesterification was used to replace the acetyl group with hydroxyl group to afford the quinoline-indole compound **17** [114].

This compound exhibited prominent antioxidant effects (ORAC-FL = 5.0), biometal chelation (e.g., Cu^{2+}, Zn^{2+}, Fe^{2+}, and Fe^{3+}), neuronal cell proliferation, neuroprotective capacity ($EC_{50} = 0.10$), BBB permeability ($P_e = 10.2 \times 10^{-6}$ cm/s), as well as inhibitory activity toward the self- or Cu^{2+}-induced Aβ42 aggregation (71.6% and 85.8%, respectively) and disaggregation of self-induced and Cu^{2+}-associated Aβ42 fibrils (72.7% and 83.3%, respectively) [114]. Moreover, the inhibitory effects of **17** was further elucidated by TEM studies, which showed less dense and thinner fibrils and lower order species of Aβ42 aggregates for self-induced aggregation and smaller-sized nonfibrillar Aβ42 aggregates for Cu^{2+}-induced aggregation. The in vitro cell proliferation of **17** was assessed using undifferentiated PC12 neuronal cells via MTT assay. In the in vivo assays, the efficacy of **17** in the hippocampal cell proliferation of living adult C57BL/6 mice was evaluated by the immunohistochemical assay by intracerebroventricular (icv) injection with osmotic minipumps (compound **17** at 10 μM) and oral dose (compound **17** at 30 mg/kg). The neuroprotection properties were tested at 1−0.01 μM using SH-SY5Y cells. The in vitro metabolic stability of **17** in the liver microsomes of SD rats was determined ($t_{1/2} = 116.8$ min). The acute toxicity studies of **17** in adult C57BL/6 mice (12 weeks old) were evaluated, demonstrating that it was well-tolerated and no acute toxicity was seen at a dose up to 2000 mg/kg by intragastric infusion.

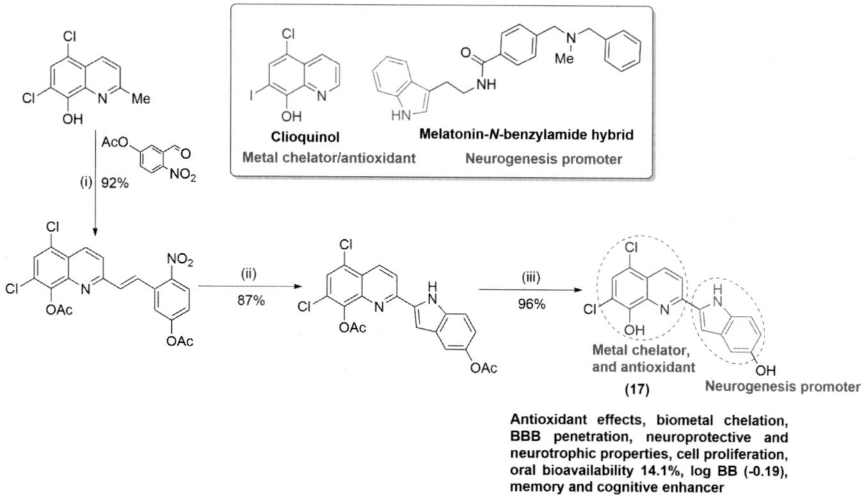

FIGURE 11.18

Synthesis and activity of quinoline-indole MTDL **17**. (i) Acetic anhydride, 150°C, 12 h; (ii) DMF, 1,10-phenanthroline•H_2O, CO, Pd(OAc)$_2$, 80°C, 24 h; (iii) MeOH, NaOCH$_3$, rt, 1 h.

Pharmacokinetic (PK) profile of **17** in SD rats was carried out through two routes of administration including intravenous and oral, which indicated a rational PK property and oral bioavailability (F = 14%), and BBB penetration (log BB = −0.19). In vivo studies using the Morris water maze assay demonstrated that long-term oral intake of **17** (Fig. 11.18) was not toxic and the cognitive and memory dysfunction in double transgenic APP/PS1 AD mice (6 months old) significantly improved, as well as the cerebral Aβ aggregation substantially reduced (30 mg/kg/day oral dose). These results confirmed that the quinoline-indole derivative **17** to be an excellent drug candidate in the treatment of AD.

11.3 Conclusions and future directions

AD has a very complex pathophysiology which involves a numbers of factors that makes it extremely challenging to develop small molecules to prevent and cure this disease. In the last decade, tremendous advances have been achieved and currently we have a better understanding of AD. However, a permanent cure remains elusive. In this regard, the strategy of using small molecule MTDLs to treat AD holds great potential as both small molecule and biological therapies based on the idea of targeting only one of the factors involved in AD pathogenesis have failed. This further supports the development of MTDLs. Another aspect to consider while developing small molecule MTDLs, is the critical need to optimize and develop standards or assay protocols to compare research data across academia and industry since currently, assays to evaluate multifunctional properties are not standardized. Furthermore, other anticipated challenge in MTDL approach includes obtaining vast amount of research data for regulatory bodies such as the US Food and Drug Administration (FDA) or European Medicines Agency (EMA) to provide evidence that they are capable of exhibiting multiple functionality and are safe and effective. This can lead to significant increases in the cost and duration to bring these drugs to the market. Championing the research and development of MTDLs for AD treatment requires the collective efforts of academic and industrial researchers, policy makers, regulatory agencies, stakeholders, patients, and patient advocacy groups which can pave way to the discovery of novel MTDLs that can prevent AD progression and potentially cure this devastating disease. Ultimately, the success of MTDL therapy is dependent on the availability of innovative diagnostic tools for early detection of AD in order to start intervention with MTDLs during the early stages of AD.

Acknowledgments

PPNR would like to thank the University of Waterloo, NSERC-Discovery RGPIN 2014 and 2019, and Ministry of Research and Innovation, Government of Ontario, Canada for an Early Researcher Award (ERA) for the financial support.

References

[1] H.W. Querfurth, F.M. LaFerla, Alzheimer's disease, N. Engl. J. Med. 362 (4) (2011) 329–344.

[2] P. Scheltens, K. Blennow, M.M. Breteler, B. de Strooper, G.B. Frisoni, S. Salloway, W.M. Van der Flier, Alzheimer's disease, Lancet 388 (10043) (2016) 505–517.

[3] M.W. Bondi, E.C. Edmonds, D.P. Salmon, Alzheimer's disease: past, present and future, J. Int. Neuropsychol. Soc. 23 (9–10) (2018) 818–831.

[4] L. Colucci, M. Bosco, A.M. Fasanaro, G.L. Gaeta, G. Ricci, F. Amenta, Alzheimer's disease costs: what we know and what we should take into account, J. Alzheimer's Dis. 42 (4) (2014) 1311–1324.

[5] L. Jonsson, P.J. Lin, A.S. Khachaturian, Special topic section on health economics and public policy of Alzheimer's disease, Alzheimers Dement 13 (3) (2017) 201–204.

[6] R. Cimler, P. Maresova, J. Kuhnova, K. Kuca, Predictions of Alzheimer's disease treatment and care cost in European countries, PLoS One 14 (1) (2019) e0210958.

[7] GBD 2016 Dementia Collaborators, Global, regional, and national burden of Alzheimer's disease and other dementias, 1990-2016: a systematic analysis for the global burden of disease study 2016, Lancet Neurol. 18 (1) (2019) 88–106.

[8] T. Dua, K.M. Seeher, S. Sivananthan, N. Chowdhary, A.M. Pot, S. Saxena, World health organization's global action plan on the public health response to dementia 2017–2025, Alzheimers Dement 13 (7) (2017) P1450–P1451.

[9] World Health Organization WHO. 2017. https://www.who.int/mental_health/neurology/dementia/hscprovider_infosheet.pdf.

[10] J. Birks, Cholinesterase inhibitors for Alzheimer's disease, Cochrane Database Syst. Rev. 25 (1) (2006) CD005593.

[11] M. Bond, G. Rogers, J. Peters, R. Anderson, M. Hoyle, A. Miners, T. Moxham, S. Davis, P. Thokala, A. Wailoo, M. Jeffreys, C. Hyde, The effectiveness and cost-effectiveness of donepezil, galantamine, rivastigmine and memantine for the treatment of Alzheimer's disease (review of technology appraisal no. 111): a systematic review and economic model, Health Technol. Assess. 16 (21) (2012) 1–470.

[12] W.V. Graham, A. Bonito-Oliva, T.P. Sakmar, Update on Alzheimer's disease therapy and prevention strategies, Annu. Rev. Med. 68 (2017) 413–430.

[13] D.J. Selkoe, The amyloid hypothesis of Alzheimer's disease at 25 years, EMBO Mol. Med. 8 (6) (2016) 595–608.

[14] E. Karran, M. Mercken, B. De Strooper, The amyloid cascade hypothesis for Alzheimer's disease: an appraisal for the development of therapeutics, Nat. Rev. Drug Discov. 10 (9) (2011) 698–712.

[15] I.W. Hamley, The amyloid beta peptide: a chemist's perspective. Role in Alzheimer's and fibrillization, Chem. Rev. 112 (10) (2012) 5147–5192.

[16] T. Mohamed, A. Shakeri, P.P.N. Rao, Amyloid cascade in Alzheimer's disease: recent advances in medicinal chemistry, Eur. J. Med. Chem. 113 (2016) 258–272.

[17] T.V. Huynh, D.M. Holtzman, In search of an identity for amyloid plaques, Trends Neurosci. 41 (8) (2018) 483–486.

[18] A.K. Ghosh, H.L. Osswald, BACE1 (β-secretase) inhibitors for the treatment of Alzheimer's disease, Chem. Soc. Rev. 43 (19) (2014) 6765–6813.

[19] B. De Strooper, T. Iwatsubo, M.S. Wolfe, Presenilins and secretase: structure, function, and role in Alzheimer disease, Cold Spring Harb. Perspect. Med. 2 (1) (2012) a006304.

[20] X.L. Bu, P.P.N. Rao, Y.J. Wang, Anti-amyloid aggregation activity of natural compounds: implications for Alzheimer's drug discovery, Mol. Neurobiol. 53 (6) (2016) 3565–3575.

[21] H. Liu, L. Wang, W. Su, X.Q. Xie, Advances in recent patent and clinical trial development for Alzheimer's disease, Pharm. Pat. Anal. 3 (4) (2014) 429–447.

[22] J. Godyn, J. Jonczyk, D. Panek, B. Malawska, Therapeutic strategies for Alzheimer's disease in clinical trials, Pharmacol. Rep. 68 (1) (2016) 127–138.

[23] M.H. Baig, K. Ahmad, G. Rabbani, I. Choi, Use of peptides for the management of Alzheimer's disease: diagnosis and inhibition, Front. Aging Neurosci. 10 (2018) 21.

[24] D. Goyal, S. Shuaib, S. Mann, B. Goyal, Rationally designed peptides and peptidomimetics as inhibitors of amyloid-β (Aβ) aggregation: potential therapeutics of Alzheimer's disease, ACS Comb. Sci. 19 (2) (2017) 55–80.

[25] R. Samo, Peptides as potential therapeutics for Alzheimer's disease, Molecules 23 (2) (2018) E283.

[26] Y.J. Wang, Alzheimer's disease: lessons from immunotherapy for Alzheimer disease, Nat. Rev. Neurol. 10 (4) (2014) 188–189.

[27] S. Reardon, Alzheimer antibody drugs show questionable potential, Nat. Rev. Drug Discov. 14 (9) (2015) 591–592.

[28] L.M. Gaya, S. Villegas, Immunotherapy for neurodegenerative diseases: the Alzheimer's disease paradigm, Curr. Opin. Chem. Eng. 19 (2018) 59–67.

[29] J. Cao, J. Hou, J. Ping, D. Cai, Advances in developing novel therapeutic strategies for Alzheimer's disease, Mol. Neurodegener. 13 (1) (2018) 64.

[30] J. Cummings, G. Lee, A. Ritter, K. Zhong, Alzheimer's disease drug development pipeline, Alzheimers Dement 4 (2018) (2018) 195–214.

[31] R.S. Doody, R. Raman, M. Farlow, T. Iwatsubo, B. Vellas, S. Joffe, K. Kieburtz, F. He, X. Sun, R.G. Thomas, P.S. Aisen, A phase 3 trial of semagacestat for treatment of Alzheimer's disease, N. Engl. J. Med. 369 (4) (2013) 341–350.

[32] B. De Strooper, L. Chavez Gutierrez, Learning by failing: ideas and concepts to tackle secretases in Alzheimer's disease and beyond, Annu. Rev. Pharmacol. Toxicol. 55 (2015) 419–437.

[33] M.F. Egan, J. Kost, P.N. Tariot, P.S. Aisen, J.L. Cummings, B. Vellas, C. Sur, Y. Mukai, T. Voss, C. Furtek, E. Mahoney, L. Harper Mozley, R. Vandenberghe, Y. Mo, D. Michelson, Randomized trial of verubecestat for mild-to-moderate Alzheimer's disease, N. Engl. J. Med. 378 (18) (2018) 1691–1703.

[34] F. Panza, M. Lozupone, V. Solfrizzi, R. Sardone, C. Piccininni, V. Dibello, R. Stallone, G. Giannelli, A. Bellomo, A. Greco, A. Daniele, D. Seripa, G. Logroscino, B.P. Imbimbo, BACE inhibitors in clinical development for the treatment of Alzheimer's disease, Expert Rev. Neurother. 18 (11) (2018) 847–857.

[35] R.M. Anderson, C. Hadjichrysanthou, S. Evans, M.M. Wong, Why do so many clinical trials of therapies for Alzheimer's disease fail? Lancet 390 (10110) (2017) 2327–2329.

[36] D. Mehta, R. Jackson, G. Paul, J. Shi, M. Sabbagh, Why do trials for Alzheimer's disease drugs keep failing? A discontinued drug perspective for 2010–2015, Expert Opin. Investig. Drugs 26 (6) (2017) 735–739.

[37] K. Herrup, The case for rejecting the amyloid cascade hypothesis, Nat. Neurosci. 18 (6) (2015) 794–799.

[38] C.R. Jack, P. Vemuri, Amyloid-β — a reflection of risk or a preclinical marker, Nat. Rev. Neurosci. 14 (6) (2018) 319–320.

[39] F. Kametani, M. Hasegawa, Reconsideration of amyloid hypothesis and tau hypothesis in Alzheimer's disease, Front. Neurosci. 30 (12) (2018) 25.

[40] K. Iqbal, F. Liu, C.X. Gong, I.G. Iqbal, Tau in Alzheimer disease and related taupathies, Curr. Alzheimer Res. 7 (8) (2010) 656–664.

[41] F.P. Chong, K.Y. Ng, R.Y. Koh, S.M. Chye, Tau proteins and taupathies in Alzheimer's disease, Cell. Mol. Neurobiol. 38 (5) (2018) 965–980.

[42] V.L. Villemagne, M.T. Fodero-Tavoletti, C.L. Masters, C.C. Rowe, Tau imaging: early progress and future directions, Lancet Neurol. 14 (1) (2015) 114–124.

[43] Y. Wang, E. Mandelkow, Tau in physiology and pathology, Nat. Rev. Neurosci. 17 (1) (2016) 5–21.

[44] M.R. Khanna, J. Kovalevich, V.M. Lee, J.Q. Trojanowski, K.R. Brunden, Therapeutic strategies for the treatment of taupathies: hopes and challenges, Alzheimers Dement 12 (10) (2016) 1051–1065.

[45] J. Gotz, G. Halliday, R.M. Nisbet, Molecular pathogenesis of the taupathies, Annu. Rev. Pathol. 14 (2019) 239–261.

[46] P.J. Dolan, G.V. Johnson, The role of tau kinases in Alzheimer's disease, Curr. Opin. Drug Discov. Dev 13 (5) (2010) 595–603.

[47] L. Martin, X. Latypova, C.M. Wilson, A. Magnaudeix, M.L. Perrin, C. Yardin, F. Terro, Tau protein kinases: involvement in Alzheimer's disease, Ageing Res. Rev. 12 (1) (2013) 289–309.

[48] B. Bulic, M. Pickhardt, E. Mandelkow, Progress and developments in tau aggregation inhibitors for Alzheimer disease, J. Med. Chem. 56 (11) (2013) 4135–4155.

[49] C.M. Wischik, C.R. Harrington, J.M. Storey, Tau-aggregation inhibitor therapy for Alzheimer's disease, Biochem. Pharmacol. 88 (4) (2014) 529–539.

[50] M. Medina, J. Avila, New perspective on the role of tau in Alzheimer's disease. Implications for therapy, Biochem. Pharmacol. 88 (4) (2014) 540–547.

[51] M.E. Pickhardt, J. Biernat, S. Hubschmann, F.J.A. Dennissen, T. Timm, A. Aho, E.M. Mandelkow, E. Madelkow, Time course of tau toxicity and pharmacologic prevention in a cell model of tauopathy, Neurobiol. Aging 57 (2017) 47–63.

[52] E. Giacobini, G. Gold, Alzheimer disease therapy — moving from amyloid-β to tau, Nat. Rev. Neurol. 9 (12) (2013) 677–686.

[53] S. Gautheir, H.H. Feldman, L.S. Schneider, G.K. Wilcock, G.B. Frisoni, J.H. Hardlund, H.J. Moebius, P. Bentham, K.A. Kook, D.J. Wischik, B.O. Schelter, C.S. Davis, R.T. Staff, L. Bracoud, K. Shamsi, J.M. Storey, C.R. Harrington, C.M. Wischik, Efficacy and safety of tau-aggregation inhibitor therapy in patients with mild or moderate Alzheimer's disease: a randomized, controlled, double-blind, parallel arm, phase 3 trial, Lancet 388 (10062) (2016) 2873–2884.

[54] F. Panza, V. Solfrizzi, D. Seripa, B.P. Imbimbo, M. Lozupone, A. Santamato, Z. Zecca, M.R. Barulli, A. Bellomo, A. Pilotto, A. Daniele, A. Greco, G. Logroscino, Tau-centric targets and drugs in clinical development for the treatment of Alzheimer's disease, Biomed. Res. Int. 2016 (2016) 3245935.

[55] M. Medina, An overview on the clinical development of tau-based therapeutics, Int. J. Mol. Sci. 19 (4) (2018) 1160.

[56] A. Boutajangout, T. Wisniewski, Tau-based therapeutic approaches for Alzheimer's disease — a mini-review, Gerontology 60 (5) (2014) 381–385.

[57] E.E. Congdon, E.M. Sigurdsson, Tau-targeting therapies for Alzheimer disease, Nat. Rev. Neurol. 14 (7) (2018) 399–415.

[58] A.S. Pithadia, M.H. Lim, Metal-associated amyloid-β species in Alzheimer's disease, Curr. Opin. Chem. Biol. 16 (1−2) (2012) 67−73.

[59] M.G. Savelieff, A.S. DeToma, J.S. Derricj, M.H. Lim, The ongoing search for small molecules to study metal-associated amyloid-β species in Alzheimer's disease, Acc. Chem. Res. 47 (8) (2014) 2475−2482.

[60] P.I. Moreira, C. Carvalho, Z. Zhu, M.A. Smith, G. Perry, Mitochondrial dysfunction is a trigger of Alzheimer's disease pathophysiology, Biochim. Biophys. Acta 1802 (1) (2010) 2−10.

[61] P. Picone, D. Nuzzo, L. Caruana, C. Scafidi, M. Di Carlo, Mitochondrial dysfunction: different routes to Alzheimer's disease therapy, Oxid. Med. Cell Longev. 2014 (2014) 780179.

[62] M. Audano, A. Schneider, N. Mitro, Mitochondria, lysosomes, and dysfunction: their meaning in neurodegeneration, J. Neurochem. 147 (3) (2018) 291−309.

[63] R. Wang, P.H. Reddy, Role of glutamate and NMDA receptors in Alzheimer's disease, J. Alzheimer's Dis. 57 (4) (2017) 1041−1048.

[64] M.T. Heneka, M.J. Carson, E.I. Khoury, G.E. Landreth, F. Brosseron, D.L. Feinstein, A.H. Jacobs, T. Wyss-Coray, J. Vitorica, R.M. Ransohoff, K. Herrup, S.A. Frautschy, B. Finsen, G.C. Brown, A. Verkhratsky, K. Yamanaka, J. Koistinaho, E. Latz, A. Halle, G.C. Petzold, T. Town, D. Morgan, M.L. Shinohara, V.H. Perry, C. Holmes, N.G. Bazan, D.J. Brooks, S. Hunot, B. Joseph, N. Deigendesch, O. Garaschuk, E. Boddeke, C.A. Dinarello, J.C. Breitner, G.M. Cole, D.T. Golenbock, M.P. Kummer, Neuroinflammation in Alzheimer's disease, Lancet Neurol. 14 (4) (2015) 388−405.

[65] A. Ardura-Fabregat, E.W.G.M. Boddeke, A. Boza-Serrano, S. Brioschi, S. Castro-Gomez, K. Ceyzeriat, C. Dansokho, T. Dierkes, G. Gelders, M.T. Heneka, L. Hoejimakers, A. Hoffmann, L. Iaccarino, S. Jahnert, K. Kuhbandner, G. Landreth, N. Lonnemann, P.A. Loschmann, R.M. McManus, A. Paulus, K. Reemst, J.M. Sanchez-Caro, A. Tiberi, A. Van der Perren, A. Vautheny, C. Venegas, A. Webers, P. Weydt, T.S. Wijasa, X. Xiang, Y. Yang, Targeting neuroinflammation to treat Alzheimer's disease, CNS Drugs 31 (12) (2017) 1057−1082.

[66] D. Kim, S.H. Baik, S. Kang, S.W. Cho, J. Bae, M.Y. Cha, M.J. Sailor, I. Mook-Jung, K.H. Ahn, Close correlation of monoamine oxidase activity with progress of Alzheimer's disease in mice, observed by in vivo two-photon imaging, ACS Cent. Sci. 2 (12) (2016) 967−975.

[67] S. Claevsen, J. Bockaert, P. Giannoni, Serotonin: a new hope in Alzheimer's disease? ACS Chem. Neurosci. 6 (7) (2015) 940−943.

[68] C. Sims-Robinson, B. Kim, A. Rosko, E.L. Feldman, How does diabetes accelerate Alzheimer disease pathology? Nat. Rev. Neurol. 6 (10) (2010) 551−559.

[69] S.E. Arnold, Z. Arvanitakis, S.L. Macauley-Rambach, A.M. Koenig, R.S. Ahima, S. Craft, S. Gandy, C. Buettner, L.E. Stoeckel, D.M. Holtzman, D.M. Nathan, Brain insulin resistance in type 2 diabetes and Alzheimer disease: concepts and conundrums, Nat. Rev. Neurol. 14 (3) (2018) 168−181.

[70] P. Maheshwari, G.D. Eslick, Bacterial infection and Alzheimer's disease: a meta-analysis, J. Alzheimer's Dis. 43 (3) (2015) 957−966.

[71] Y. Wang, Y. Shi, H. Wei, Calcium dysregulation in Alzheimer's disease: a target for new drug development, J. Alzheimers Dis. Parkinsonism 7 (5) (2017) 374.

[72] A. Agis-Torres, M. Solhuber, M. Fernandez, J.M. Sanchez-Montero, Multi-target-directed ligands and other therapeutic strategies in the search of a real solution for Alzheimer's disease, Curr. Neuropharmacol. 12 (1) (2014) 2–36.

[73] R.E. Hughes, K. Nikolic, R.R. Ramsay, One for all? Hitting multiple Alzheimer's disease targets with one drug, Front. Neurosci. 25 (10) (2016) 177.

[74] M. Rosini, E. Simoni, R. Caporaso, A. Minarini, Multitarget strategies in Alzheimer's disease: benefits and challenges on the road to therapeutics, Future Med. Chem. 8 (6) (2016) 697–711.

[75] A. Cavalli, M.L. Bolognesi, S. Capsoni, V. Andrisano, M. Bartolini, E. Margotti, A. Cattaneo, M. Recanatini, C. Melchiorre, A small molecule targeting the multifactorial nature of Alzheimer's disease, Angew. Chem. Int. Ed. Engl. 46 (20) (2007) 3689–3692.

[76] M.D. Shultz, Two decades under the influence of the rule of five and the changing properties of approved oral drugs, J. Med. Chem. 62 (4) (2019) 1701–1714.

[77] M.L. Bolognesi, G. Chiriano, M. Bartolini, F. Mancini, G. Bottegoni, V. Maestri, S. Czvitkovich, M. Windisch, A. Cavalli, A. Minarini, M. Rosini, V. Tumiatti, C. Melchiorre, Synthesis of monomeric derivatives to probe memoquin's bivalent interactions, J. Med. Chem. 54 (24) (2011) 8299–8304.

[78] Z. Luo, J. Sheng, Y. Sun, C. Lu, J. Yan, A. Liu, H.B. Luo, L. Huang, X. Li, Synthesis and evaluation of multi-target-directed ligands against Alzheimer's disease based on the fusion of donepezil and ebselen, J. Med. Chem. 56 (22) (2013) 9080–9099.

[79] T. Schewe, Molecular actions of ebselen-an anti-inflammatory antioxidant, Gen. Pharmacol. 26 (6) (1995) 1153–1169.

[80] M.A. Ansari, S.W. Scheff, Oxidative stress in the progression of Alzheimer disease in the frontal cortex, J. Neuropathol. Exp. Neurol. 69 (2) (2010) 155–167.

[81] J. Aaseth, J. Alexander, G. Bjorklund, K. Hestad, P. Dusek, P.M. Roos, U. Alehagen, Treatment strategies in Alzheimer's disease: a review with focus on selenium supplementation, Biometals 29 (5) (2016) 827–839.

[82] T. Ma, M.S. Tan, J.T. Yu, L. Tan, Resveratrol as a therapeutic agent for Alzheimer's disease, Biomed. Res. Int. 2014 (2014) 350516.

[83] A.R. Neves, M. Lucio, J.L. Lima, S. Reis, Resveratrol in medicinal chemistry: a critical review of its pharmacokinetics, drug-delivery, and membrane interactions, Curr. Med. Chem. 19 (11) (2012) 1663–1681.

[84] C. Lu, Y. Guo, J. Yan, Z. Luo, H.M. Luo, M. Yan, L. Huang, X. Li, Design, synthesis and evaluation of multitarget-directed resveratrol derivatives for the treatment of Alzheimer's disease, J. Med. Chem. 56 (14) (2013) 5843–5859.

[85] M. Pickhardt, Z. Gazova, M. von Bergen, I. Khlistunova, Y. Wang, A. Hascher, E.M. Mandelkow, J. Biernat, E. Mandelkow, Anthraquinones inhibit tau aggregation and dissolve Alzheimer's paired helical filaments in vitro and in cells, J. Biol. Chem. 280 (5) (2005) 3628–3635.

[86] M. Convertino, R. Pellarin, M. Catto, A. Carotti, A. Calflisch, 9,10-Anthraquinone hinders beta-aggregation: how does a small molecule interfere with abeta-peptide amyloid fibrillation? Protein Sci. 18 (4) (2009) 792–800.

[87] E. Viayna, I. Sola, M. Bartolini, A. De Simone, C. Tapia-Rojas, F.G. Serrano, R. Sabate, J. Juarez-Jimenez, B. Perez, F.J. Luque, V. Andrisano, M.V. Clos, N.C. Inestrosa, D. Munoz-Torrero, Synthesis and multitarget biological profiling of a novel family of rhein derivatives as disease-modifying anti-Alzheimer agents, J. Med. Chem. 57 (6) (2014) 2549–2567.

[88] O.M. Bautista-Aguilera, A. Samadi, M. Chioua, K. Nikolic, S. Filipic, D. Agbaba, E. Soriano, L. de Andres, M.I. Rodriguez-Franco, S. Alcaro, R.R. Ramsay, F. Ortuso, M. Yanez, J. Marco-Contelles, *N*-methyl-*N*-((1-methyl-5-(3-(1-(2-methylbenzyl)piperidin-4-yl)propoxy)-1*H*-indol-2-yl)methyl)prop-2-yn-1-amine, a new cholinesterase and monoamine oxidase dual inhibitor, J. Med. Chem. 57 (24) (2014) 10455−104563.

[89] C. Binda, F. Hubalek, M. Li, Y. Herzig, J. Sterling, D.E. Edmondson, A. Mattevi, Inactivation of purified human recombinant monoamine oxidases A and B by rasagiline and its analogues, J. Med. Chem. 47 (7) (2004) 1767−1774.

[90] O.M. Bautista-Anguilera, S. Hagenow, A. Palomino-Antolin, V. Farré-Alins, L. Ismaili, P.L. Joffrin, M.L. Jimeno, O. Soukup, J. Janočková, L. Kalinowsky, E. Proschak, I. Iriepa, I. Moraleda, J.S. Schwed, A.R. Martínez, F. López-Muñoz, M. Chioua, J. Egea, R.R. Ramsay, J. Marco-Contelles, H. Stark, Multitarget-directed ligands combining cholinesterase and mono-amine oxidase inhibition with histamine H3R antagonism for neurodegenerative disease, Angew. Chem. Int. Ed. 56 (41) (2017) 12765−12769.

[91] O.M. Bautista-Anguilera, J. Budni, F. Mina, E.B. Mediros, W. Deuther-Conrad, J.M. Entrena, I. Moraleda, I. Iriepa, F. López-Muñoz, J. Marco-Contelles, Contilisant, a tetratarget small molecule for Alzheimer's disease therapy combining cholinesterase, monoamine oxidase inhibition, and H3R antagonism with S1R agonism profile, J. Med. Chem. 61 (15) (2018) 6937−6943.

[92] S. Tordjman, S. Chokron, R. Delorme, A. Charrier, E. Bellissant, N. Jaafari, C. Fougerou, Melatonin: pharmacology, functions and therapeutic benefits, Curr. Neuropharmacol. 15 (3) (2017) 434−443.

[93] M. Shukla, P. Govitrapong, P. Boontem, R.J. Reiter, J. Satayavivad, Mechanism of melatonin in alleviating Alzheimer' disease, Curr. Neuropharmacol. 15 (7) (2017) 1010−1031.

[94] B. Lopez-Iglesias, C. Perez, J.A. Morales-Garcia, S. Alonso-Gil, A. Perez-Castillo, A. Romero, M.G. Lopez, M. Villarroya, S. Conde, M.I. Rodriguez-Franco, New melatonin-*N,N*-dibenzyl(*N*-methyl)amine hybrids: potent neurogenic agents with antioxidant, cholinergic, and neuroprotective properties as innovative drugs for Alzheimer's disease, J. Med. Chem. 57 (24) (2014) 3773−3785.

[95] D. Srikrishna, C. Godugu, P.K. Dubey, A review on pharmacological properties of coumarins, Mini Rev. Med. Chem. 18 (2) (2018) 113−141.

[96] R. Farina, L. Pisani, M. Catto, O. Nicolotti, D. Gadaleta, N. Denora, R. Soto-Otero, E. Mendez-Alvarez, C.S. Passos, G. Muncipinto, C.D. Altomare, A. Nurisso, P.A. Carrupt, A. Carotti, Structure-based design and optimization of multi-target-directed 2*H*-chromen-2-one derivatives as potent inhibitors of monoamine oxidase B and cholinesterases, J. Med. Chem. 58 (24) (2015) 5561−5578.

[97] B. Varga, A. Csonka, A. Csonka, J. Molnar, L. Amaral, G. Spengler, Possible biological and clinical applications of phenothiazines, Anticancer Res. 37 (11) (2017) 5983−5993.

[98] S. Darvesh, K.V. Darvesh, R.S. McDonald, D. Mataija, R. Walsh, S. Mothana, O. Lockridge, E. Martin, Carbamates with differential mechanism of inhibition toward acetylcholinesterase and butyrylcholinesterase, J. Med. Chem. 51 (14) (2008) 4200−4212.

[99] G. Tin, T. Mohamed, N. Gondora, M.A. Beazely, P.P.N. Rao, Tricyclic phenothiazine and phenoselenazine derivatives as potential multi-targeting agents to treat Alzheimer's disease, Med. Chem Comm. 6 (11) (2015) 1930–1941.

[100] X. Li, H. Wang, Z. Lu, X. Zheng, W. Ni, J. Zhu, Y. Fu, F. Lian, N. Zhang, J. Li, H. Zhang, F. Mao, Development of multifunctional pyrimidinylthiourea derivatives as potential anti-Alzheimer agents, J. Med. Chem. 59 (18) (2016) 8326–8344.

[101] T. Mohamed, X. Zhao, L.L. Habib, J. Yang, P.P.N. Rao, Design, synthesis and structure-activity relationship (SAR) studies of 2,4-disubstituted pyrimidine derivatives: dual activity as cholinesterase and Aβ-aggregation inhibitors, Bioorg. Med. Chem. 19 (7) (2011) 2269–2281.

[102] T. Mohamed, J. C. Yeung, M. S. Vasefi, M. A. Beazely, P. P. N. Rao, Development and evaluation of multifunctional agents for potential treatment of Alzheimer's disease: application to a pyrimidine-2,4-diamine template, *Bioorg. Med. Chem. Lett* **22** (14) 4707–4712.

[103] T. Mohamed, P.P.N. Rao, 2,4-Disubstituted quinazolines as amyloid- aggregation inhibitors with dual cholinesterase inhibition and antioxidant properties: development and structure-actviity relationship (SAR) studies, Eur. J. Med. Chem. 126 (27) (2017) 823–843.

[104] T. Mohamed, M.K. Mann, P.P.N. Rap, Application of quinazoline and pyrido[3,2-*d*] pyrimidine templates to design multi-targeting agents in Alzheimer's disease, RSC Adv. 7 (2017) 22360–22368.

[105] C.B. Mishra, S. Kumari, M.T. Thiazole, A promising heterocycle for the development of potent CNS active agents, Eur. J. Med. Chem. 92 (2015) 1–34.

[106] M. Shidore, J. Machhi, K. Shingala, P. Murumkar, M.K. Sharma, N. Agrawal, A. Tripathi, Z. Parikh, P. Pillai, M.R. Yadav, Benzylpiperidine-linked diarylthiazoles as potential anti-Alzheimer's agents: synthesis and biological evaluation, J. Med. Chem. 59 (12) (2016) 5823–5846.

[107] S.T. Forgue, B.E. Patterson, A.W. Bedding, C.D. Payne, D.L. Phillips, R.E. Wrishko, M.I. Mitchell, Tadalafil pharmacokinetics in healthy subjects, Br. J. Clin. Pharmacol. 61 (3) (2006) 280–288.

[108] A. García-Osta, M. Cuadrado-Tejedor, C. García-Barroso, J. Oyarzábal, R. Franco, Phosphodiesterases as therapeutic targets for Alzheimer's disease, ACS Chem. Neurosci. 3 (11) (2012) 832–844.

[109] F. Mao, H. Wang, W. Ni, X. Zheng, M. Wang, K. Bao, D. Ling, X. Li, Y. Xu, H. Zhang, J. Li, Design, synthesis, and biological evaluation of orally available first-generation dual-target selective inhibitors of acetylcholinesterase (AChE) and phosphodiesterase 5 (PDE5) for the treatment of Alzheimer's disease, ACS Chem. Neurosci. 9 (2) (2018) 328–345.

[110] W. Ni, H. Wang, X. Li, X. Zheng, M. Wang, J. Zhang, Q. Gong, D. Ling, F. Mao, H. Zhang, J. Li, Novel tadalafil derivatives ameliorates scopolamine-induced cognitive impairment in mice via inhibition of acetylcholinesterase (AChE) and phosphodiesterase 5 (PDE5), ACS Chem. Neurosci. 9 (7) (2018) 1625–1636.

[111] X.Y. Jiang, T.K. Chen, J.T. Zhou, S.Y. He, H.Y. Yang, Y. Chen, W. Qu, F. Feng, H.P. Sun, Dual GSK-3β/AChE inhibitors as a new strategy for multitargeting anti-Alzheimer's disease drug discovery, ACS Med. Chem. Lett. 9 (3) (2018) 171–176.

[112] A. Wieckowska, T. Wichur, J. Godyń, A. Bucki, M. Marcinkowska, A. Siwek, K. Wieckowski, P. Zareba, D. Knez, M. Gluch-Lutwin, G. Kazek, G. Latacz, K. Mika, M. Kolaczkowski, J. Korabecny, O. Soukup, M. Benkova, K. Kiéc-

Kononowicz, S. Gobec, B. Malawska, Novel multitarget-directed ligands aiming at symptoms and causes of Alzheimer's disease, ACS Chem. Neurosci. 9 (5) (2018) 1195−1214.

[113] A. Wieckowska, M. Kolaczkowski, A. Bucki, J. Godyń, M. Marcinkowska, K. Wieckowski, P. Zareba, A. Siwek, G. Kazek, M. Gluch-Lutwin, P. Mierzejewski, P. Bienkowski, H. Sienkiewicz-Jarosz, D. Knez, T. Wichur, S. Gobec, B. Malawska, Novel multitarget-directed ligands for Alzheimer's disease: combining cholinesterase inhibitors and 5-HT$_6$ receptor antagonist. Design, synthesis, and biological evaluation, Eur. J. Med. Chem. 124 (2016) 63−81.

[114] Z. Wang, J. Hu, X. Yang, X. Feng, X. Li, L. Huang, Design, synthesis, and evaluation of orally bioavailable quinoline-indole derivatives as innovative multitarget-directed ligands: promotion of cell proliferation in the adult murine hippocampus for the treatment of Alzheimer's disease, J. Med. Chem. 61 (5) (2018) 1871−1894.

Index

'*Note*: Page numbers followed by "f" indicate figures, "t" indicate tables and "b" indicate boxes.'

C

Carbohydrate chemical mimics, 200–201
Carbohydrate processing enzymes, 197
Carboxypeptidases A (CPA) enzyme, 9
CDPs. *See* Consensus Diversity Plots (CDPs)
Cellular thermal shift assay (CETSA), 146
Centrocountins
  biologically active complex natural products, 248, 248f
  cellular targets
    acentrosomal spindle poles, 259–261
    affinity-based approach, 257
    chemical structure, 257, 258f
    dissociation constants, 259
    fluorescence lifetime imaging microscopy (FLIM), 259b
    fluorescence polarization (FP), 259b
    mitotic modulators, 257
    nuclear transport process, 259–261
    nucleophosmin (NPM), 258–259, 260f
    peptide sequence tags, 258
  genomics applications, 247–248
  high-throughput screening (HTS), 247–248
  intellectual property rights, 247–248
  molecular scaffolds, 247–248
  phenotypic screening
    anticancer drugs target mitosis, 254
    chromosome congression, *(R)*-18a, 256–257, 256f
    cyclin-dependent kinases (CDKs), 253–254
    high-content cell-based assay, 254–255
    immunoblotting/Western blotting, 255b
    immunostaining, 255b
    mitotic inhibitors, 253, 254f
    structure-activity relationship (SAR), 253
  polycyclic indoles, 248
  structural and molecular properties, 247–248
  tetrahydroindolo[2,3-*a*]quinolizines). *See* Tetrahydroindolo[2,3-*a*]quinolizines
Chalcone isomerase (CHI), 24–25
Chalcone synthase (CHS), 24–25
Chemical diversity analysis, 83
  alignment analysis, 90–91
  chemical libraries, 84, 85f
  consensus diversity analysis, 92–93, 93f
  evaluation methods, 85
  molecular diversity, 88–89, 88f
  pharmacophore, 84
  R-group decomposition, 85–87, 87f
  scaffold diversity, 91, 92f
  structural fingerprints, 89–90, 90f
  types, 84, 86t
  virtual screening, 84
  visual representation
    biology-oriented synthesis (BIOS), 94
    components, 94
    molecular diversity and chemical space, 94, 95f
    principal component analysis (PCA), 94
    principal moment of inertia (PMI) plot, 94
    self-organizing maps (SOMs), 94
Chemical genetics, 2, 4, 11
Chemical reactions
  "biotech" drugs, 35–36
  click chemistry, 35–36
  cross-coupling reactions, 35–36
    N-allenamides, 49–51
    copper-mediated process, 46–47
    enantioselective process, 37–38
    environmental and economic advantages, 47–48
    gold catalytic properties, 48–51
    Hiyama-Denmark coupling, 46–47
    iron catalysis, 46–49
    nickel-mediated cross-couplings, 43–46
    olefin metathesis, 43–44
    organometallic reagent, 37–38
    palladium-mediated coupling, 37–38
    see also Palladium-mediated cross-couplings
    sesquiterpene analogues, 48–50
    Stille-type coupling, 46–47
    Suzuki-Miyaura coupling, 46–47
    unactivated alkenes, 46–48
  cycloadditions, 35–36
    alkenols, 52
    asenapine synthesis, 54–56
    diversity-oriented synthesis, 54–55
    fragment-based drug discovery, 53–54
    Huisgen reaction, 52–53
    MCR/CuAAC approach, 52–54
    metalloprotease (MMP) inhibitors, 53–54
    microwave-assisted cross-coupling reactions, 52–53
    propargyltrimethylsilanes, 54
    retrocyclization, 52
    scaffold-divergent synthesis, 54–55
    solid phase organic synthesis (SPOS), 50–52
    triazoles, 52–53
  diversity-oriented synthesis, 35–36, 37f
  high-throughput screening (HTS), 35–36
  late-stage functionalization (LSF) methods, 35–36. *See also* Late-stage functionalization (LSF) methods
  molecular biology, 35–36